Practical Hazops, Trips and Alarms

Other titles in the series

Practical Data Acquisition for Instrumentation and Control Systems (John Park, Steve Mackay)

Practical Data Communications for Instrumentation and Control (Steve Mackay, Edwin Wright, John Park)

Practical Digital Signal Processing for Engineers and Technicians (Edmund Lai)

Practical Electrical Network Automation and Communication Systems (Cobus Strauss)

Practical Embedded Controllers (John Park)

Practical Fiber Optics (David Bailey, Edwin Wright)

Practical Industrial Data Networks: Design, Installation and Troubleshooting (Steve Mackay, Edwin Wright, John Park, Deon Reynders)

Practical Industrial Safety, Risk Assessment and Shutdown Systems for Instrumentation and Control (Dave Macdonald)

Practical Modern SCADA Protocols: DNP3, 60870.5 and Related Systems (Gordon Clarke, Deon Reynders)

Practical Radio Engineering and Telemetry for Industry (David Bailey)

Practical SCADA for Industry (David Bailey, Edwin Wright)

Practical TCP/IP and Ethernet Networking (Deon Reynders, Edwin Wright)

Practical Variable Speed Drives and Power Electronics (Malcolm Barnes)

Practical Centrifugal Pumps (Paresh Girdhar and Octo Moniz)

Practical Electrical Equipment and Installations in Hazardous Areas (Geoffrey Bottrill and G. Vijayaraghavan)

Practical E-Manufacturing and Supply Chain Management (Gerhard Greef and Ranjan Ghoshal)

Practical Grounding, Bonding, Shielding and Surge Protection (G. Vijayaraghavan, Mark Brown and Malcolm Barnes)

Practical Industrial Data Communications: Best Practice Techniques (Deon Reynders, Steve Mackay and Edwin Wright)

Practical Machinery Safety (David Macdonald)

Practical Machinery Vibration Analysis and Predictive Maintenance (Cornelius Scheffer and Paresh Girdhar)

Practical Power Distribution for Industry (Jan de Kock and Cobus Strauss)

Practical Process Control for Engineers and Technicians (Wolfgang Altmann)

Practical Telecommunications and Wireless Communications (Edwin Wright and Deon Reynders)

Practical Troubleshooting Electrical Equipment (Mark Brown, Jawahar Rawtani and Dinesh Patil)

Practical Hazops, Trips and Alarms

David Macdonald, BSc (Hons) Inst. Eng., Senior Engineer,
IDC Technologies, Cape Town, South Africa

Series editor: Steve Mackay

ELSEVIER

AMSTERDAM • BOSTON • HEIDELBERG • LONDON
NEW YORK • OXFORD • PARIS • SAN DIEGO
SAN FRANCISCO • SINGAPORE • SYDNEY • TOKYO

Newnes is an imprint of Elsevier

Newnes

Newnes
An imprint of Elsevier
Linacre House, Jordan Hill, Oxford OX2 8DP
200 Wheeler Road, Burlington, MA 01803

First published 2004

British Library Cataloguing in Publication Data
Macdonald, D. M
  Practical hazops, trips and alarms – (Practical Professional)
  1. Machinery – Safety appliances 2. Machinery – Monitoring
  I. Title
  621.3'0289

Library of Congress Cataloguing in Publication Data
A catalogue record for this book is available from the Library of Congress

ISBN   0 7506 6274 3

For information on all Newnes Publications
visit our website at www.newnespress.com

Typeset and edited by Integra Software Services Pvt. Ltd, Pondicherry, India
www.integra-india.com
Printed and bound in The Netherlands

Contents

Preface

Introduction to the book

This introduction maps out the reasons why this book has been prepared and introduces the ideas behind it. This section also provides a guide to the contents and uses of the manual.

What is the book about?

This book is about some of the critical activities involved in making sure that a manufacturing process is:

- Safe for people to work with
- Safe against damage to the environment
- Secure against failures that could result in major asset losses in the business.

It concentrates on the application of hazard study methods and the actions that follow from them for providing protection against hazards. The book seeks to provide training in three basic steps that form part of the overall risk management framework for industries such as chemicals, oil and gas, pharmaceuticals and food processing. The steps can be seen in Figure P.1.

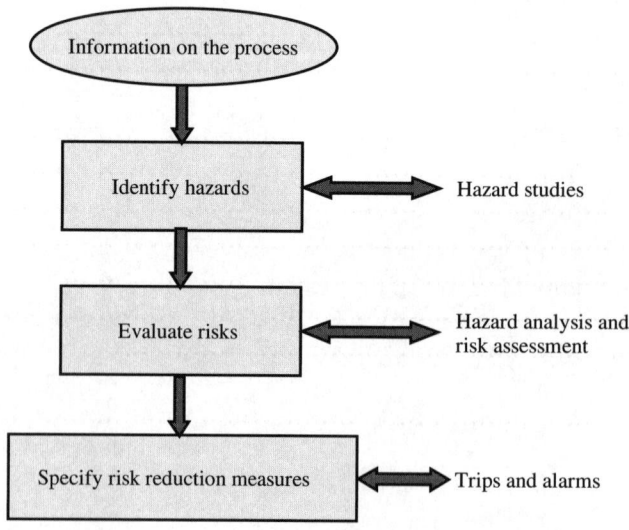

Figure P.1
Scope of the book

- Hazards of a process plant or an activity are identified through the systematic application of hazard studies based on the best possible information available.
- Hazards may create risks to people, environments and property. The risks may or may not be acceptable. This requires evaluation through the techniques of risk assessment and hazard analysis.
- Whenever risks are found to be unacceptable, solutions have to be found either by fundamental design changes or by providing protection measures. Protection measures may be mechanical or organizational or they may be provided by safety-related control systems employing alarm devices or automatic controls. For simplicity in this book we call them trips and alarms.

For protection measures to succeed they must be based on three key factors:

1. There must be correct and up-to-date knowledge of the hazardous situations that are to be controlled including knowledge of the possible causes.
2. There must be a clearly defined course of action to be taken in response to the approach to a hazardous event.
3. The protection systems must be *appropriate* for the problem and be correctly designed and maintained.

Appropriate implies that the protection systems must be practical to use, will carry out the correct actions to restore safe conditions and will be engineered to a level of reliability sufficient to match the degree of risk reduction demanded by the hazardous event. Furthermore, the protection systems must not impede productivity or impact negatively on production volumes through its complexity nor create unacceptable production losses through unreliable operations.

These points appear to be simple and obvious. However, there is a lot of evidence to show that it is not unusual for risk reduction measures to be out of touch with the original problem they were intended to deal with. For example, Figure P.2 shows the results of a survey of control and safety system failures by the United Kingdom Health and Safety Executive (HSE), the body responsible for administration and control of occupational and home safety in the UK. The survey classified the causes of 34 accidents involving failures of control systems, which were supposed to protect against such incidents.

HSE summary of findings

The Summary of the problems of failed safety systems found by HSE included some interesting paragraphs.

Analysis of incidents showed:

- The majority of incidents could have been anticipated if a systematic risk-based approach had been used throughout the life of the system.
- Safety principles are independent of the technology.
- Situations often missed through lack of systematic approach.

Quoting from the report:

The analysis of the incidents shows that the majority were not caused by some subtle failure mode of the control system, but by defects which could have been anticipated if a systematic risk-based approach had been used throughout the life of the system. It is also clear that despite differences in the underlying technology of control systems, the safety principles needed to prevent failure remain the same.

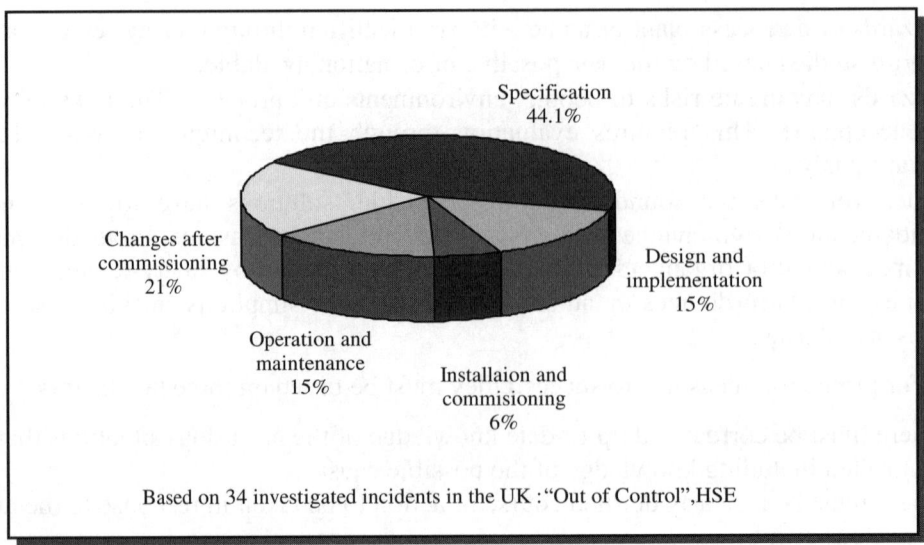

Figure P.2
Summary of causes for failures of safety-related control systems

A review of design issues showed:

- Need to verify that the specification has been met
- Overdependence on single channel of safety
- Failure to verify software
- Poor consideration of human factors.

A review of specification issues included this comment:

The analysis shows that a significant percentage of the incidents can be attributed to inadequacies in the specification of the control system. This may have been due either to poor hazard analysis of the equipment under control, or to inadequate assessment of the impact of failure modes of the control system on the specification.

The HSE report is a very useful guide to safety system project engineers and it is available through HSE Books at www.hse.gov.org.

Hazards of replacement and modification

Another indicator of the weak links between protection systems and the problem they are supposed to solve is in the situation that has occurred when a process plant is faced with the task of replacing an obsolete item of safety instrumentation. Typically the plant may have an electrical trip and interlock control panel based on relays or old solid-state logic devices and they want to replace it with a new package, perhaps based on safety PLC equipment. The question arises: How do we specify the exact requirements for the new safety system? Where is the original specification? Does it still meet the needs of the plant as it stands today?

This often leads to a search for the original hazard or Hazop study reports. Sometimes these are fully up to date and in good shape. Sometimes they cannot even be found. Often they are available but slightly out of date.

In the USA a wide-ranging investigation of chemical plant accidents by James Belke for the US Dept of Labor included many telling comments on the failure of companies to make adequate provisions for hazard studies and the design of safety systems.

Regarding change control the following comment is noteworthy:

Recurring causes of these accidents include inadequate process hazards analysis, use of inappropriate or poorly-designed equipment, inadequate indications of process condition, and others. Of particular note, installation of emissions or pollution control equipment has preceded several significant accidents, highlighting the need for stronger systems for management of change.

Continuity and validation

Today's standards for safety-related control systems (they are also called safety instrumented systems or functional safety systems) demand fully traceable links from their performance specifications and their functional testing all the way back to the current hazard study reports and records for the process. This is the only way to ensure that the safety systems are valid for the plant. No safety system can be considered to be correct for the plant without 'Validation'.

All of these factors point strongly to the need for continuity to be assured through the steps from identification of hazards into the delivery of safety systems. This book sets out to support this approach by introductory level training linking the three fields shown in Figure P. 1. Namely:

- Hazard studies
- Hazard analysis
- Design of safety instrumentation and alarm systems.

Safety projects will benefit if all contributors understand the context and methods of hazard studies whilst having a good understanding of the principles of alarm and trip systems.

Objectives

The objective of this book is to provide an introductory level of training in the methods of hazard studies and in the associated risk reduction methods achieved by using alarm and trip systems. The hazard study methods are based on those explained in internationally supported manuals and engineering standards. The alarm and trip system principles include introductory training on the new internationally accepted standard IEC 61508 and all design principles expressed in the manual are intended to be consistent with this standard.

Through feedback from an existing IDC training course in safety instrumentation it became apparent that there is a need for a training course that covers the basics of hazard studies and bridges the gap between hazard studies and the delivery of safety-related control systems. In simple terms:

"If you work with hazard studies it helps to know how trips and alarms are supposed to be built.

If you work with trips and alarms it helps to know how hazard studies are supposed to be done."**

The course is intended to be useful for:

- Process plant engineers, technicians and supervisors involved in new plant projects or in the modification or upgrading of existing plants
- Loss prevention officers, trainee Hazop team leaders
- Plant managers, project managers and planners seeking an awareness of the role of Hazops in overall safety management
- Instrument and electrical engineers, process control engineers and system integrators who are likely to be participants in Hazops or who will be asked to engineer safety control systems
- Commissioning engineers and plant supervisors, process maintenance technicians.

   This book consists of a set of nine chapters each on a particular topic. A set of appendices have been placed after Chapter 9 to carry information that may be useful to support the reader in future use. Questions for practical exercises can be found after the appendices. Suggested answers follow after the set of questions. References, together with a list of suggested further reading are given in Appendix A to this module.

Useful websites

Websites are naturally available for any significant authority or supplier for safety studies. Some of the most useful web sites relevant to this workshop are listed in Appendix B to this module. In some places, the book includes extracts from material obtained from these web sites.

1

Introduction to hazard studies

1.1 Scope and objectives of this chapter

This chapter introduces the basic terms and concepts underlying risk management and describes well-established methods of quantifying risk. It then describes the concepts of tolerable risk and the principles of risk reduction. These include an introduction to the principles of risk ranking and the development of risk matrices. Then follows an overview of the typical legal requirements for implementation of safety principles in USA and in Europe. With this background in place, the chapter concludes by describing the range of hazard identification techniques, commonly used in industry, according to the application and the stage of the project work.

The objective of this chapter is to provide grounding in risk management principles and help participants to see the relevance of hazard studies to safety management.

When you have completed study of this chapter you should be able to:

- Provide an outline of the principles of risk management, risk assessment and risk reduction.
- Define what is generally meant by the terms hazard and risk, and explain the meaning of the term Alarp.
- Know how to design and use a risk matrix.
- Identify typical features of regulatory frameworks for risk management.
- Explain the differences between hazard identification and hazard analysis.
- Know of several methods to identify hazards.

1.2 Introduction to hazards and risk management

There is a common saying in the control systems world 'If you want to control something, first make sure you can measure it.' We need to control the risks of harm or losses in the workplace due to hazards of all forms. So what we need to measure is RISK. Here we need to be clear on the terms 'Hazard' and 'Risk'.

1.2.1 What is hazard and what is risk?

A hazard is 'an inherent physical or chemical characteristic that has the potential for causing harm to people, property, or the environment'.

In chemical processes, 'It is the combination of a hazardous material, an operating environment, and certain unplanned events that could result in an accident'.

Risk

'Risk is usually defined as the combination of the severity and probability of an event. In other words, how often can it happen and how bad is it when it does happen? Risk can be evaluated qualitatively or quantitatively.'

Roughly,

$$\text{Risk} = \text{Frequency} \times \text{Consequence of hazard}$$

Risk reduction

Risk reduction can be achieved by reducing either the frequency of a hazardous event or its consequences or by reducing both of them. Generally, the most desirable approach is to first reduce the frequency since all events are likely to have cost implications, even without dire consequences. Figure 1.1 illustrates how this principle applies in cricket.

If we can't take away the hazard we shall have to reduce the risk

This means: Reduce the frequency and / or reduce the consequence

Example:
Glen McGrath is the bowler. He is the Hazard
You are the batsman. You are at risk
Frequency = 6 times per over.   Consequence = ouch!

Risk = 6x ouch!

Risk reduction: Limit bouncers to 2 per over. Wear more pads.

Risk = 2x ouch!

Figure 1.1
An example for risk reduction in sport

Safety systems are all about risk reduction. If we can't take away the hazard we shall have to reduce the risk.

1.2.2   Safety management principles

It helps to look at the principles of risk management because they can be applied directly to safety management. Understanding risk management will show us how hazard studies and risk analysis activities fit in to the overall task of managing risk in a company. We shall then look at the principles of hazard identification, risk assessment and risk reduction, knowing how they all come together under risk management.

Why is this important?

Because you can do a better job with hazard studies and extract more value from the studies if you can see the point of them.

1.2.3 The meaning of safety management

What does safety management mean for a manufacturing plant or large item of equipment?

Safety management involves the provision of a safe working environment for all persons involved in the manufacturing process. It extends to cover the safety of the environment and the security of the business from losses.

The fundamental components of safety management will include:

- Having a systematic method of identifying and recording all hazards and risks presented by the subject plant or equipment
- Ensuring that all unacceptable risks are reduced to an acceptably low level by recognized and controllable methods that can be sustained throughout the life cycle of the plant
- Having a monitoring and review system in place that monitors implementation and performance of all safety measures
- Ensuring all departments and personnel involved in safety administration are aware of their individual responsibilities
- Responding to regulatory requirements from national and local authorities for the provision of adequate safeguards against harm to persons and the environment
- Maintaining a risk register and a safety case report that demonstrate adequate safety measures are in place and are being maintained at all times.

Safety management is effectively the same as the more general term, risk management, but applied specifically to risks associated with harm to persons, property or environment. Let us take a closer look at risk management principles to see what we can learn from them.

1.2.4 Risk management defined

Risk management is a very broadly used term and it is typically applied to business and organizational activities. The broad scope of this term can be seen in the definition of risk management taken from the Australian/New Zealand standard AS/NZ 4360 (Ref. 2 as listed in Appendix A) clause 1.3.24.

'Risk management – The culture, process and structure, which come together to optimize the management of potential opportunities and adverse effects.'

Industrialized countries encourage a culture of risk management in all enterprises and at all levels within an organization. The culture and processes of risk management are typically applied to business decision-making to achieve a logical balance between opportunities for growth and improvement on the one side and the potential for losses and failure on the other.

The application of risk management to occupational health and safety is just one of the many areas where the techniques are used. Let us look at a few basic processes in risk management to show how they match up to established or emerging methods in engineering systems.

The following notes are based on the guidance provided in the guideline document 'A basic introduction to managing risk' published as an Australian guideline HB 142-1999 by Standards Australia (Ref. 2 as listed in Appendix A).

Managing risk

- Requires rigorous thinking. It is a logical process, which can be used when making decisions to improve the effectiveness and efficiency of performance.
- Encourages an organization to manage pro-actively rather than reactively.
- Requires responsible thinking and improves the accountability in decision-making.
- Requires balanced thinking . . . 'Recognizing that a risk-free environment is uneconomic (if not impossible) to achieve, a decision is needed to decide what level of risk is acceptable'.
- Requires understanding of business operations carried on, where conformity with process will alleviate or reduce risk.

Hazard studies are part of the disciplined approach to managing risks in plant operations and they must be conducted in accordance with the principles shown here.

1.2.5   The process for managing risk

It turns out that the models suggested for managing risk are the same as those we find in the procedural models described for safety life cycle (SLC) activities that we shall be looking at later in this book. This is encouraging since it means that one procedural model fits all circumstances and no specialities are involved for safety. If the company recognizes risk management in its business, it should have no problem in understanding safety management.

Here is a diagram of a general risk management model (Figure 1.2) based on the version published in AS/NZS4360: 1999.

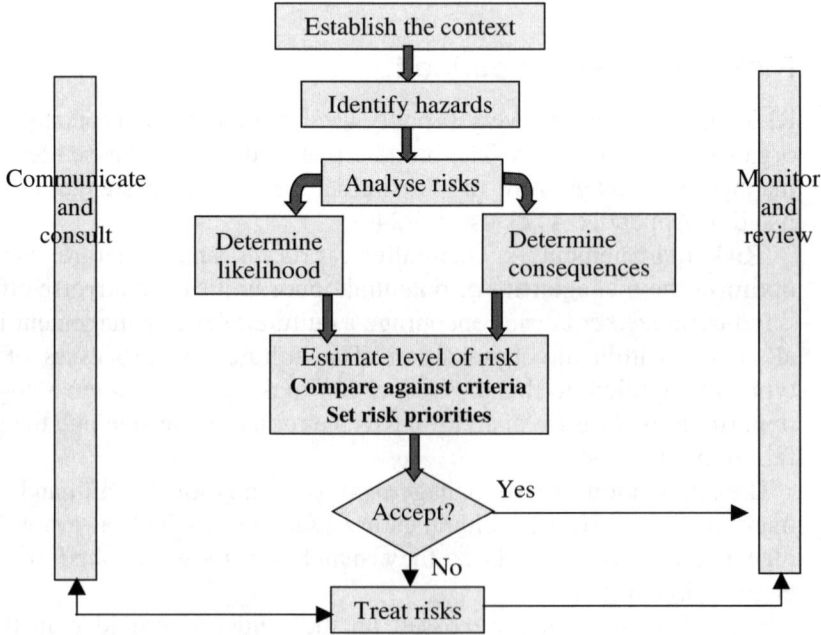

Figure 1.2
The process for managing risk

This model is intended to serve for all risk management activities within a company. These begin with strategic risk management applicable to the corporate planning levels where key business decisions can be subjected to risk evaluation and treatment. There are close parallels with the management of engineering risks and the management of functional safety. Let us examine the meaning of each step of the process.

Establishing the context

In business risk management, this first step involves wider issues that are out of our scope in this book. The context includes:

- *Strategic context:* The relationship between the organization and its environment, including financial, operational, competitive, political, social, cultural and legal aspects. In our field of work this would be typically defined by the organization's overall Safety Health and Environment (SHE) policy. It would also define the legal framework or regulatory compliance needs for the plant in question.
- *Organizational context:* Requires an understanding of the organization and its capabilities.
- *Risk management context:* Defining which part of the organization or which activities are in the scope. This would be the specific manufacturing plant or process under consideration.
- *Risk evaluation criteria:* Defines the criteria against which any risk is to be evaluated. We shall see that in our field this includes the so-called tolerable risk criteria for risks of harm to persons, environment and asset losses. Risk management and risk reduction cannot be conducted without some reference points for what is acceptable.
- *Structural context:* Deals with how the risk management process is to be handled and documented within the organization. Expect this to lead to a definition of who is responsible for the supply of information, conducting studies and managing the documentary records. In the case of SHE risk management, the documentary records are of critical importance and will require a quality management system.

Stakeholder identification

One of the problems experienced in hazard study work is that the participants are not always aware of all the parties that may be affected by the results of the studies. Obviously, decisions made by hazard study teams must be followed up by actions by the affected parties. Knowing who all the parties are and having their support is essential. What is a stakeholder? AS/NZS 4360 1999 1.3.31 'Stakeholders – Those people and organizations who may affect, be affected by, or perceive themselves to be affected by, the decision or activity'.

The risk management study must identify the stakeholders at the earliest opportunity. In process hazard studies the stakeholders are likely to include:

- Project managers, representing the business and shareholders
- Local authority regulators. Environmental protection officers. The NGOs for public participation processes
- Design engineers of relevant disciplines. Frequently, these will be process engineers, control and instrumentation specialists, electrical engineers

- Process and environmental safety officers. Fire prevention officers
- Commissioning engineers, Production managers
- Union or staff representatives. Safety officers
- Design contractors and equipment suppliers
- Risk insurance companies.

The challenge for a person managing a hazard study project must surely be to establish the relationships with stakeholders, such that all parties are satisfied with their level of involvement, whilst ensuring the study work avoids 'too many cooks'.

Identify hazards (alternative is risks)

With the context in place, the risk management model says, 'identify the risks'. The HB 142 guide raises the issue of 'perceptions of risk' and points out that 'perceptions of risk can vary significantly between technical experts, project team members, decision makers and stakeholders'.

In this book we have to take the 'technical experts' route to risks, as we shall see below. It is instructive to note that the layperson sees risk on a more personal and subjective scale.

'Lay persons are less accepting of risk over which they have little or no control (e.g. public transport versus driving one's own car), where the consequences are dreaded or the activity is unfamiliar.'

This is the stage where hazard studies are performed to answer the questions: 'What can happen, how can it happen.' The result is a list of hazards with the possible causes.

Analyze risks

The next step is 'Analyze the risk'. You can see from Figure 1.3 that it is necessary to establish a level of risk based on the criteria we mentioned earlier. The likelihood and the consequences must be found and multiplied together and applied to a scale of risk used to set priorities. In our field of work, this is the activity we call 'Hazard Analysis'. We are going to look at that again in a moment.

Evaluate risks

The next step is to compare the risk level with certain reference points (Figure 1.3) to decide if the risk level is acceptable or not.

If the risks are unacceptable the choice is to treat the risks or decide to avoid the risks altogether by doing something else.

The diagram introduces the concept of 'tolerable risk' or 'acceptable risk'. In business practice, the reference point for acceptable risks may depend on the company and its senior management. When it comes to safety and operability there is less room for flexibility. We are concerned with what is acceptable to society and our workers as a 'tolerable risk'. We are going to take a closer look at tolerable risk concepts in a few moments. Before that, let us look at the general-purpose model for risk treatment.

Figure 1.4 is informative for us in safety management because it demonstrates the options and decision that have to be considered during a hazard analysis and after a Hazard and operability (Hazop) study. In fact, this diagram covers all stages in the life cycle of the situation being considered. We shall see this theme recurring throughout the book. Let us consider the terms on the left-hand side of the diagram.

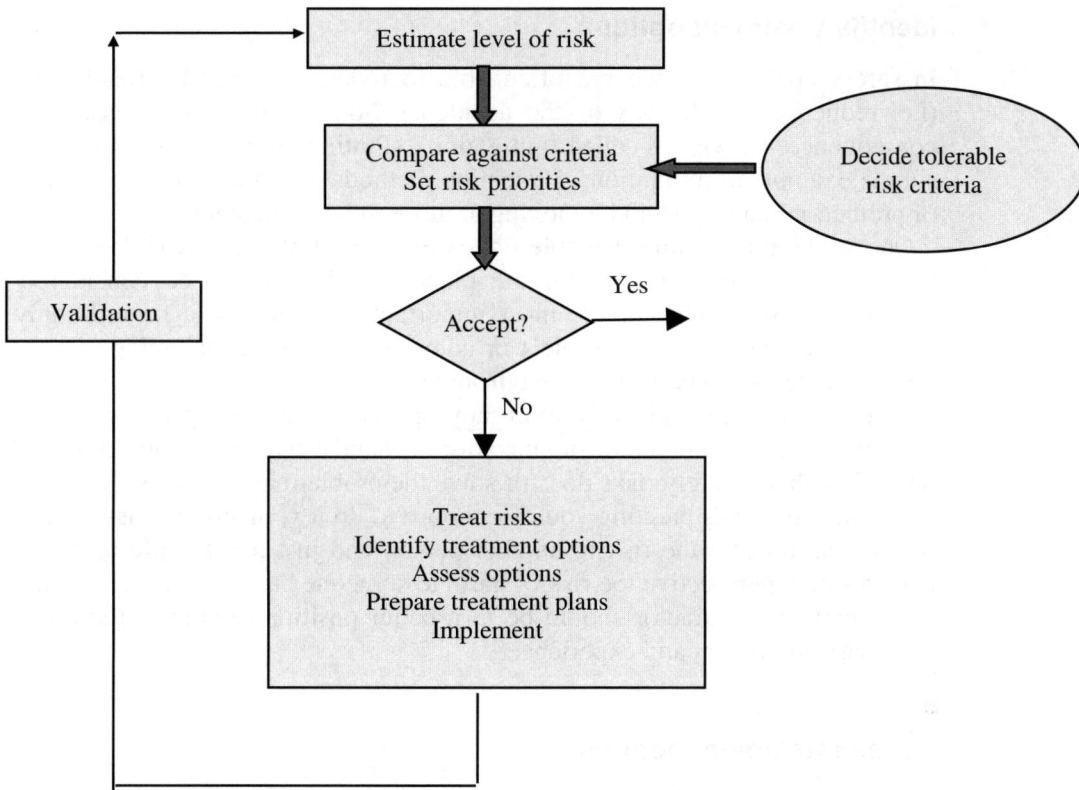

Figure 1.3
Evaluation of risk and the treatment stages

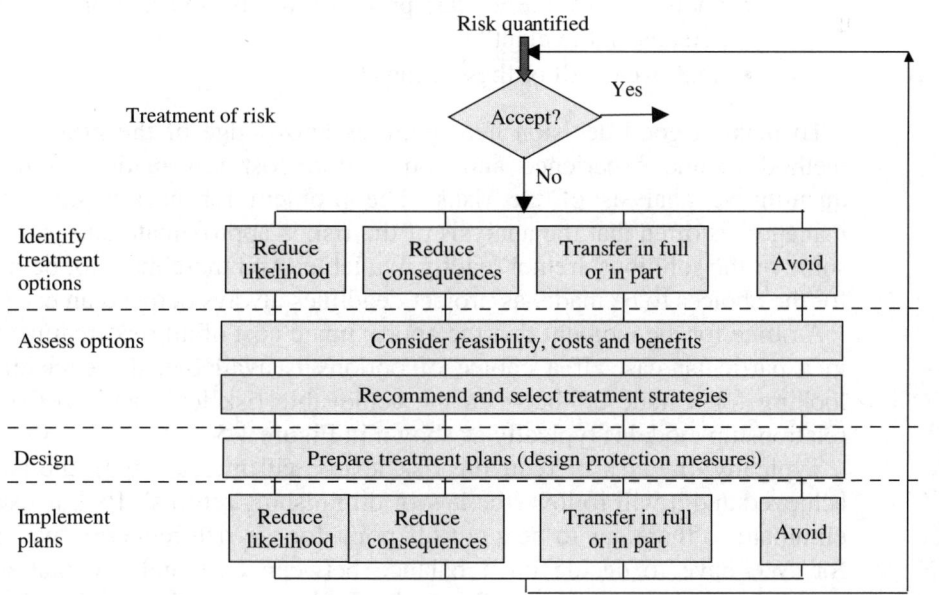

Figure 1.4
Details of treatment of risks (based on AS/NZS 4360: 1999 but modified for hazard studies)

Identify treatment options

In safety applications we are often able to reduce the risk by treating the likelihood (i.e. reducing the chances of the accident). Sometimes it is necessary to reduce the consequences by what is called 'mitigation'. (Putting on gas masks after a gas escape is a simple example of mitigation.) Protection methods to reduce risk are described as 'layers of protection' and we shall be looking at those in later chapters.

One solution to an unacceptable risk is to avoid it altogether. Unfortunately, this route sometimes implies not building the plant and this has to be considered along with all other options. One of the most important outcomes of a hazard study can be the decision to abort the whole project or adopt an alternative technology on the grounds of unacceptable risk to persons and environment.

Transfer of risk is generally more appropriate to business processes where it may be attractive to find some other organization to handle the risk in question. Subcontracting out of health and safety risks doesn't solve the problem in our cases. For example:

If your roof needs painting you can choose to do it yourself and risk falling off the roof or you can transfer the risk to another person and just pay for life insurance. However, from a safety perspective the risk of harm to someone has not been removed.

The painting contractor should be in a better position to control the risk if he has the necessary equipment and experience.

Assess treatment options

This is a very interesting stage of risk analysis. We have to consider feasibility, costs and benefits of the possible risk treatment options.

In the case of an engineering project the choices typically come down to:

- Shall we redesign the process to minimize hazard?
- Shall we provide alarms and trips to shutdown the process when the hazardous condition approaches?
- Shall we provide a blast-proof room and evacuation facilities to protect the persons on the plant?
- Shall we do all of these things?

To make a good decision here requires knowledge of the process and the protection methods, some experience and some good cost information. Someone has to do a quantitative analysis of the risks. The problem for hazard study teams and project managers is often that the analysis of the risk is approximate and the cost implications of some of the solutions are not readily available. And there may not be much time available for the choices to be made, as project deadlines always demand an early decision.

Assume for the moment that the approximate cost of all risk treatment options is known in a particular case. If a choice of options is available, the decision can be made by looking for a trade off between the achievable risk level and cost of achieving it. The relationship model is typically as shown in Figure 1.5.

Typically, the cost of reducing risk levels will increase with the amount of reduction achieved and it will follow 'the law of diminishing returns'. Risk is usually impossible to eliminate so there has to be a cut off point for the risk reduction we are prepared to pay for. We have to decide on a balance between cost and acceptable risk. This is the principle of Alarp, that we shall examine in the next section. Before that, let us look at the general-purpose model for risk treatment as shown in Figure 1.6 which is based on a diagram included in AS/NZ 4360:1999.

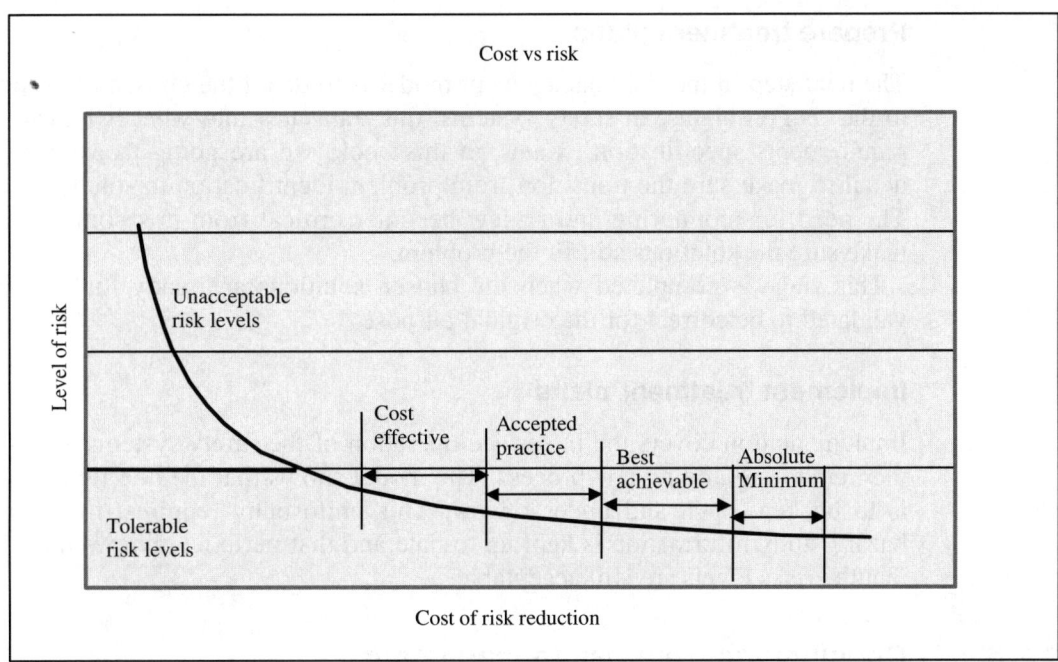

Figure 1.5
Risk reduction vs cost

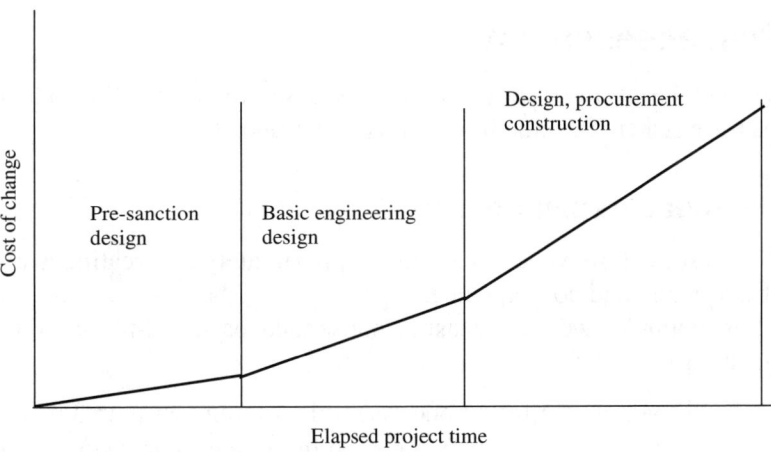

Figure 1.6
Cost of design changes against project time

The second factor that will influence the hazard study work is the relationship between design changes and their impact on project costs. There are heavy cost penalties involved in late design changes. Hence it pays to design the hazard study program to identify critical safety and operability problems at an early stage. This is where preliminary hazard study methods are valuable. Preliminary studies can often identify major problems at the early stage of design, where risk reduction measures or design changes can be introduced with minimum costs.

Prepare treatment plans

The next step in the risk management model is to detail the chosen or proposed solutions to the risk problems. In safety systems, this translates into what is known as the 'safety requirements specification'. Later, in this book, we are going to examine this stage in detail to make sure the transition from problem identification to solution works properly. The need for monitoring and review becomes critical from this point on, as we seek to make sure the solutions still fit the problem.

This stage is completed when the chosen solutions are ready for use and have been validated to be correct for the original purpose.

Implement treatment plans

Implementation covers the in-service operation of the safety systems and is supported by the monitoring and review process. The model shows that the question of acceptable risk is to be kept open and under review. This philosophy requires, for example, that the hazard study information is kept up to date and that periodic reviews must be held to see that the risks levels are still acceptable.

1.2.6   Conclusions from risk management

We have seen how the generalized models for risk management are directly applicable in safety management. When we look at the new application standards for safety instrumented systems (SIS) and for alarms, we shall recognize the same principles being applied. Hazard studies are an integral part of the process.

1.3   Risk assessment

Let us take a closer look now at risk assessment and find out how we may be able to measure risk and decide if it needs to be reduced.

1.3.1   The measurement of risk

Risk is something we can measure approximately by creating a scale based on the product of frequency and consequence.

For example, we can measure consequences in terms of injury to persons. Here is a quantitative scale:

- Minor – Injury to one person involving less than 3 days absence from work
- Major – Injury to one person involving more than 3 days absence from work
- Fatal consequences for one person
- Catastrophic – Multiple fatalities and injuries.

Likewise, the frequency or likelihood of an event causing injury can also be placed on a scale. For example here is a qualitative scale (descriptive but does not define numbers):

- Almost impossible
- Unlikely
- Possible
- Occasionally
- Frequently
- Regularly.

Alternatively, frequency can be placed on a quantitative scale. This would simply be the event frequency in events per year. For example:

- One hazardous event occurring on the average once every 10 years will have an event frequency of 0.1 per year.
- A rate of 10^{-4} events per year means that an average interval of 10 000 years can be expected between events.

Another alternative is to use a semi-quantitative scale or band of frequencies to match up words to frequencies. For example:

- Possible = Less than once in 30 years
- Occasionally = More than once in 30 years but less than once in 3 years
- Frequently = More than once in 3 years
- Regularly = Several times per year.

Once we have these types of scales agreed, the assessment of risk requires that for each hazard we are able to estimate both the likelihood and the consequence. For example:

- Risk item no. 1 – 'Major' injury likely to occur 'Occasionally'
- Risk item no. 2 – 'Minor' injury likely to occur 'Frequently'.

Whilst both of the above items are undesirable, we cannot yet tell which of them is the most important problem in need of risk reduction, or even if they need any reduction at all. What we need is a system of comparative values for risk.

1.3.2   Introducing the risk matrix

From the above, it is clear that a scale of risk can be created from the resulting products of frequency and consequence. One popular way to represent this scale is by means of a simple chart that is widely known as a risk matrix. Here are some examples.

Risk matrix example 1

Figures 1.7 and 1.8 are examples of simple risk matrices where frequency of the possible event is ascending on the *Y*-axis and the consequence categories are ascending on the *X*-axis.

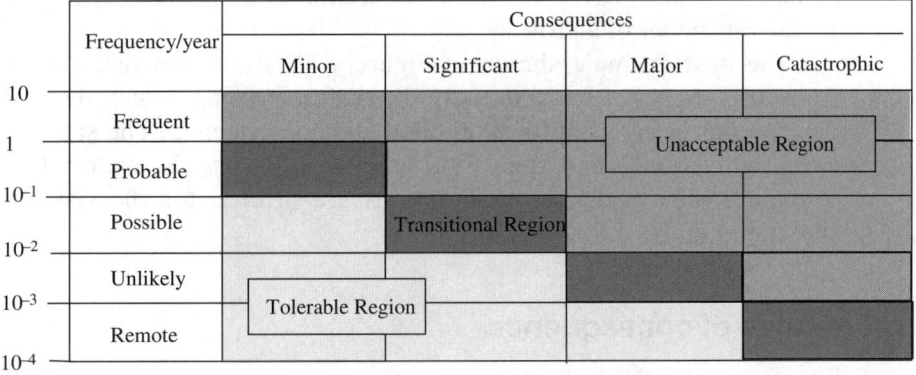

Figure 1.7
Risk matrix tolerability bands

When the product of frequency and consequence is high, the risk is obviously very high and is unacceptable. The unacceptable region extends downwards towards the acceptable region of risk as frequencies and/or consequences are reduced. The transitional region, as shown in the diagram, is where difficult decisions have to be made between further reduction of risk and the expenditure or complexity needed to achieve it. Our diagram shows some attempt at quantifying the frequency scale by showing a range of frequencies per year for each descriptive term. This is usually necessary to ensure some consistency in the understanding of terms used by the hazard analysts.

Some companies go a step further and assign scores or values to the descriptions of frequency and consequence. This has the advantage of delivering risk ranking on a numbered scale, allowing some degree of comparison between risk options in a design. Figure 1.8 shows a possible score and ranking values on the same risk matrix.

Frequency/year	Consequences			
	Minor: 1	Significant: 3	Major: 6	Catastrophic: 10
Frequent: 10	6	10	60	100
Probable: 8	8	24	48	80
Possible: 4	4	12	24	40
Unlikely: 2	2	9	12	20
Remote: 1	1	3	6	10

Frequency scale (left axis): 10, 1, 10^{-1}, 10^{-2}, 10^{-3}, 10^{-4}

Figure 1.8
Risk matrix showing ranking and tolerability bands

The scoring system adopted in Figure 1.8 is an arbitrary scheme devised to suit the tolerability bands as best as possible. Each company and each industry sector may have its own scoring system that has been developed by experience to provide the best possible guidelines for the hazard study teams working in their industry. There does not appear to be any consensus on a universally applicable scoring system but the ground rules are clear. The scales must be proportioned to yield consistently acceptable results for a number of typical cases. Once the calibration of a given system is accepted, it will serve for the remainder of a project.

In the next example shown in Figure 1.10, there is a risk classification chart taken from IEC 61508 (the standard we shall be using later). Here we see qualitative descriptions being used for likelihood and consequence. The standard uses this diagram as a sample only and does not suggest any specific values be adopted from this example. Some specimen descriptions are offered for the risk rank classes I–IV as shown in Figures 1.9 and 1.10.

1.3.3   Scales of consequence

The format of the risk matrix allows companies to set down their interpretations of consequences in terms of losses to the business as well as harm to the environment and harm to persons.

Frequency	Consequences			
	Catastrophic	Critical	Marginal	Negligible
Frequent	I	I	I	II
Probable	I	I	II	III
Occasional	I	II	III	III
Remote	II	III	III	IV
Improbable	III	III	IV	IV
Incredible	IV	IV	IV	IV

Figure 1.9
Risk classification of accidents: Based on table B1 of IEC 61508-5

Risk class	Interpretation
Class 1	Intolerable risk
Class II	Undesirable risk. Tolerable only if risk reduction is impracticable or if the costs are grossly disproportionate to the improvement gained
Class III	Tolerable risk if the cost of risk reduction would exceed the improvement gained
Class IV	Negligible risk

Figure 1.10
Risk classification of accidents: Table B2 of IEC 61508-5 (interpretation of risk classes)

Thus, a 'significant' consequence may equate to a financial loss due to:

- Loss of quality or contamination of product
- Lost production time
- Damage to plant and repair cost
- Failure to deliver/loss of market.

The company will set up a scale of loss in terms of plant damage and lost production
For example:

Minor	Critical	Severe	Catastrophic
Short-term loss of production	Damage to machines. Repairable in short time	Damage to plant. Major repair costs. Serious loss of production	Substantial damage to plant. Potential loss of overall plant

A similar table can carry a scale of environmental damage:

Minor	Critical	Severe	Catastrophic
Temporary excursion in emission levels	Significant release. Effluent clean up required	Ecological damage for up to 1 year. Risk of penalties	Ecological damage for more than 1 year. Pressure to cease business

Let us add a scale of harm to personnel:

Minor	Critical	Severe	Catastrophic
Reportable but non-disabling injuries causing over 3 days absence	Disabling injury or severe injury requiring extensive recovery. 1 in 10 chance of fatality	Critical injuries, and possibly 1 fatality	One or more fatalities

Integrating these scales will be helpful for comparative purposes but sometimes leads to unwanted conclusions. The values of business loss, loss of life and severe damage to the environment do not really have direct equivalence. The scales may be supported by a price tag, for example:

Minor	Critical	Severe	Catastrophic
Less than $US 10 000	Up to $US 100 000	Up to $US 1 M	More than $US 1M

Warning note

Integrated scales can be informative, but we must be very careful how the scale is used in risk assessment work. Remember that many hazardous events will have the potential for consequence on two or more of the categories, persons, environment and assets. Hence, each event must be recorded for its effect on all three categories individually. We may succeed in reducing the risk in one category, but we must always check the risk level in the other categories.

Sometimes, it requires three risk matrices for each hazard.

Progress check

Let us stop at this point and recall what we were looking for at the start of this exercise. We wanted to achieve a comparative scale of risk so that we can know which problems need the most attention for risk reduction.

Risk matrix does this for us. In fact, it does three valuable things:

- It tells us how each risk compares with another, so we can find the highest priorities for attention.
- It tells us which risks are totally unacceptable and shows which risks may be acceptable.
- It guides us to how much risk reduction will be needed to make the risk tolerable.

However, there seem to be two problems here that need to be sorted out:

- *Problem 1:* Where are the boundaries for the tolerable risk zone? Who defines the risk graph? Who defines the tolerable risk band?
- *Problem 2:* How far down risk the scale is good enough for my application?

These problems bring us to issues of tolerable risk and deciding how much risk reduction is justified.

1.4 Concepts of Alarp and tolerable risk

To deal with the above problems, let us first recall the principles of risk reduction, as discussed earlier. It helps to use Figure 1.11 to show what we are doing in safety systems.

The reduction of risk is the job of protection measures. In some cases, this will be an alternative way of doing things or it can be a protection system such as an SIS. When we set out to design a protection system, we have to decide how good it must be. We need to decide how much risk reduction we need (and this can be one of the hardest things to agree on). The target is to reduce the risk from the unacceptable to at least the tolerable. The concept of tolerable risk is part of the widely accepted principle of Alarp.

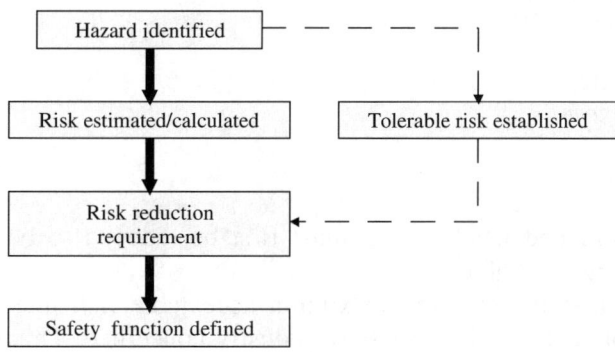

Figure 1.11
Risk reduction design principles

The Alarp (as low as reasonably practicable) principle recognizes that there are three broad categories of risks:

- *Negligible risk:* Broadly accepted by most people as they go about their everyday lives, these would include the risk of being struck by lightning or of having brake failure in a car.
- *Tolerable risk:* We would rather not have the risk but it is tolerable in view of the benefits obtained by accepting it. The cost in inconvenience or in money is balanced against the scale of risk, and a compromise is accepted.
- *Unacceptable risk:* The risk level is so high that we are not prepared to tolerate it. The losses far outweigh any possible benefits in the situation.

Represented by the Alarp diagram in Figure 1.12.

The width of the triangle represents risk, and hence, as it reduces, the risk zones change from unacceptable through to negligible. Clearly this is following the same principle that we saw earlier in the risk management section. The hazard study and the design teams for a hazardous process or machine have to find a level of risk that is as low as reasonably practicable in the circumstances or context of the application. The problem here is 'How do we find the Alarp level in any application?'

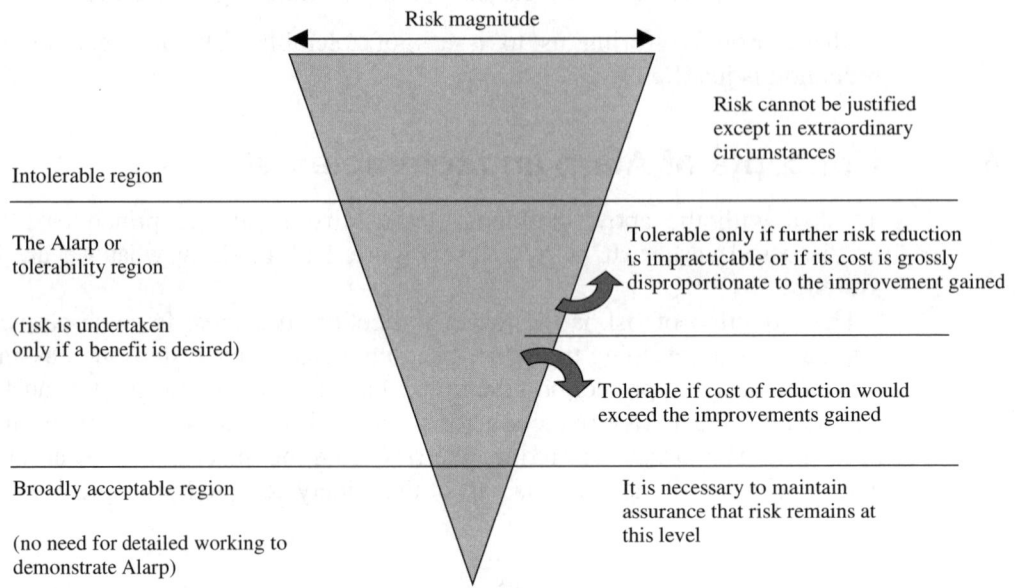

Risk magnitude

Intolerable region

Risk cannot be justified except in extraordinary circumstances

The Alarp or tolerability region

Tolerable only if further risk reduction is impracticable or if its cost is grossly disproportionate to the improvement gained

(risk is undertaken only if a benefit is desired)

Tolerable if cost of reduction would exceed the improvements gained

Broadly acceptable region

(no need for detailed working to demonstrate Alarp)

It is necessary to maintain assurance that risk remains at this level

Figure 1.12
Alarp diagram

Step 1

The estimated level of risk must first be reduced to below the maximum level of the Alarp region at all costs.

This assumes that the maximum acceptable risk line has been set as the maximum tolerable risk for the society or industry concerned. This line is hard to find, as we shall see in a moment.

Step 2

Further reduction of risk in the Alarp region requires cost benefit analysis to see if it is justified. This step is a bit easier and many companies define cost benefit formulae to support cost justification decisions on risk-reduction projects.

The principle is simple 'If the cost of the unwanted scenario is more than the cost of improvement the risk reduction measure is justified'.

The tolerable risk region remains the problem for us. How do we work out what is tolerable in terms of harm to people, property and environment?

1.4.1   Establishing tolerable risk criteria

This is a complex question that has exercised many minds and companies over the years. There is much debate on the issue. The UK HSE has published material and reports on this subject over the years and in this book we can take some guidelines from a published paper by Edward Marzal at Exida.com called 'Tolerable Risk Guidelines'.

Marzal has investigated the different ways in which risk can be expressed and has developed some interesting conclusions. He identifies two types of risk criteria that are best suited to risk reduction design in industries, such as the process industry. These are 'Individual risk' and 'Risk integrals'.

Individual risk

This is the frequency at which an individual may be expected to sustain a given level of harm from the realization of specified hazards. The most commonly used example is:

> *Probability of fatality per year:*   Usually this is applied to the most exposed individual on a plant. It does not indicate how many persons will die in an accident, but it does provide a comparative value for risk in any situation.

Risk integrals

These calculations are the sum of risk indices obtained from 'Frequency of accident × Consequence'. These provide a measure of the total risk presented by a given plant, taking into account all the risks it presents.

$$\text{Risk Integral} = \sum_{i=1}^{n} C_i F_i \qquad \text{(meaning that all for a given plant are summed)}$$

Examples are:

- *Probable Loss of Life (PLL):*   Number of fatalities × frequency of event
- *Fatal accident rate (FAR):*   Number of fatalities per 10^8 h worked at the site where the hazard is present.

All of the above measures can be used for assisting with Alarp decisions and for relating them to risk reduction design. Generally, in trip and alarm (or SIS) systems, we are concerned with achieving risk reduction by reducing the frequency component of risk. Hence, if the risk assessments and tolerable risk criteria lead to frequency targets, they are easy for us to use.

1.4.2   Using individual risk as a guide to tolerable risk

Marzal examined risk management guidelines in several countries and found that some countries have government-based risk tolerance criteria but most countries do not have them. In general, the risk criteria are defined by individual companies.

The evidence from government material examined by Marzal included the following individual risk values:

- **Lower ALARP boundary for a worker in UK: 1×10^{-5}**
- **Lower ALARP boundary for a worker in UK: 1×10^{-6}**
- **Lower ALARP boundary for public in Netherlands: 1×10^{-8}**
- **Upper ALARP boundary for a worker in UK: 1×10^{-3}**
- **Upper ALARP boundary for public in UK: 1×10^{-4}**
- **Upper ALARP boundary for public in Hong Kong: 1×10^{-5}**
- **Upper ALARP boundary for public in Netherlands and NSW, Australia: 1×10^{-6}**

To illustrate the meaning of these scales:

Assume the target for worst-case risk to a member of the public in UK is 1×10^{-4}. Take the example of an accident with a 10% chance of causing a fatality. Calculate the highest event frequency that would be considered tolerable.

$$F_{max} = \frac{\text{Max.fatal risk frequency}}{C} = \frac{1 \times 10^{-4}}{0.1} = 10^{-3} \text{ events/year}$$

i.e. It should not be tolerated if its event frequency is higher than 1 in 1000 years.

On the same basis, if the single event were to cause 10 fatalities, this criteria would require the target frequency to be below 1 in 100 000 years.

Edward Marzal points out that the USA is specifically opposed to setting tolerable risk guidelines arguing that they are open to misapplication due to uncertainty about the nature of risks and the population numbers exposed to each risk, amongst other factors. The USA achieves a very good safety record and this is attributed to 'the flexibility to apply capital where it will produce the most benefit and the unrestricted ability of the free market to determine third party liability costs'.

Marzal concludes that many companies are now finding that where a financial basis for the risk-reduction project is calculated, the results always justify the greatest amount of risk reduction. For example, a 1995 report by Mudan found that 'risk due to third party liability of personnel injury is insignificant when compared to other losses such as property damage and business interruption'.

It is not clear whether this philosophy would be successful in the case of environmental damages but it is a significant point to keep in mind.

1.4.3   Using injury statistics as a guide to tolerable risk

Another way to arrive at tolerable risk targets is to examine accident statistics and see how various countries and types of industries normally perform.

If we can obtain a consistent method of measuring accident rates then we can see how we are doing in comparison to others. The following data from European studies show how different countries measure up on an approximately consistent scale of measurement. The scale of measurement for injury rates is based on relating accident records to the number of workers employed in an industry.

Workplace injury in Europe and the USA in 1996

This table shows the rates of fatal and over 3-day injury per 100 000 workers or employees.

Country	Rate of Fatal Injury	Rate of Over 3-Day Injury	Employed People Covered
EU average	3.6	4200	
Great Britain	1.9	1600	Workers
Sweden	2.1	1200	Workers
The Netherlands	2.7	4300	Employees
USA	2.7	3000	Workers
Germany	3.5	5100	Workers
Italy	4.1	4200	Workers
Spain	5.9	6700	Workers
Portugal	9.6	6900	Employees

The above and following data are taken from an HSE report issued 27 September 2000 and is based on an Eurostat publication 'Accidents at Work in the EU in 1996 – Statistics in FOCUS, Theme 3–4/2000'. HSE supplied data on USA and the Netherlands from their own studies.

Industry sectors for the EU average and Great Britain in 1996

This table shows the rates of fatal and over 3-day injury per 100 000 workers.

Industry Sector	EU Average		Great Britain	
	Fatal	Over 3 days	Fatal	Over 3 days
Construction	13.3	8000	5.6	2700
Agriculture	12.9	6800	10.8	2000
Transport, storage and communication	12.0	6000	1.2	2400
Electricity, Gas and Water	5.7	1600	1.4	1700
Manufacturing	3.9	4700	1.4	2200
Within manufacturing				
Food, beverage, tobacco	4.7	6600	0.9	Na
Non-metallic mineral products	8.1	6500	1.4	Na
Basic metals and metal products	7.7	8500	2.1	Na
Wood and wood products	8.5	10 800	4.9	Na
Wholesale, retail trade and repairs	2.5	2400	0.4	1300
Financial, real estate, business	1.6	1600	0.3	700
Hotels and restaurants	1.1	3500	0.3	1500

Eurostat's results and a study by HSE show that the rate of fatal injury in Great Britain is one of the lowest in Europe, and is lower than the USA. Other charts in this series allow comparisons between individual countries.

Note that the data here identifies significant injuries causing over 3 days of absence from work. This data may be useful for evaluating tolerable rates for accidents of lower severity levels.

There are two issues to be considered here:

- Are these accident rates to be considered tolerable as targets for risk? Or do they represent the result of setting targets well below these achieved figures?
- How do we covert these figures into practical target rates for risk reduction?

The figures can be related to the scale of 'Probability of fatality per year'. If the EU average fatality rate per 100 000 workers is 3.6, it converts to 3.6×10^{-5} per average worker. How many times greater risk is faced by the most exposed worker?

If we compare this with a PLL of 1×10^{-3} for a worst-case site target in UK, we find that the worst case is approximately 30 times more risky than the average worker experiences. If we drop the target to 1×10^{-4} it will be 3 times worse than the achieved rate. This begins to look like a reasonable risk target for the most exposed worker in a hazardous process plant.

1.4.4   Fatal accident rate

This is another 'risk integral' method of setting a tolerable risk level. If a design team is prepared to define what is considered to be a target FAR for a particular situation it becomes possible to define a numerical value for the tolerable risk. Whilst it seems a bit brutal to set such targets, the reality is that certain industries have historical norms and have targets for improving those statistical results.

The generally accepted basis for quoting FAR figures is the number of fatalities per one hundred million hours of exposure. This may be taken as the fatalities per 10^8 worked hours at a site or in an activity, but if the exposure is limited to less than all the time at work, this must be taken into account.

Very roughly, 1 person working for 50 years will accumulate 10^5 working hours or 50 people will be working for 1 year. So, 1000 people over 50 years will accumulate 10^8 h of exposure.

You can see from the following table that this scale of measurement allows some comparisons to be made between various activities.

Activity	FAR per 10^8 Exposure Hours
Travel	
Bus	4
Car	50–60
Occupation	
Chemical Industry	4
Manufacturing	8
Shipping	8
Coal-mining	10
Agriculture	10
Boxing	20 000
Rock-climbing	4000
Staying at Home	1–4

Fatal accident rate can be used as basis for setting the tolerable rate of occurrence for a hazardous event. For example:

Suppose a plant has an average of 5 persons on site at all times and suppose that 1 explosion event is likely to cause 1 person to be killed. The site FAR has been set at 2.0×10^{-8}/hr.

We can calculate the minimum average period between explosions that could be regarded as tolerable, as follows:

$$\text{Fatality rate per year} = (\text{FAR/hour}) \times (\text{hours exposed/year})$$
$$= (2 \times 10^{-8}) \times (5 \times 8760)$$
$$= 8.76 \times 10^{-4}$$
$$\text{Average years per explosion} = \frac{1}{8.76} \times 10^{-4} = 1140 \text{ years}$$

If, for example, the risk analysis predicts an explosion rate of 1 event per 30 years, the risk will be unacceptable and the frequency of explosion will have to be reduced by at least 1140/30 = 38 times lower to be tolerable. This amount of risk reduction therefore defines the minimum requirements for the safety system in this application.

1.4.5 Tolerable risk conclusion

The indications are that many companies determine tolerable risk targets using consensus from the types of statistics we have been looking at. Marzal concluded that the range of PLL values in industry is still a wide one from 10^{-3} to 10^{-6} for the upper level. We must also remember to allow for the effect of multiple hazard sources. It appears that financial cost benefit analysis often justifies greater risk reduction factors than the personal or environmental risk criteria. We shall revisit this issue when we come to safety integrity level (SIL) determination practices later in this book.

1.4.6 Practical exercise

Now is good time to try practical Exercise No. 1, which is set out towards the back of the manual after the appendices. This exercise demonstrates the calculation of individual risk and FAR, and uses these parameters to determine the minimum risk reduction requirements.

1.5 Regulatory frameworks and examples from EU and USA

In the early stages of any project involving potential hazards the question of regulations is bound to arise: Where do we stand with regard to the legal requirements for safety? What does the law require us to do? Are there any safety targets that we are legally required to meet? The simple answer is that most industrialized counties have legal frameworks in place that are similar in nature and have been substantially improved in the past 10 years. Safety regulation now emphasizes the need for a complete safety management system to try to deal with the fact that around 90% of accidents can be traced back to failures to manage the various aspects of safety from identification of hazards through to training and continued monitoring of safety performance.

1.5.1 Legal requirements for hazard studies

The scope of legal requirements is far too large in this book but we are particularly interested in knowing the following:

- What are the legal requirements for hazard studies to be done?
- How often must they be done?
- What sort of study is acceptable?
- What sort of reporting is required?

Our approach in this book is to pick out the general principles that are commonly seen in regulations in the USA and in Europe. These provide a good indication of what one should expect to be doing to satisfy good practices anywhere. Each project will need to decide at the outset which occupational health and safety (OHS) legislation is applicable and then decide on the scope of hazard studies and reporting requirements needed to satisfy the authorities.

In the hazard study life cycle these questions must be answered in the first of the preliminary hazard studies (hazard study 1, see next chapter). In this book, we outline some of the generally observed principles whilst, in Appendix C we have added some information specific to some countries as it becomes available.

1.5.2 Legal requirements for safety instrumentation

Similar questions arise with the provision of safety solutions. Where there are well-established safety standards available (e.g. for many common types of machines), the regulations usually require compliance or will accept conformity to an approved standard. This is where Harmonized European Standards are used.

When it comes to process safety systems such as alarms and trips the solutions cannot be directly prescribed and very few direct application standards exist beyond boiler and furnace safety measures. However, the new international standards for functional safety, IEC 61508 and IEC 61511, provide a comprehensive method of applying instrumented safety systems. These standards are beginning to be used by legislators as references for demonstrating that risks have been reduced to an acceptable level and that a suitable regime is being maintained for functional safety.

1.5.3 International trends in safety practices

In most countries OHS regulations lay down basic requirements for employers to safeguard their workers and public from harm. The overall OHS requirements are typically supplemented by additional regulations that target particular sectors of industry where significant problems with hazards are known. Figure 1.13 shows the characteristic structure seen in USA, Europe, South Africa, Australia and other countries.

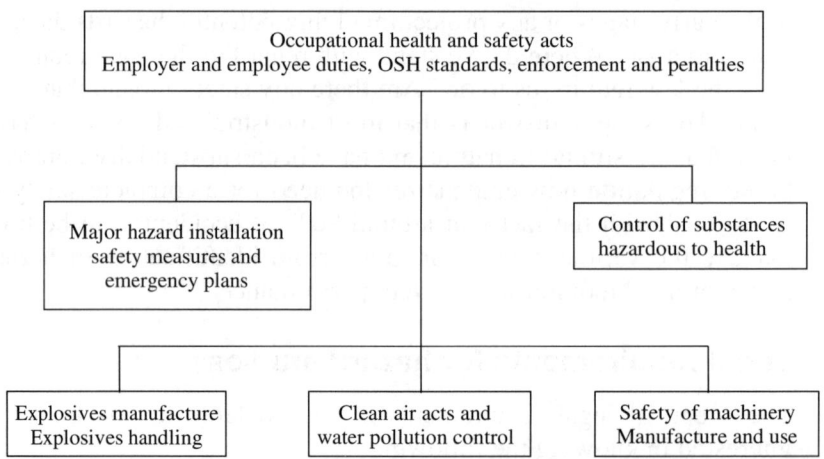

Figure 1.13
Typical structure of health and safety regulations for industry

The OHS types of regulations usually require that a risk assessment be carried out on the occupations and processes at the place of work. They normally require a reporting and review system to assist regulatory oversight.

Specific regulations have been generated for particular types of industry that supplement the basic OHS requirements. For example, the principal regulations affecting the chemical industries in the USA are:

- OSHA regulations for 'Process Safety Management of Highly Hazardous Chemicals and Blasting Substances' (known as the PSM rule) (29 CFR 1910.119)
- USA: Clean Air Act: EPA-40 CFR Part 68: Accidental Release Prevention Requirements. Risk Management Program. Referred to as the 'RMP Rule'.

The PSM rule was an improvement over earlier safety regulations and was driven by the realization that major hazard potentials at plants were not being managed to adequate standards in some areas. The main driving force was said to be the Pasadena, Texas incident.

Pasadena incident

- Petrochemical plant producing polyethylene at Pasadena, Texas
- *October 1989:* Release of isobutene, ethylene and catalyst carrier during routine maintenance on a reactor. Vapor cloud ignited after 1 minute with equivalent explosion energy of 10 t of TNT
- *Consequence:* 23 killed, 130 injured. Damage costs: approximately $750 million.

The widespread application of the PSM and RMP rules means that Process Hazard Analysis (PHA) is an essential technique for very many companies in the USA. In particular, the more critical process plants will be most likely to employ detailed Hazop procedures as the routine method for assisting them to comply with the regulations.

1.5.4　European regulations

In Europe, the major hazard regulations are derived from the Seveso II directive (96/82/EEC) and its amendments. The directive originates from the Seveso 1 directive that was introduced following the disastrous events at Seveso in Northern Italy.

Seveso 1

- Plant manufacturing pesticides and herbicides. On 10 July 1976 a dense vapor cloud containing tetrachlorodibenzoparadioxin (TCDD) was released from a reactor, used for the production of trichlorofenol. Commonly known as dioxin, this was a poisonous and carcinogenic by-product of an uncontrolled exothermic reaction.
- Kilogram quantities of the substance lethal to man, even in microgram doses, were widely dispersed which resulted in an immediate contamination of some ten square miles of land and vegetation. More than 600 people had to be evacuated from their homes and as many as 2000 were treated for dioxin poisoning.

The first Seveso directive was later revised and extended, again stimulated by accidents such as Bhopal, India 1984 and Basel, Switzerland, 1986. The current version is known as the Seveso II directive. The following information is from the EU's Major Accident Hazards Bureau:

'The Seveso II Directive has fully replaced its predecessor, the original Seveso Directive. Important changes have been made and new concepts have been introduced into the Seveso II Directive. This includes a revision and extension of the scope, the introduction of new requirements relating to safety management systems, emergency planning and land-use planning and a reinforcement of the provisions on inspections to be carried out by Member States!'

Outline of Seveso II

The Seveso II Directive sets out basic principles and requirements for policies and management systems, suitable for the prevention, control and mitigation of major accident hazards.

Establishments that have the potential for major accidents are required to comply with the requirements of the directive in the form of national laws that are passed to enact the EU directives. The establishments are classed into 'lower tier' and 'upper tier' according to the size of inventories and the size of the plant.

Lower tier establishments are to draw up a Major Accident Prevention Policy (MAPP), designed to guarantee a high level of protection for man and the environment by appropriate means, including appropriate management systems, taking account of the principles contained in Annex III of the Directive.

Upper tier establishments (covered by Article 9 of the Directive and corresponding to a larger inventory of hazardous substances) are required to demonstrate in the 'safety report' that an MAPP and a Safety Management System (SMS) for implementing it have been put into effect, in accordance with the information set out in Annex III of the Directive.

Here is a very abbreviated description of some of the requirements of the directive but for detailed information we suggest reference to the website and guidance source listed in Appendix B. Guidance material is freely available from the EU website and in the form of free leaflets and purchased books from the HSE books website.

1.5.5   Requirements for a safety management system

The Seveso II directive emphasizes the need for an SMS to protect against major hazard installations. Text from the directive:

'The safety management system should include the part of the general management system which includes the organizational structure, responsibilities, practices, procedures, processes and resources for determining and implementing the major accident prevention policy.'

It is recognized that the safe functioning of an establishment depends on its overall management. Within this overall management system, the safe operation of an establishment requires the implementation of a system of structures, responsibilities and procedures, with the appropriate resources and technological solutions available. This system is known as the SMS.

Development of an MAPP

The directive states:

'The major accident prevention policy should be established in writing and should include the operator's overall aims and principles of action with respect to the control of major accident hazard.'

Activities in support of the SMS are defined in the directive. These include:

- *Organization and personnel:* Roles and responsibilities of personnel, identification of training needs and the provision of training. The operator should identify the skills and abilities needed by such personnel, and ensure their provision.
- *Hazard identification and evaluation:* Includes procedures to systematically identify and evaluate hazards, define measures for the prevention of incidents and mitigation of consequences.
- *Operational control:* Documented procedures to ensure safe design and operation of the plant. Safe working practices should be defined for all activities relevant for operational safety.
- *Management of change:* Operating company should adopt procedures for planning and controlling all changes in people, plant, processes and process

variables, materials, equipment, procedures, software, design or external circumstances which are capable of affecting the control of major accident hazards.

- *Planning for emergencies:* An emergency plan is required.
- *Monitoring performance:* The operator should maintain procedures to ensure that safety performance can be monitored and compared with the safety objectives defined.
- *Audit and review:* Independent audit of the organization and its processes. Management to keep its SMS under review for essential correction or changes.

The above principles have been transferred into national laws in member states of the EU. So in the UK, for example, the directive is implemented as the Control of Major Accident Hazards (COMAH) regulations and has been in force since February 1999. The two-tier reporting requirements are defined, as per the directive. Additionally, all hazardous chemical and other substances used in industry are subject to the Control of Substances Hazardous to Health (COSHH) regulations, 1994.

1.5.6 What can we conclude from this overview?

- It should be clear that an SMS incorporating all the above measures is the underlying principle for delivering safety in the workplace.
- We find that the same principles and practices reappear in most of the recent safety legislations.
- In industries or plants where lower levels of accident potential are to be found, the rules become simplified and the organizational burden is reduced but the principles remain basically the same.
- In all cases, where risk to persons are involved, the regulations require an identification of hazards and an assessment of risk.

1.5.7 Regulations summary

In Appendix B of this module we have compiled a brief summary of some of the key regulatory instruments in various countries that may entail the use of hazard study methods. IDC will continue to add to this summary, as workshop experience dictates.

The trend in international safety practices is to move away from prescribed solutions to safety problems in favor of allowing individuals to carry out assessments of risk followed by risk-reduction measures appropriate to the problem. Independent but approved assessment bodies are available to carry out conformity assessments. Their reports are then used to show to the authority that a company is complying with the requirements of the law.

1.5.8 Conclusion on hazard study legal requirements

Following up on the questions raised at the start of this section:

What are the legal requirements for hazard studies to be done? They are part of the SMS procedures. These must be done to satisfy the major accident hazard regulations in USA, UK and other countries where similar laws are applied.

How often must they be done? Hazard studies are done at the design and planning stages and then kept under review. Periodic reviews of existing hazard studies are part of the mandatory review procedures built into SMS. Some countries define mandatory review intervals. The PSM rule in USA requires companies to update or revalidate their PHAs at least every 5 years, the period in RSA has been set as every 3 years.

What sort of study is acceptable? This is not clearly defined but the overall requirement remains that the method used must be sufficient to ensure confidence that most hazards have been identified and the study must be good enough to satisfy an independent auditor.

What sort of reporting is required? Reporting of results must be suitable for auditing and for revalidation studies. For example, exception-only reporting may not be acceptable to show an auditor that all possibilities have been considered.

1.6 Methods of identifying hazards

Now that we have established the widespread requirements for hazard study and analysis in many sectors of industry, we should catalog the best-known methods and get to know where they can be most effectively used.

This section outlines some of the methods available for the identification of hazards as distinct from the assessment or analysis of hazards. We separate these two activities as they are usually performed in separate steps. One person often does risk analysis after a team has identified the problems; sometimes the analysis is done ahead of the Hazop when the hazards are obvious.

1.6.1 Terminologies for hazard studies

Process hazard analysis is a general term used (particularly in USA) to describe the tasks of identification of hazards and the evaluation of risks in the process industries. Within the range of PHA activities we find two main stages: Hazard Identification and Hazard Assessment. The second stage is also known as risk analysis.

In the UK, HAZAN is a term that applies to the technique of quantitative assessment of particular risks. (The likelihood or frequency of the event and the severity of the consequence.) This is often combined with the analysis of proposed risk reduction (or protection) measures to provide a risk assessment report (Ref. 4).

Tevor Kletz points out that 'if someone asks you to carry out a Hazop or hazan on a design you should first make sure that they are clear on the difference'.

- Hazop is a particular type of hazard analysis that finds out the problems of operability in the design and identifies any hazards involved in operating the plant.
- Hazan works out the risks for a particular hazard once it has been identified. Alternative names often used for this work are 'hazard assessment' and 'hazard analysis'.
- The term 'hazard analysis' has come to mean any or all of the above activities and is therefore a very general term.

1.6.2 Hazard analysis techniques

The requirements for risk assessment cover all fields of industrial activity and we find that the hazard analysis methods and the risk assessment techniques used in process design have equivalents in other industries. For example, in safety of machinery, the requirements for hazard analysis are laid down in the European Standard EN 1050.

In EN 1050 Annex B there are descriptions of several techniques for hazard analysis. The notes there make an important distinction between two basic approaches. These are called deductive and inductive. This is how the standard describes them:

'In the deductive method the final event is assumed and the events that could cause this final event are then sought.

In the inductive method the failure of a component is assumed. The subsequent analysis identifies the events which this failure could cause.'

Deductive method

A good example of a deductive method is Fault tree analysis or FTA. We shall study FTA in Chapter 7. The technique begins with a top event that would normally be a hazardous event. Then all combinations of individual failures or actions that can lead to the event are mapped out in a fault tree. This provides a valuable method of showing all possibilities in one diagram and allows the probabilities of the event to be estimated. As our practical will show, this also allows us to evaluate the beneficial effects of a protection measure.

Deductive methods are useful for identifying hazards at earlier stages of a design project where major hazards such as fire or explosion can be tested for feasibility at each section of plant. We start with a hazard event such as 'Fire' and search for possible causes. Its like a cause and effect diagram where you start with the effect and search for causes. We shall see examples of this in Chapter 2.

Inductive method

So-called 'what if' methods are inductive because the questions are formulated and answered to evaluate the effects of component failures or procedural errors on the operability and safety of the plant or a machine. For example, 'What if the flow in the pipe stops?' This category includes:

- Failure Mode and Effects Analysis or FMEA
- Hazop studies
- Machinery concept hazard analysis (MHCA).

1.6.3   Summary of hazard-identification methods

Here is a summary of the hazard-identification methods to be found in guide manuals (Refs 8 and 10) and standards (Ref. 7) for hazard studies. It is useful to have this list because many companies will have preferences for certain methods or will present situations that require a particular approach. We need to have a choice of tools for the job and to be aware of their pros and cons. It is also apparent that similar methods will have a variety of names. All guides agree that Hazop provides the most comprehensive and auditable method for identification of hazards in process plants but that some types of equipments will be better served by the alternatives listed here.

Name of Method	Type/Procedure	When	Advantages	Disadvantages
Hazard study level 1 Alternative name: Concept stage hazard review	Inductive. Identifies potential hazard types from list of raw materials and operations	Concept stage before flow sheet	Provides database. Assists layout and siting. Legal obligations identified	Based on minimal info

Name of Method	Type/Procedure	When	Advantages	Disadvantages
Hazard study level 2 Alternative names: Preliminary PHA (Pr HA). Also called Screening level, Risk Analysis	Inductive. Finds potential hazards at each system or unit. Uses guideword tables applied to plant sections to prompt for hazards and possible causes and consequences. See Chapter 2	Flow sheets and materials known	Used on new facilities or previously untested facilities to get an overall view of where major hazards exist. Early detection offers chance to design out the hazards. Allows protection measures to be designed in	Does not find detailed hazards. Still requires Hazop later
Hazop	Deductive and Inductive.Structured analysis tool. Takes small sections or nodes of plant and applies all conceivable deviations to design intent. Searches for both cause and consequence	Can be used at any stage where detailed equipment or piping and instruments diagram (P&IDs) are available best used in design at latest stage possible	Very thorough, systematic. Provides high level of confidence in detection of hazards. Improves operability. The most widely used methodology for hazards identification	Very time-consuming and costly. If not set up correctly and managed, it can be unreliable. Requires experienced leadership
What-if analysis	Deductive. Similar to Hazop but uses team of experienced persons to test for hazards by asking relevant 'what-if' questions	Any stage of a project for new or existing plants	Easy to use. Faster than Hazop, best used with checklists	Much less systematic than Hazop. Depends on experience of team and of the process/ equipment. Hence requires justifying
Checklist	Deductive. Divide plant into nodes as for Hazop. Apply previously developed or published checklists for known failure and deviations. Record consequences, safeguards and actions	Any stage, provided the checklist has been made available by experienced staff	Useful where only one or two persons are available to study the plant	Requires time to obtain good checklists. Depends on checklists and lacks creative thinking. Hence not thorough, especially for new designs

Name of Method	Type/Procedure	When	Advantages	Disadvantages
What-if + Checklist	Combination of the above two methods	Any stage	Easy to use. Better than basic 'what-if' or checklist	Needs an experienced team and good checklists. Not thorough. Not easy to audit and prone to short cuts
FMEA	Deductive method. Starts with components of system or process and presumes failures. Results are then deduced to see if they cause a hazard	Final design stages or for evaluation of reliability	Good for electronic systems and mechanical equipment. Good for complex equipment	Not suited to processes because deviations and hazards may not be due to any failure of components. Does not detect common cause failures
Component functional analysis	As for FMEA but divides a process into functional objects. Uses list of typical functional failures and deduces the effects	Instead of Hazop or for deriving deviations for Hazop or failure modes for FMEA	Good for deriving plant equipment failures	Does not cover for all process deviations. Requires to be used with Hazop
Fault tree analysis	Inductive. Structuring the consequence back to the causes	Usually to quantify risks after the hazard has been identified	Graphical views of the causes and effects. Good for quantifying risks and seeing the primary dominant causes	Not suitable for initial identification of hazards. Not structured
Hazardous human error analysis	Deductive. Tests for effects of human activities on a machine or plant	Evaluation of a man–machine interface or task	Useful for hazards of machinery usage or maintenance tasks	Limited to human tasks

Name of Method	Type/Procedure	When	Advantages	Disadvantages
MHCA	Deductive. Structures a machine into functional parts and operating phases. Reviews each phase for possible malfunctions and deuces hazards. Incorporates HHEA as above	Safety review of a completed machinery design	Good method for proving the overall safety of a machine or assembly line	Not suitable for process plants

1.6.4   Conclusions

We have looked at the principles of risk management and seen how they apply to SMS. Hazard identification and risk ranking are part of risk assessment and risk reduction. Risk reduction requires an understanding of tolerable risk concepts and the measurement of risk. We have seen how a risk matrix or risk profile supports quantification and ranking of risks.

Safety regulations recognize these methods as part of the system needed to manage and ensure safety. Companies therefore need to develop competencies in these subjects and be able to develop the skills in hazard identification. Hazard studies and Hazop, in particular, have emerged as essential tools for the tasks.

Hazard studies lead to the requirements for safeguards and we shall see that alarm and trip systems are one of the key means of providing those safeguards. Firstly, we need to know how to do hazard studies, and then we need to know how to define the safeguards.

2

Hazard studies at levels 1 and 2

Objectives

This chapter begins by describing the use of hazard studies at critical stages of a project. Different styles of study are deployed at each stage to ensure the continuity of the studies throughout the life of the project. The first or concept stage is then presented in some detail with guideline material. This stage seeks to provide a database and context for the plant being designed.

The second level hazard study is particularly important for the identification and treatment of potentially severe risks. A methodology for this study is also examined. This chapter also identifies the relevance of preliminary hazard analysis to the specification of protection measures involving instrument trips and alarms.

The objective of this chapter is to provide grounding in risk management principles and help participants to see the relevance of hazard studies to safety management.

When you have completed this chapter you should have a good understanding of the role of hazard studies in the life of a plant or machine.

You should have a foundation of knowledge that will assist you to participate usefully in preliminary hazard studies specific to your project.

2.1   Introduction

2.1.1   History

The technique of conducting hazard studies appropriate to the stage of design, construction or operation of a process plant became established as a standard methodology in the mid-1970s through the efforts of people such as Trevor Kletz and SB Gibson working at ICI. Ref. 5 (see Appendix A) is a paper by SB Gibson (1975) to I Chem E describing the six levels of hazard study recommended by ICI at that time. By the late 1980s, this procedure was in common use.

Trevor Kletz has become a prolific writer on the subject of hazards and safety in general and his material is entertaining and informative. See Ref. 4 for one of his many publications on the subject.

In compliance with local OHSA requirements, chemical manufacturing companies have incorporated hazard studies into their codes of practice for all capital projects. Hazard studies are also standard practice for modifications to existing plants and operations. Unfortunately, these are sometimes omitted and now again this leads to dire consequences.

IEC has recently published a very useful standard on Hazop (IEC 61882 see Ref. 7). This standard incorporates well-established basic practices in Hazop Studies; it covers a wide range of application areas, including continuous and batch processes, electronic control systems and emergency planning.

2.1.2   Planning for quality assurance in hazard studies

At the start of a hazard study program, it is essential to develop a clear understanding of the scope and purpose of the studies. As we have seen, the process industries, in particular, are required to develop comprehensive safety management systems (SMS in the UK, and PSMs and RMPs in the USA). The success and credibility of safety plans in regard to the plant equipment and processes will be dependent on the quality of the hazard studies and on the effectiveness of the protection measures.

It follows that we should take steps to ensure high quality in the study procedures. It also follows that we require good quality descriptions of the hazard problems and accurate specifications for the instrumented safety system requirements.

2.1.3   Evidence of quality problems

Here is a good description of typical quality problems in PHA extracted from a paper by David A Moore published by Accutech (see www.acutech.com).

Quality problems with PHA studies

'Since the OSHA requirements are performance-based and the guidelines are somewhat vague, a wide disparity exists in how PHA studies are performed in different companies. Problems have included:

- Inadequate process safety information used as a basis of the study
- Unclear scope and objectives
- Lack of a set of assumptions for the study
- Lack of a common basis for acceptable risk decision-making
- Varying degrees of completeness (number of hazards discussed, thoroughness of documentation, hazards vs operability problems discussed)
- Varying amounts of time invested in the study.

As a result, in some cases, the original objectives that OSHA established are not met. If this happens, employees and contractors cannot rely on the information in the PHA as complete. Management may not be presented with a consistent and thorough analysis from which to make risk decisions. The employer may not be in compliance, but more importantly, unnecessary risks may be taken as a result.

Employers have an opportunity to improve the quality of their studies to minimize these problems. Setting new guidelines for PHAs is very timely and prudent right now because not all initial PHA studies are complete, facility and management of change PHAs are on-going, and the revalidation requirement is just now being addressed.'

2.1.4   The case for a systematic approach

Moore recommends that 'all employers should have a prescribed approach for conducting PHAs. In addition, a protocol for auditing the PHAs should be developed. These standards will encourage continuous improvement in process safety, and will, no doubt, greatly contribute to the entire PSM effort.'

The need for quality assurance and a systematic approach to hazard studies was recognized by the developers of hazard studies at ICI Ltd in the UK. Each study verifies that the actions of previous studies have been carried out. This should lead to continuity in the design and application of measures to protect health, safety and environmental issues throughout the life of the plant.

In this book, we call this the 'life cycle approach' to hazard studies and it has a very close parallel with the latest techniques for design and implementation of SIS that we are going to study in later modules. The safety instrumentation or 'Functional safety' requirements of the plant are to be managed by what is known as the 'Safety life cycle' and we shall look closely at this in Chapter 4.

Here we use the six-level model for hazard studies as an example of the quality assurance approach.

2.1.5 The process hazard study life cycle

A typical process SLC model comprises six levels, as shown in Figure 2.1.

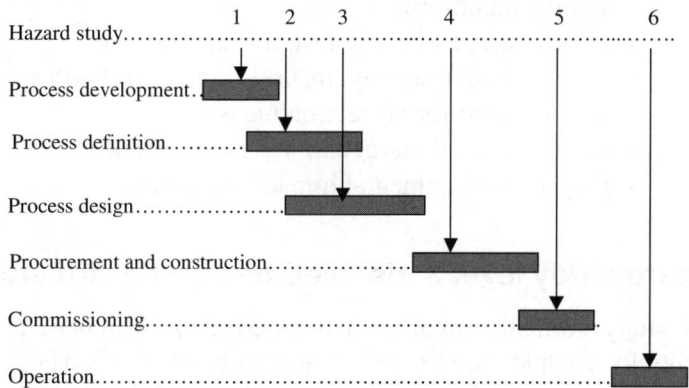

Figure 2.1
Process hazard studies at project stages

The six levels correspond to the appropriate stages in the progress of a new process plant design. Here is a summary of the scope and purpose of each type of study.

2.1.6 Hazard study level 1

- Basic hazards of the materials and the operations are identified.
- Collects information (if any) on previous hazardous experiences with the processes.
- SHE criteria are defined (e.g. critical exposure levels for toxic materials, atmospheric emission targets, effluent constraints).
- Identifies what information is needed and what studies are needed to address safety, health and environmental issues.
- Identifies relevant legislation and defines any constraints arising from them.
- Affirms the use of inherently safe design principles wherever possible.
- Evaluates suitability of proposed sites and their constraints.

In terms of the risk management model we saw in Chapter 1, this study establishes the 'Context' of the risks. On a new project this study would be expected to take place during the 'conceptual design' stage. The project will be in the planning phase and no significant engineering design will have been started.

For an existing plant this study would perhaps be replaced by a review of the baseline information available on the process and its hazards to provide a reference for the detailed studies.

Later we shall examine the methodology for this study.

2.1.7   Hazard study level 2

This study takes place during the project definition stage or 'Front engineering design', as it is sometimes known. It corresponds to a typical 'Preliminary PHA' in USA terms.

The study is usually performed on the draft flow sheets for the design at a stage where design changes and the introduction of additional safety measures will not incur penalties of wasted design effort.

- Examine plant items and equipment on process flow sheet and identify significant hazards.
- Identify areas where redesign is appropriate.
- Assess plant design against relevant hazard criteria.
- Identify need for protection measures.
- Identify critical alarm and trip requirements.
- Prepare environmental impact assessment.

2.1.8   Hazard study level 3: detailed design hazard study

This study normally involves a detailed examination of a firm design that should be essentially complete and nearly ready to be 'frozen'. The control system functions and operability features of the design as well as its critical safety functions should also be in place based on the guidance obtained from the level 2 studies.

As Trevor Kletz explains in his guidebook (Ref. 4, Appendix A) 'It brings hazards and operating problems to light at a time when they can be put right with an India-rubber rather than with a welding torch'.

The review at this stage expects to confirm that the detailed flow sheet or P&ID design is indeed safe and operable but will very often find additional and hopefully minor hazard and operability problems that will require some actions. This desirable state will most likely be attained if the previous level 2 studies have been properly done and if the design team has followed up the action items.

In summary:

- Critical 'line by line' examination of plant operations on completed design
- Searches for detailed hazard, control and operability problems
- Reviews existing safety measures
- Often uses Hazop method
- May use FMECA (Failure mode, effect and criticality analysis)
- Should be completed before detailed design/procurement begins.

Completion of the level-3 hazard study releases the 'basic engineering package' for detailed design and procurement to go ahead. It follows that the design package should then be 'frozen' as far as this is possible. All changes to the approved design will now come under Management of Change (MOC) procedures.

After this study, any new design changes affecting the plant equipment must be evaluated for their impact on safety and operability. Very often, the best way to do this is to repeat part of the detailed hazard study work in the area of change. The same applies to changes made in response to actions raised by the study. The essential point here is to ensure that for the life of the plant the hazard study record must be valid for the version of the design actually planned or already in use at the plant. This requirement often poses problems for the operating company, and we need to look at this again in Chapter 3.

2.1.9 Hazard study level 4: construction/design verification

This is a design/construction review. It is performed at the end of the construction stage to ensure the equipment has been built as intended and that there are no violations of the design intent.

In summary:

- The hardware is checked to ensure that it has been built as intended and that no violations of design intent have occurred.
- Checks that all actions raised during earlier studies have been incorporated into the design.
- Checks operating and emergency procedures are documented and comply with previous hazard study actions.
- Reviews commissioning and operating instructions for safety.

In a new plant project it is likely that handover from the design/construction team to the operating team will take place on completion of this study.

2.1.10 Hazard study level 5: pre-startup safety review

This review is carried out to examine the preparedness of the operating team to carry out the final commissioning of the plant.

The subjects to be covered include:

- Final operating procedures documented and applied in training
- Training of operators completed as far as practicable
- Preparation and readiness for startup including function testing cleanliness and purging
- SIS and alarms fully operational and validated. In USA this means pre-startup acceptance tests (PSAT) completed
- Confirmation of compliance with company and legislative standards.

Under the new functional safety standard IEC 61508, validation of SIS includes the completion of functional tests to prove the logic and response rates of the trip systems. It also requires that third party functional safety assessments (FSA) of the design and implementation have been completed with satisfactory results.

2.1.11 Hazard study level 6: post-startup review

Hazard study guides recommend that this study be carried out a few months into the production phase. Its purpose is:

- To check that all outstanding issues from the previous studies have been completed
- To examine whether operating experience so far matches the assumptions made about the process or its operability during the previous studies

- To look for any significant deviations from the intended operating procedures
- To capture experience from the plant for future use.

This is particularly important for the functional safety systems, where the estimates of likely hazard rates need to be checked periodically against experience, such as alarm frequencies.

2.1.12   Life cycle models for hazard studies

What we have seen so far is a model for ensuring continuity and quality in the execution of hazard studies from design to startup. This version is likely to have many variations according to need. The model we have seen so far is based on ICI's original methods and is also described by the other guides such as the 'Hazops guide to best practice' published by the UK I Chem. Eng. It is also similar to the descriptions given in IEC 61822 Hazop studies – Application Guide.

Here are some additional points to consider:

- The application of Hazops now extends far beyond its original application to chemical processes. The chemical process model may not be appropriate to other industries but the principle of progressively detailed studies and verification of each stage is likely to be needed in all applications.
- Timing of studies is clearly important. They must synchronize with other project activities so that:
  - The relevant design information is available for study.
  - The actions arising will cause minimum disturbance to the design progress.
- Hazard studies take up time and personnel resources and hence can be expensive. They must be used cost-effectively by customizing their scope to the essential needs of safety and operability.
- If hazard study records are carefully planned and maintained the quality and efficiency of the method will be improved. Software products can be of great help here.
- Hazard study plans must be designed to support revalidation over the life of the plant. This is now essential where regulations such as COMAH and PSM require audits and periodic reviews.

Later in this book, we are going to look at the relationship between hazard studies and the specification of safety-instrumented protection systems. The quality management issues are very similar for trips and alarms. Hence we shall try to produce an alignment model in Chapter 6.

Figure 2.2 considers that the core requirement is for a risk assessment table where every risk found in the plant is registered and evaluated over the life of the plant. The relationship to hazard studies and to the SLC of the trip and alarm system can then be seen.

In Figure 2.2:

- Risks are identified by hazard studies and are placed into a risk management listing (often known as a Risk Register). Hazop actions that concern operability and do not involve hazards will not go into this list.
- Initial risk rating is stated against a profile or risk matrix, and risk reduction needs are defined.
- This leads to the specification of overall safety requirements for the risk reduction measures.

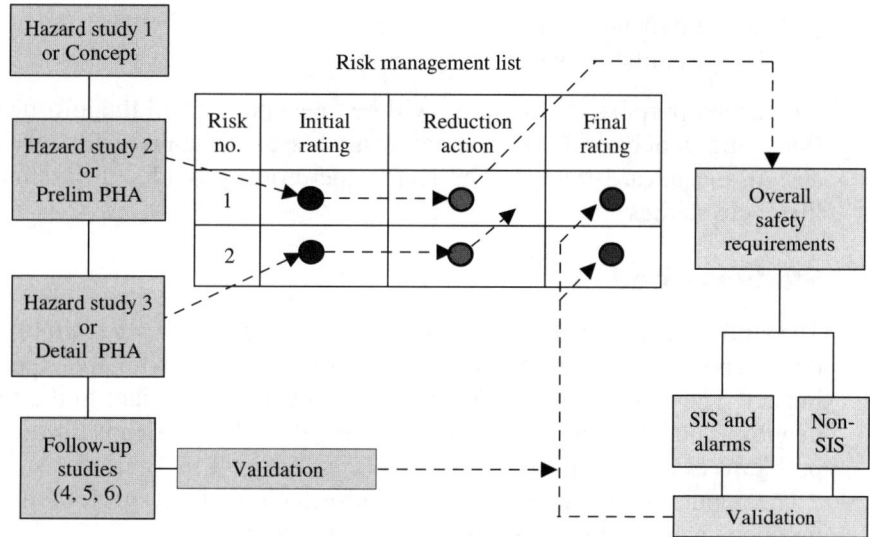

Figure 2.2
Relationship of hazard studies, risk management list and safety systems

- Safety requirements are allocated to SIS and/or to non-SIS functions.
- Validation of the safety system performance confirms the desired risk reduction has been achieved.
- Validation of the completed Hazop actions is achieved by follow-up hazard studies.
- The risk management list is maintained and reviewed throughout the life of the plant.

These general-purpose models for risk management can be used as the basis for planning for a particular project.

2.2 Methodologies for hazard study 1

Here is a brief expansion on some of the functions and features of the preliminary or concept phase hazard review we have described as level 1. This method is described under various names such 'concept and definition phase hazard study' or 'Screening level risk analysis'. For example, Screening level risk analysis formulates a list of hazards and situations by considering characteristics such as materials processed, operating environment, material inventories and plant layout.

2.2.1 Purpose

The purpose of a preliminary hazard review is to carry out a critical review of the proposed plant, machine or project in terms of SHE issues. The participants in the study need to be satisfied that all SHE matters that could affect the feasibility and progress of the project have been identified, and actions listed, where necessary.

The SHE topics will include:

- Identification of all materials being handled or generated
- Identification of all types of potential hazard including, fire explosion, toxic, environmental
- Process characteristics and previous incidents history
- Citing proposals and consideration of adjacent plants

- Environmental impact
- Regulatory issues.

A further purpose of this study will be to ensure that all the information brought into the study and generated by it remains available as a database for the succeeding stages of design and hazard review. The study, therefore, provides a baseline for the subsequent life cycle studies.

2.2.2   Study method

The study is conducted in one or more formal meetings by a team of persons consisting of those who know the project scope and intent and those who are experienced in its areas of impact. The study team examines the 'Concept' of the plant in the form of a simple flow diagram and outlines descriptions to identify all significant hazards and environmental problems as early as possible.

The inputs to the study are basic information of the project, the materials used and produced and the proposed process or equipment.

A checklist method is most likely to be employed because this study covers a number of basic issues that are common to all projects. Hence a reference template can be set up by the company.

The outputs from the study will be a list of the topics examined, with actions and the conclusions arising. These may form the basis of the SHE statement for the whole project.

For example, a chemical engineer from the design team will describe the process and any potential it has for releasing substances to atmosphere or to ground water. An environmental advisor or pollution control specialist will then question the designer on the risks of pollution targets being exceeded. This may relate to control systems and containment philosophy and the feasibility of avoiding serious releases. The result would be agreement on what performance targets are acceptable and what control measures should be considered in the design. The study would also comment on how critical is the risk to the project.

In some cases, it may be advisable to prepare a preliminary risk assessment chart, such as the one shown in Chapter 1:

- The study would also be expected to approve the basis of the environmental impact statement and the criteria for safety, health and environment matters.

These should include:

- Toxic gas emission criteria
- Tolerable risk targets for employee risk and off-site public risk
- Major incident target frequency
- Effluent limits for gaseous, aqueous and solid wastes
- Noise level limits
- Occupational exposure limits.

These criteria should enable the study team to approve the risk matrix profiles for the project that can be used in the succeeding risk assessment and reduction studies.

2.2.3   Preparation

The participants will need to establish a thorough understanding of the proposed plant or equipment. For the study to proceed efficiently and quickly (and so at lower cost) the best possible information should be assembled before the formal meeting. It should be made available to the team members.

Some suggested items are:

- Draft project definition
- Process or equipment description with outline diagrams or flow sheets
- A listing of known SHE issues and incidents with similar projects (if any)
- Chemical or material hazard data sheets
- A hazards checklist for the type of activities in the process
- List the applicable legislation for compliance (e.g. Comah, Cossh)
- Draft environmental statement
- Draft occupational health statement (OHS).

2.2.4 Team members

Most companies will seek to involve their most experienced persons in the team relevant to the technology being applied. Representation should be available for the following:

- Hazard study leader to chair discussions and ensure all records are made and issued
- Project manager and/or business manager
- Design team leader and any key specialist engineers
- Operations or end-user representative
- Occupational health representative
- Environmentalist
- Construction representative or manufacturer.

2.2.5 Scope and outputs of the study

The scope of the study can be limited to basic hazard and environmental issues or it can be extended to cover a number of related topics such as organizational and human factors. For some companies, this study constitutes a comprehensive review of all safety, health and environmental issues related to the plant. In other applications, the study will result in a listing of all hazards and a ranking of all risks.

Here is a suggested minimum scope for the final report when applied at the start of a project life cycle:

- Plant and process description
- A database of all substances stored as well as processed at the facility, this will include the physical properties such a toxicities, flash points, upper and lower explosive limits, etc.
- Inventories of materials and where they are contained
- Interactive points such as sources of ignition
- Vulnerable locations such as office blocks or canteens
- Possible release scenarios from vessels, storage tanks or pipelines
- Consequence potentials in terms of severity of impact on people, plant and environment
- Estimated likelihood of any severe events with risk ranking
- Hazards or risks register for the plant
- Reference to or statement of tolerable risk targets for the site
- Identification of known safeguards and outline proposals for any additional safeguards to be considered by the design team

- Environmental impact summary to include the position on air pollution, water pollution, solid waste disposal, odor, noise, aesthetics and any special characteristics of the plant
- Impact on existing plants at the site. In particular, the potential for hazardous interactions
- Issues of transport to and from the site and any hazardous material transportation risks
- Review and approval of the OHS for the project
- Listings of regulatory compliance issues and consent levels.

2.2.6   Conclusion

Whilst the hazard study at this level may be little more than a long checklist, it has great value for decision-making at the early stages of project. It also provides a convenient reference point for all subsequent safety-related design activities and for the next levels of hazard study.

From experience, it is surprising how difficult it can be to trace back from later stages of the project to find reliable baseline data on the plant safety parameters. This level of study has great potential value to the project in its ability to highlight key issues, whilst there is scope to tackle the problems in the most effective manner.

2.3   Process hazard study 2

2.3.1   Introduction

Key phrases: Systematic search for hazards, preliminary hazard analysis, risk reduction decisions.

The level 2-hazard study is typically a study of a unit or combinations of units of plant intended to establish knowledge of the main hazard and operability problems at an early stage in the design. This allows design improvements to be made with the minimum of disruption to the design work. It also has a significant impact on the introduction of safety-instrumented trips and alarms, as well as helping to establish the best operating philosophies for the plant. (We shall be looking into these aspects a bit later.)

The method is also used to evaluate proposed modifications to the plant, again at the stage when the plans can be changed without undue losses.

2.3.2   Scope of studies

During this study, team members look at the potential hazards of the main plant systems that may include the following:

- Material hazards of feeds, intermediates, products and any other substances generated by the process
- Process equipment and packaged plant units (for example, filters, compressors, absorbers)
- Plant layout, in terms of physical hazards and proximities
- External factors such as environmental or the impact of upstream and downstream deviations from expected conditions
- Operational environment, for example, shift operations, seasonal peaks, long-term storage
- Management systems, impact on skills and staffing.

2.3.3   Study methods

Study methods are adapted to suit the applications but they usually come down to:

- Preliminary PHA method: Uses prompts for typical hazards and searches for possible causes
- Checklist/What-if analysis: Considering feasible deviations in the intended operations against a checklist. Supported by experienced staff posing 'What-if questions'.

Here we take a look at a preliminary PHA method that has been used by ICI and by AECI in South Africa.

2.3.4   Outline

Method

- Study team from all role players
- Systematic search for specific chemical/physical hazards against plant preliminary flow sheet or block diagram
- Select operational blocks for study. Apply keywords from a guide diagram using a sequence of questions (see diagrams)
- For each hazard, carry out assessment of consequences and frequency
- Use fault tree analysis where appropriate
- Record results in a chart form.

Inputs

- Process flow sheet or equipment descriptions
- Block diagrams of control systems
- Hazard study 1 records and data sheets.

Outputs

- Identification of critical hazards and design constraints
- Hazard summary table
- Risk assessment listing and requirements for risk reduction
- Confirmation or modification of overall control systems
- List of items requiring further action or study
- Major project decisions.

Related design actions

- Pressure relief requirements
- Leads to specification of critical alarms, interlocks and trips
- Supports determination of SIL requirements
- Feeds information and documentation into the SLC (see Chapter 3).

2.3.5   Hazard study 2: systematic procedure

An example of a practical method of searching for hazards, evaluating the consequences and outlining corrective actions is given here; details are based on methods used in the AECI Process Safety Manual.

Preparation

Requires that the plant design has reached the stage of having completed flow sheet designs so that all major equipment items have been defined on the diagrams. The intended operating procedures for the process are known, and the volumes, rates and materials involved are all known to approximate values.

Most of the data described for hazard study level 1 should be accessible to participants before the study. If there was no level 1 study, then, at least the material hazard data listed for that study should still be assembled. Otherwise, very little preparation is needed.

Procedure

The engineer with the best knowledge of the design describes the essential features of the plant items and explains the intended methods of operation. This should be supported by a written description passed out to each team member.

The study team should first review the hazard study 1 report or baseline data, as available. It should check the actions listed in the level 1 report to establish which of these have been implemented at this stage and should note outstanding key points.

The study team leader should seek agreement on the scope of plant to be covered in each particular study. It is essential to ensure that no misunderstanding occurs on the limits of plant being reviewed.

Flow sheets are divided into major operational stages so that a unit or stage can be considered individually. The chosen area should constitute a complete process operation or unit so that all factors affecting its status can be seen. All interconnected upstream and downstream plants and services should be identified during the presentation.

The selected area or system will then be subjected to a systematic review of hazards using the sequence illustrated in Figure 2.3.

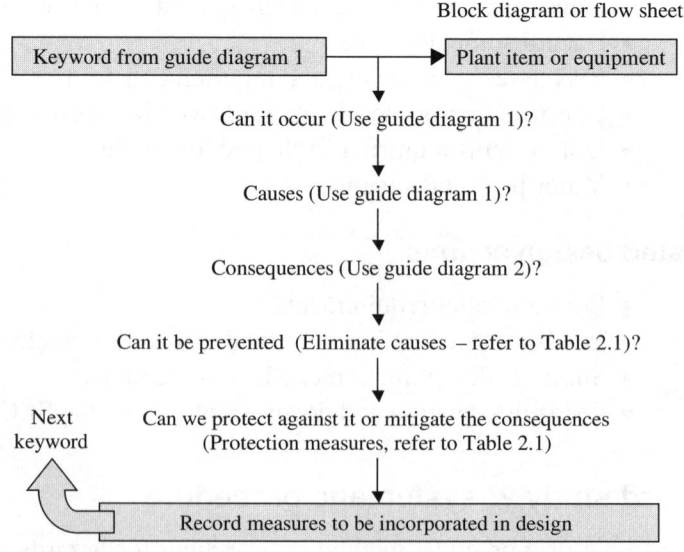

Figure 2.3
Search procedure for hazardous events

- Begin with the first item in 'Hazard/Event' column in Guide Diagram No. 1 (see Figure 2.4) to check the possibility of this hazard occurring.
- The second column supplies the essential prompts for feasibility, the third column contains basic checklists of possible sources of the hazard. Only the sources are listed. The study team will discuss and decide the possible causes, as each feasible hazard is examined.

Hazardous Event/ Situation	Prompts	
External fire	Fuel	Flammable gas, vapour, solid, metal, wood, waste material, pyrophoric material
	Release mechanism	LOC, poor housekeeping
	Ignition	Sparks, flares, static, friction, vehicles, hot spots, welding, lightning, auto-ignition, furnaces
Internal fire (in equipment)	Flammable mixture	Flammable gas, vapour, liquid, solid, metal, dust, residue, pyrophoric material, oxygen, halogen
	Ignition	Sparks, static, friction, welding, decomposition
Internal explosion (in equipment)	Physical over pressure	LOC (Burst-Physical overpressure), head pressure, liquid filling, testing, purging
	Uncontrolled reaction	Runaway reaction, decomposition, polymerization, contamination
	Flammable mixture	Flammable gas, vapour, liquid, solid, dust, mist, oxygen, halogen, NO_{13}, explosive/unstable compound, polymerization, loss of ignition/re-ignition
	Ignition	Sparks, static, friction, hot spots, welding, decomposition
Confined explosion (in building)	Flammable mixture	Flammable gas, vapour, dust, mist, oxygen enrichment, halogen, explosive/unstable compound
	Release mechanism	LOC, storage, handling
	Ignition	Sparks, static, friction, welding, machines, vehicles, hot spots
Unconfined explosion	Fuel	Flammable gases, vapours, dusts, mists, explosives/unstable compounds
	Release mechanism	LOC, storage
	Ignition	Sparks, flares, static, friction, vehicles, hot spots, welding, lightning, furnaces, pylons
Acute harmful /noxious exposure	Acute harmful/noxious	Toxic gases, vapours, mists, liquids, dusts, fumes, acids, alkalis, biological
	Exposure mechanism	LOC, decontamination, mechanical handling, sampling, manipulation, ventilation failure
Chronic harmful/noxious exposure	Chronic harmful/noxious	Toxic gases, vapours, mists, liquids, dusts, fumes, biological, radio-active
	Exposure mechanism	Leaks from seals, valves, charging, discharging, preparing for maintenance, loading, packing
Pollution	Pollutant	Aqueous, gaseous, ground, silts, smells, fire water, surfactants lubricants, foams, acids, algae, fuel gases, by-products, residues, dust, mists, steam
	Release mechanism	LOC, decontamination
Violent release of energy	Energy source	Electrical, potential, kinetic
	Release mechanism	Electrical explosion, LOC (burst), impct, mechanical failure
Noise	Sources	Machinery, ejectors, flares, pressure let-down, vents, reliefs, road/rail traffic, sirens, conveyors, mechanical handling, demolition, construction
Visual impact	Appearance	Building profile, stacks, layout, colour, location, smoke, steam, plumes, flares, flashes
Major financial effect	Factors to consider	Loss of business, business interruption, downtime, overhauls, loss of services, spares, major downtime, e.g. computers, single stream equipment, etc., major operating costs

Figure 2.4
Hazard study 2 guide diagram 1

- Identify any feasible hazard and record it and its possible causes in the Hazard Summary Table (see Figure 2.6).
- For the same hazard event consider the 'consequences' using Guide Diagram No. 2 (see Figure 2.5) and record them in the summary sheet.
- For 'prevention' the team will use possible ideas from experience or will look for prevention layers such as those indicated in Table 2.1 or any other prevention layer technique based on their own experience.
- For 'protection against consequence' the team will look for mitigation layers of protection such as those indicated in Table 2.2.
- Now the team must agree on 'what must be done' and 'who is responsible'. The actions required must be placed on an individual to see that they are implemented.

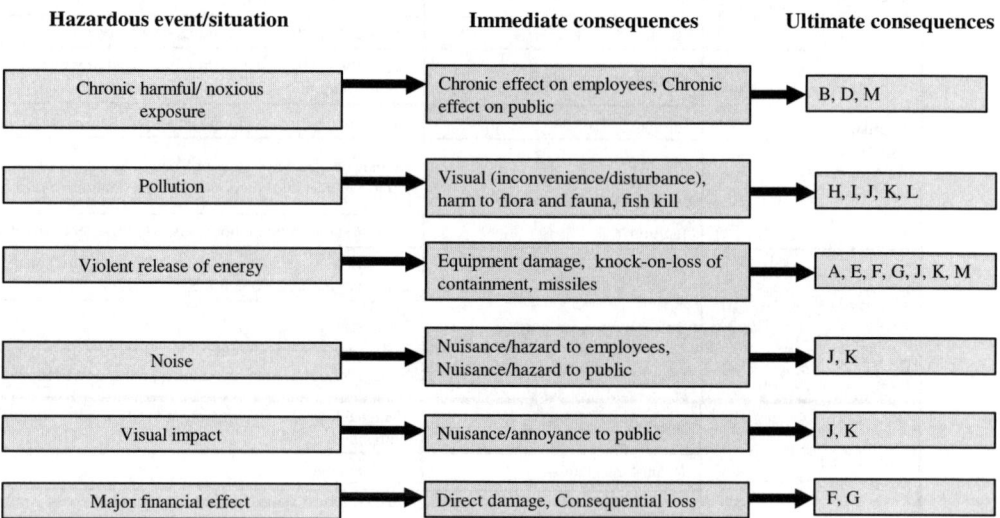

Code	Group	Consequences
A	Employees	Injuries/fatalities
B		Ill health/long-term fatalities
C	Public	Injuries/fatalities
D		Ill health/long-term fatalities
E	Fire fighters	Injuries/fatalities
F	Plant damage	Damage to plant and equipment
G		Loss of production
H	Environmental damage	Harm to Flora and Fauna
I		Fish kill
J	Publicity/media	Bad publicity
K		Public/product concern/site licence
L	Authorities	Environmental protection
M		Industrial incidents/accident investigators
N	Other effects	Evacuation of site
O		Evacuation of public
P		Obnoxious odor

Figure 2.5

Hazard study 2 guide diagram 2 sheet 2

Hazard Study 2 Summary Form	Project:			Drawing number	Rev nos.
Team Members:				Date:	Sheet no.
Hazard ref no.	Hazardous event or situation	Caused by/ sequence of events	Consequences immediate/ ultimate	Estimated likelihood. Suggested measures to reduce likelihood	Emergency measures (to reduce consequences)	Action required by

Figure 2.6
Hazard study 2 summary table

2.3.6 Worksheet for detail records

The most important task of the team leader at this point is to ensure that these actions are properly recorded in the minutes or records of the study. There are a number of ways of doing this but generally, the format will consist of a worksheet for all notes that should be filed.

The study team will need to have a check list where each guideword, deviation or 'what-if' question is noted down for the record to show that these points were raised and dealt with and not just overlooked. The problem with a hazard summary table is that it only reports action items and leaves a presumption that all other possibilities were considered and dismissed as 'not a problem'. Worksheet records are easy to keep on computer, using a spreadsheet. Typically, the format will be as follows:

Plant Area / Description	Area 05/ Sterilizer			
Flow sheet no.	05-001 rev 1			
Hazard Type	Reviewed	Possible hazard		Hazard ref. no.
		Yes	No	
External fire	√	√		05/1
Internal fire	√		√	
Internal explosion	√		√	
Environmental Pollution	√	√		05/2
Special items raised	√		√	

Optionally, the reasons for stating 'no possible hazard' can also be referenced.

The results of the worksheet records should always lead to a formal report version such as the Hazard Summary Table. A basic format for the summary table is shown in Figure 2.7.

2.3.7 Guideword diagrams and hazard summary table

Before we comment any further on the outputs or results of the hazard study 2 activities, we can run through Figures 2.4–2.6 and then try out the method on a practical example.

Figure 2.6 captures the results of a hazard study level 2. It serves to record the main hazards anticipated for the plant found from the high-level hazard study. A similar but more detailed type of form is recommended for reporting the results of a Hazop. Examples of such forms are to be found in IEC 61822: Hazard and operability studies (Hazop studies) – Application Guide.

Notes

Hazardous Event	Each possible event is recorded as it is identified by the study team (use of guide diagram 1 or from any other prompts).
Caused by/sequence	Possible causes typically as shown in guide diagram 1. Details of the sequence of events that could create the conditions.
Consequences	Possible consequences as indicated by guide diagram 2 or from specific analysis of the event. Scale of consequences will help decide risk reduction requirements.
Estimated likelihood	Essential to have an estimate of frequency of event or an agreed descriptive term that translates to a frequency band. May be adjusted later after risk analysis.
Measures to reduce likelihood	These are the suggested layers of protection and may include an SIS function.
Emergency measures	An action to be taken once the event occurs. Also known as mitigation layers, these should reduce the ultimate consequences of the event.
Action required	Requirements for further study or for the next steps in the design of the safety measures.

This form may be expanded for wider usage into a risk assessment or risk register table. An example of this is shown below, with typical risk matrix parameters added to the table.

Hazard Study 2 Summary Form		Project:			Drawing number			Rev nos.	
Hazard ref. no.	All fields as for summary form	Risk matrix before safety measures			Risk matrix after safety measures			Decision	Action required by
		Prob.	Cons	Risk	Prob.	Cons	Risk		
G-1	Details as above	5	4	20	2	4	8	Accepted with safety measures	Inst. Eng

2.3.8   Guideword libraries and diagrams for other industries

So far, we have shown the guidewords appropriate to typical chemical process hazards. The contents of the diagrams have to be customized to particular industry sectors. For example, in machinery safety hazard studies there will be an equivalent set of more appropriate hazards.

The following guideword diagrams (Figure 2.7) were constructed using the hazard types listed in EN 292-1 1991: Safety of Machinery-Basic concepts, general principles for design. Please note that all guideword tables shown in this manual are indicative and should be verified or supported by consulting other guides and standards, where available.

Hazard guidewords are sometime published in hazard study manuals or software packages as 'Libraries'. Most companies will be able to develop their own libraries based on the experience in their field of operations.

Hazardous Event/ Situation	Prompts	
Mechanical hazards (Static Loads)	Shape/ location	Sharp objects, cutters, hand traps, eye levels, projections, loose objects, large masses
	Potential energy	Load falls, machine tipping, lack of stability,collision of machines
	Movement close to persons	Uncontrolled loading, overloading, machine tipping, uncontrolled amplitude of movement, unintended movement of load, inadequate holding devices
Mechanical hazards (dynamic)	Shape/ location	Sharp objects, cutters, hand traps, projections, work piece tables, loose objects, large masses, robot arms, gantries
	Kinetic energy/thrust/force	Elements in controlled or uncontrolled motion
	Movement close to persons	Cutting, shearing, crushing, entanglement, drawing in or trapping. Impacts stabbing or puncturing, friction or abrasion. Ejection of objects or fluids
Electrical hazards	Live parts	Panels, drives, switches , solenoids, lamps, controls, heaters, isolators
	Fault conditions	Removal of covers, failure to isolate, insulation or wiring fault. Electrostatic discharge, thermal effects of overloads and shorts.
		Approach to live parts under high voltage, maintenance, cleaners using water. Shock-induced falls
Thermal hazards	Heat sources	High temperature objects, flames, explosion , thermal radiation, steam pipes, heaters, cutting torches, inductive heating, welding
	Persons exposed	Operators, cleaners, loaders, technicians, public
Noise hazards	Noise sources	Cutting, drilling, grinding, impacts and stamping. Motors, pump fluids in valves, fluid/steam releases, compressed air releases
	Persons exposed	Operators, cleaners, loaders, technicians, public
Vibration hazards	Vibration sources	Cutting, drilling, grinding, Motors, pumps, rotating parts, impact drills and hammers
	Persons exposed	Operators, loaders. Hand-held machines, long-term exposure
Radiation hazards	Radiation sources	Low-frequency, RF and Micro waves, infra-red, visible light, ultra-violet, X-rays and gamma rays α, β, electron or ion beams, neutrons
	Persons exposed	Operators, loaders, technicians, patients, public
Hazards of materials and substances (check chemical hazards guide diagram)	Materials processed or released	Solids, liquids, gases, vapours, oils, drugs, biohazards, dusts
	Contact conditions	Contact or inhalation, exposure of skin, eyes, etc.
	Type of harm	Toxic, corrosive, irritant, long-term exposure effects
Hazards of neglecting ergonomics	Physical	Unhealthy postures, excessive repetition, overloads
	Phsycological	Mental overload, stress underload, boredom, distraction

Figure 2.7
(a) Guideword diagram for level 2 hazard study of machinery safety. (b) Continuation of guideword diagram for machinery safety

2.3.9   Risk analysis and risk reduction measures

The difficult part of the hazard study sequence is when the team comes to decide the question 'How to prevent or reduce the likelihood'. This where everyone can offer ideas and the team leader has to allow all the ideas to be sorted out hopefully into agreement.

It helps a lot here to have a systematic approach. There is a logical sequence involved in arriving at solutions to the risk problems and it follows the model we saw in chapter 1. Here is the suggested sequence for study to follow:

- Define the causes or event sequences leading to the hazard
- Estimate the likelihood (known as the hazardous event frequency)

- Estimate the consequence
- Decide the amount of risk reduction needed
- Look opfor inherent safety solutions first
- If needed, suggest measures to reduce likelihood
- Then suggest measures to reduce consequences
- Review the revised risk levels and decide if these will be acceptable.

Here are some guideline notes intended to assist with each of the above steps.

2.3.10   Event sequences leading to a hazard

There will be one or more logical sequences of events leading to the hazard. These must be recorded in a clear format. Fault trees are often the best medium to use because they:

- Clarify the logic of events leading to a hazard
- Capture the estimated event frequencies
- Model and calculate hazardous event frequencies (before protection measures)
- Help to develop and evaluate protection layers.

We are going to look at fault trees and get some practice with them later on in this book. The point to note here is that all the possible combinations of events leading to the 'top event' should be tracked down and recorded in the hazard 2 sessions or should be actioned for detailed study by an individual.

2.3.11   Hazardous event frequencies

Once the logic of an event has been determined, the 'likelihood' or 'event frequency' has to be estimated. The hazard study team should make an initial estimate, and in some cases they will call for a more detailed evaluation to be done outside of the main study. As we have seen in Chapter 1, the risk reduction requirement depends on knowing the unprotected risk frequency. Again, fault trees provide a basic platform for developing the risk frequencies.

2.3.12   Estimating the consequences

We have seen that guide diagram 2 suggests some ideas but the team needs to agree on the possible consequences.

2.3.13   Estimating the risk

It is the task of the hazard study team to estimate the level of risk associated with each hazard. This brings us directly back to the risk classification work we covered in module 1. Recall that the quantitative method of risk analysis begins with an estimate of unprotected risk frequency and an estimate of the consequences.

This is where the risk classification chart or risk matrix becomes very helpful because it serves as a guideline for the hazard 2 team to decide on the acceptability or not of any given risk. It is most important to ensure that the risk classification information is properly agreed and approved within the company structures before the start of a hazard 2 study.

2.3.14   Inherent safety solutions

Hazard study guidebooks advise that the design team should look first for inherent safety solutions, i.e. eliminate the hazard.

AVOID	Not having anything that is hazardous
REDUCE	Reduce the amount that is hazardous (inventory)
SUBSTITUTE	Replace with something that is not hazardous
SIMPLIFY	Use conditions that are not hazardous (this is not always simpler!)

It looks as if inherent safety becomes part of the process design, effectively the first layer of protection.

2.3.15 Adding more protection

Typically, at the point of deciding risk reduction needs the study team will move along to the next step of suggesting more protection. Hence, the layers of protection that we saw earlier will be added. For the SLC we are going to need to recognize and classify these layers to assist in our risk reduction models.

2.3.16 Group exercise: protection layers/risk reduction categories

Evaluate the following measures suggested for risk reduction in terms of their categories:

- Basic design
- Process control
- External risk reduction measure
- E/E/PES safety-related system
- Other technology-related risk reduction systems
- Mitigation layer.

Table of suggested risk reduction measures

Flammable/Acute Harmful Material/Pollutant Hazard	Category Code
Inventory reduction	B
Pressure/Temperature reduction	
Minimize equipment, piping, seals and joints	
Design for containing maximum pressures	
Pressure relief systems location/layout/spacing	
Containment/bunding/safe disposal	
Eliminate sources of ignition	
Rapid leak detection	
Control operating parameters within safe limits	
Interlocks for drives and valve settings	
Monitor and alarm deviations in critical parameters	
Shutdown process on critical deviations	
Rapid fire detection	
Control room/occupied building blast resistant	
Toxic refuges (gas escape rooms)	
Fire protection	
Dispersion aids – water jets, air dilution fire-fighting facilities	
Emergency procedures on/off site	
Vent/relief discharges – treatment/containment/recovery	
Chronic harmful material hazard	

Flammable/Acute Harmful Material/Pollutant Hazard	Category Code
Design for hygiene requirements as per OHS act	
Containment of low-level discharges	
Monitoring of work place atmospheres	
On-going health screening	
Building ventilation	

2.3.17   Key measures to reduce the risk

Once the general philosophy for prevention and mitigation has been established, the hazard study team can follow-up with more specific proposals. Typical key measures are listed here:

- Layout/separation distances
- Pressure relief requirements
- SIS (trips and alarms)
- Interlocks (e.g. mechanical, electrical, PLCs, computer, procedures)
- Fire prevention, protection and fire fighting
- Preventing sources of ignition
- Management systems for safe operation and maintenance
- Correct operator interaction (i.e. knowing when and how to shutdown safely)
- Spillage containment and recovery absorption systems
- Reducing exposure to harmful substances.

Let us consider some of the measures that will affect the SLC activities and the instrument engineers.

2.3.18   Process and operational safety measures

These are typical routine measures that are built-in to the operating instructions for the plant. Operators are given standard operating procedures that include such duties as closing all drain valves and perhaps locking devices off before startup of a plant unit. These measures can be suggested by the hazard study team and they will be found again when the detailed Hazop study is done.

2.3.19   Alarm functions

Many concerns raised in a Hazop will lead to alarm functions being specified to the control systems engineer. The typical alarm functions will begin with simple deviations from the normal range: e.g. high level in a tank serves as a first warning of an approaching problem.

- More serious alarms will merit a separate sensor for process conditions.
- Smarter alarms bring rate of change warnings.
- Even smarter alarms turn themselves off when not applicable.
- At some point in the discussion of a hazard, issues concerning response to alarms will be raised.
- Can we be sure the operator will respond?
- Will he be there?
- Is there enough time to respond?
- Will he do the right thing?
- How many other alarms will there be at the same time?

As we know, all large control systems include a full range of alarm facilities and the problem is to avoid a proliferation of alarms that defeat the purpose when a real upset occurs. There are two relevant articles on this subject available in this book, which stem from the huge explosion at Milford Haven, in 1994, and there are copies of these available for reference when we come to the chapter on instrumentation and hardware. Essentially, the issue is about the management of alarms. On its own, an alarm is OK, in a large crowd they are a real problem.

2.3.20   Safety-instrumented functions

When a Hazop study finds there is a situation that is likely to arise that requires a positive and, sometimes, very fast response to ensure safety, it will conclude that a safety instrumented function must be performed.

This decision should not be taken lightly because the cost of such functions is high and the degree of upkeep needed is not trivial. The SIS brings with it the additional factor of potential for nuisance tripping, reducing the overall availability of the plant.

2.3.21   Measures to prevent causes or mitigate consequences

The identification of measures to reduce risk takes place during the hazard study 2. It is useful for the study team to have a set of prompts of typical measures available. The best measures are those that prevent the causes of hazards, as given, for example, in Table 2.1.

Measure	Reduce Hazard Due to
Pressure/temperature reduction in process	High energy levels, stresses
Minimize equipment, piping, seals and joints	Leaks
Design for containing maximum pressure	Rupture/bursting
Provide pressure relief system	Rupture/bursting
Location/layout/spacing	Interactions/confined spaces
Operational alarms	Wrong operating conditions
Automatic protection systems (SIS)	Wrong operating conditions, dependency on human response

Table 2.1
Measures to prevent or eliminate causes of hazards

Measures to reduce consequences are used when the causes of a hazard cannot be further reduced. These measures accept that the hazardous event may occur but provide means of mitigating the scale of events or reduce the consequences. Examples are given in Table 2.2.

Chronic harmful material hazard

Design for hygiene standards, containment of low-level discharges, monitoring of work place, on-going health screening, building ventilation.

Measure	Mitigate Consequences of
Containment/bunding/safe disposal	Uncontrolled dispersion, contamination
Rapid leak detection	Leaks leading to gas cloud/liquid pool
Rapid fire detection	Runaway fire
Control room/occupied buildings design for pressure shocks	Injury to occupants

Measure	Mitigate Consequences of
Toxic refuge (Gas-safe room)	Toxic vapor exposure
Fire protection/dispersion aids – water jets	Spread of fire
Fire-fighting facilities	Uncontrolled fire
Off-site vent/Relief discharges	Uncontrolled emissions
Isolation of stages and units	Migration of fires Feeding of fires from other units
Emergency procedures	Uncontrolled responses Chaotic evacuation
Emergency shutdown systems (ESD)	Slow response to hazardous event Dependency on human factors

Table 2.2
Measures to mitigate or reduce consequences

2.3.22   Closing the study session

Some important things must be done before the study team can be allowed to go home!

- Check and agree the hazard summary inputs.
- The action points on the summary must be set down and agreed with the nominated person. Remember that these actions are going to be followed up under the SMS. The person nominated to carry out the actions must be willing to accept the responsibility.
- Consider/decide the risk rating for each hazard. If the hazard study is part of the risk assessment record for the plant, it will be best to include a risk rating in the summary; in which case, the study team should be asked to agree their risk matrix ratings at the time.
- Decide on follow-up meeting. If the study calls for more hazard analyses of identified problems, a report-back meeting with some or all of the participants may be needed.

The important thing here is to make sure that there is continuity and follow through from the good work done by the hazard study.

2.3.23   Hazard 2 report

The initial hazard study sessions are often followed by a development phase during which the design team will improve and further specify any hazard control measures found necessary from the systematic study. This work should lead to a closing report with the following scope:

- The hazard summary table
- Risk appraisals with quantified assessments of risks to people or the environment. Layout and spacing issues may be dealt with in this section
- Specification of key protective systems, mechanical, electrical and instrumental. This will include estimates for hazard demand rates and required risk reduction factors supported by hazard analyses or fault tree logic diagrams
- Relief systems philosophy
- Service requirements in the event of an emergency
- Copy of environmental impact assessment
- Correspondence with authorities.

All the above information matches the needs of the records for the IEC SLC and so we can link this report directly into the SLC documentary records.

We must now check the IEC requirements for Phases 3 and 4 to see that we can complete the SIS specification task with the aid of information from the hazard studies.

2.4   Practical example of hazard 2 application

Here we can try out the above method on an example of a widely used item of process equipment. This is a low-temperature sterilizer unit using ethylene oxide (ETO) gas. Ethylene oxide sterilizers are commonly used for the sterilization of medical devices and appliances where the materials of construction do not permit the use of high-temperature steam sterilizing.

In our example, a large sterilizing chamber is proposed for mass production work. It is large enough to allow several pallet trolleys to be pushed into the chamber with boxes stacked on pallets and it has an entry door at one end and an exit door at the other. Non-sterile loads are placed in the chamber and the doors are sealed. After sterilizing, the exit door is opened and the sterilized boxes are taken to a secure storage area for final de-gassing.

In plan view the chamber looks roughly like the diagram in Figure 2.8.

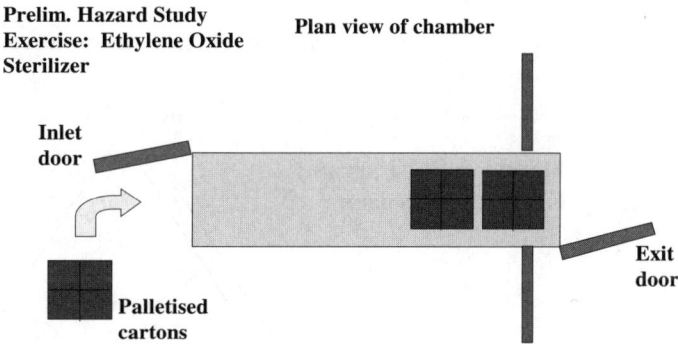

Figure 2.8
Plan view of sterilizer with dividing wall between stages

The process requires that products are impregnated with an ETO gas mixture with air or an inert gas such as nitrogen or carbondioxide. The mixture strength is typically 50% and it is infiltrated into the products by means of first vacuuming out the air from the packages and replacing it with the ETO mixture. This is done in a large sealed chamber. After several hours of exposure, all possible bacteria in the products have been destroyed and the ETO gas mixure is removed by vacuum and replaced by fresh air. The basic process sequence can be summarized as follows:

- Place goods in chamber
- Remove most of the air by vacuum pump
- Relieve vacuum with ETO mix
- Expose for several hours
- Remove ETO mix by vacuum pump
- Relieve vacuum by air

- Repeat vacuum cycle four or five times to flush out residual ETO (de-gassing cycles)
- Remove sterilized goods from chamber.

2.4.1   Preparing for hazard 2 study

Imagine we are preparing for a preliminary hazard analysis or hazard study level 2. The first task is to assemble essential data. Hopefully, a hazard 1 study has done the groundwork!

The hazard study 1 information will include the following:

- *Location:*   Integral to a pharmaceutical plant and arranged to be indoors so that goods leaving the production line can flow directly to the inlet side. The outlet side leads to a secure holding/de-gassing area where the goods are ventilated and subjected to microbiological sample testing to prove they have been sterilized.
- *Size of unit:*   H = 2.0 m, W = 2.0 m, L = 10 m, Volume = 40 m³.
- *Operating temperature:*   50 °C using double-skin walls filled with hot water in circulation.
- *ETO gas properties:*   Flammable gas mixture defined by the diagram in Figure 2.9.

Figure 2.9 indicates that ETO is a flammable gas in the range 0–95% air. Even 75% ETO: 25% pure nitrogen mixture is still flammable.

ETO sterilizer: gas mixtures

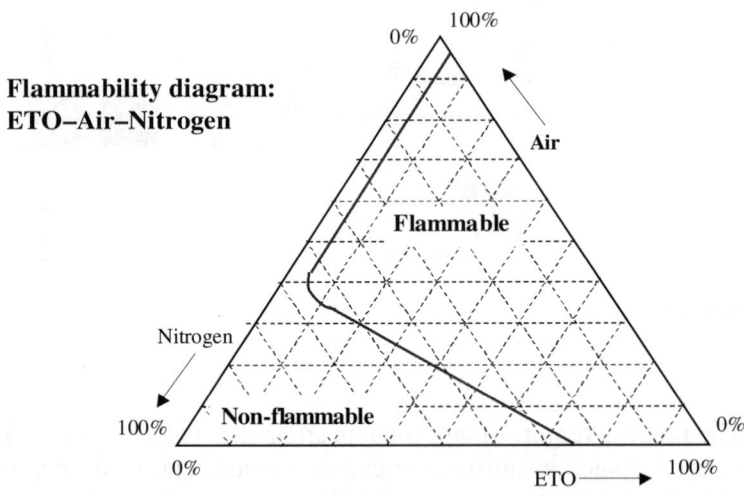

Figure 2.9
Flammability diagram: ETO–Air–Nitrogen

- ETO is a colorless gas with an ether-like odor. It is heavier than air and is readily absorbed in water where it forms ethylene glycol.
- The toxicity of ETO presents a problem of long-term exposure because it has been linked to cancer, reproductive damage and other health hazards. In the USA, OHSA has set the exposure limit, i.e. Permissible exposure limit (PEL): An 8 hour time weighted average of 1 ppm. OHSA also requires that frequent exposure monitoring be done by the employer to check the exposure levels of the workers on each shift.

- Respirators can be used to control exposure during maintenance work or during work practices where no other means exist for reducing the exposures below the PEL.
- Environmental constraints will not permit the systematic exhausting of ETO to atmosphere, hence an absorber (scrubber) system will be needed. Disposal of ethylene glycol must be done to an approved conversion process or to an approved waste disposal site. A typical environmental interface block diagram of the process was drawn up as in Figure 2.10.

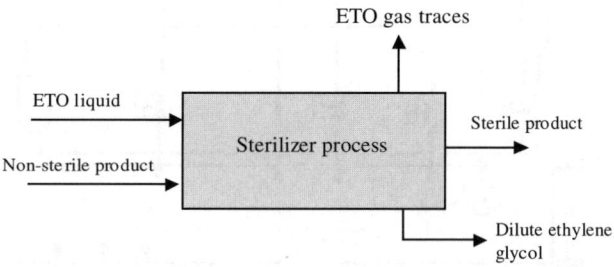

Figure 2.10
Environmental interfaces block diagram for sterilizer

2.4.2   Hazard study 2: presentation of design

The proposed design of the sterilizer system is represented in the block flow diagram shown in Figure 2.11. For this example, we have to short-cut some of the detailing but the essential features of the design are:

- The sterilizer chamber doors are closed and sealed gas-tight for vacuum by means of inflatable silicon rubber seals. Doors are operated manually.
- A PLC control system sequences valves and drives to execute the operating cycle without operator intervention.
- Evacuation of the chamber is done by a vacuum pump delivering its discharge through a water column absorber and exhausting to atmosphere.
- Ethylene oxide is supplied from an individual gas bottle with enough capacity for one or more cycles. The bottle has a dip pipe, and the liquid gas is passed through a water-heated evaporator to supply the gas into the chamber. The gas bottles and evaporator are sited in a separate room from the operating areas.
- The proposed operating cycle, see Figure 2.12, replaces air in the chamber with nitrogen before admitting ETO. The operating point and the transient gas mixture values can be seen on the flammability diagram in Figure 2.13.

In the sterilizer operating cycle in Figure 2.12:

- The gas composition in the chamber is air whilst vacuum is being pulled.
- Then nitrogen is admitted to a target pressure.
- Ethylene oxide is then admitted until the designed operating pressure is reached. This operating point is selected to be outside of the flammable region.

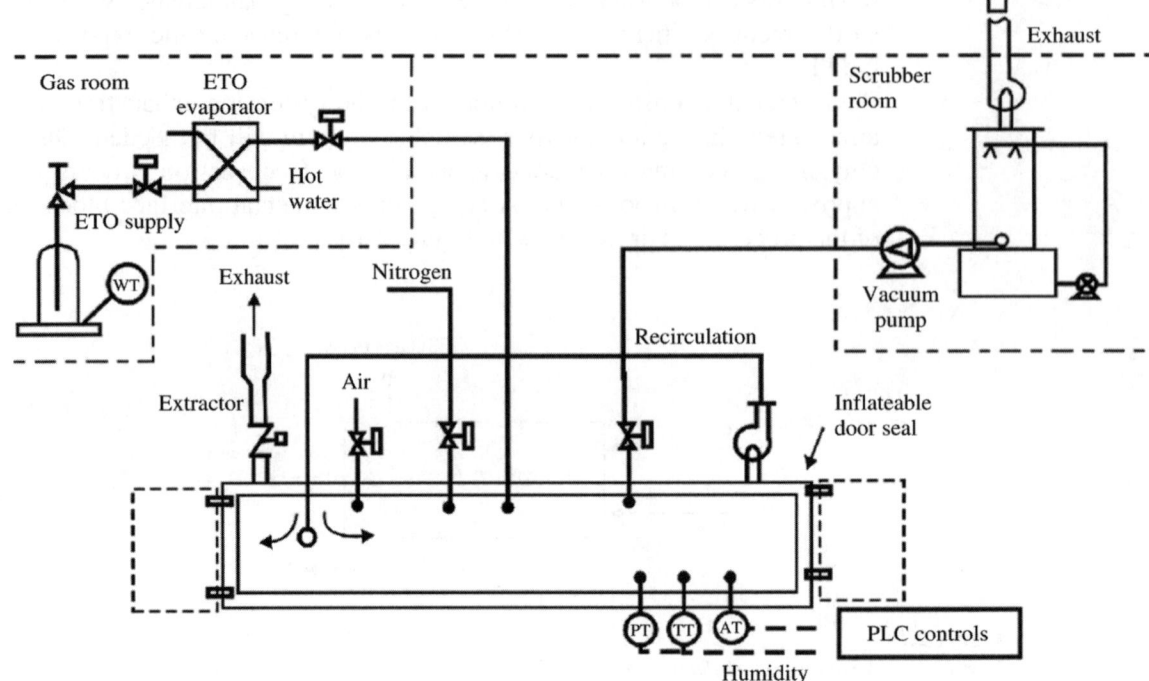

Figure 2.11
Simplified flow sheet – ETO sterilizer

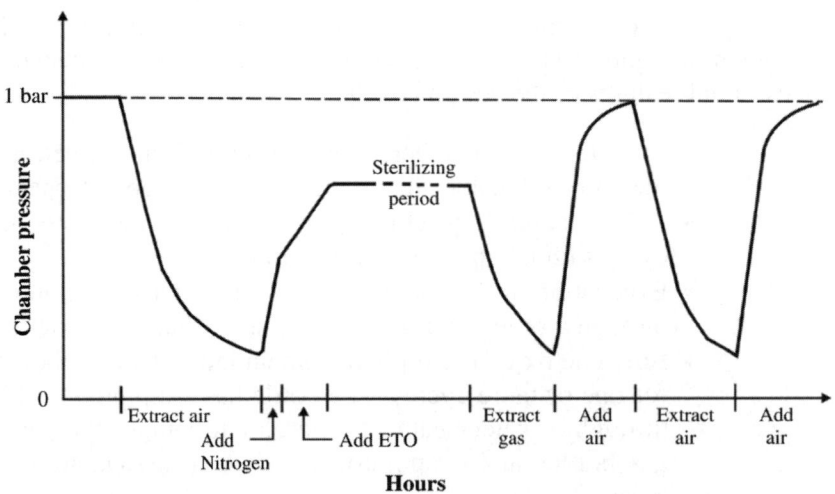

Figure 2.12
Sterilizer operating cycle

- At the end of the sterilizing period, the vacuum pump extracts the mixture.
- Air is admitted and the mixture changes to normal air as the purging takes place.

Note that this operating strategy has been designed to avoid the presence of a flammable mixture in the chamber, except for the purge stage at the end of the cycle. This is indicated by Figure 2.13.

Now we are ready to practice the hazard study 2 guideword method using the process Guideword Diagrams 1 and 2.

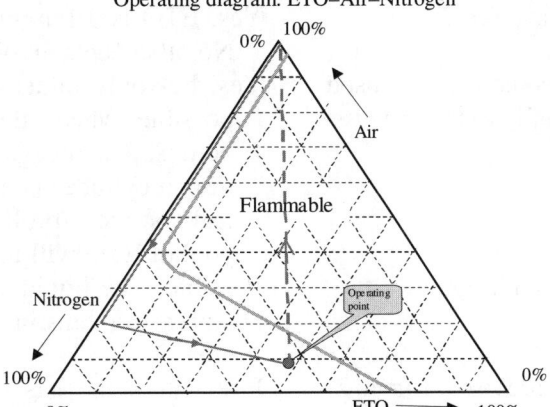

ETO sterilizer: Gas mixtures

Operating diagram: ETO–Air–Nitrogen

Figure 2.13
Gas conditsions

2.4.3 Scope

The scope of the study can be set to cover the whole sterilizer system from health, safety environment and business loss viewpoints. At this stage, we should exclude product-related hazards arising from the sterilization process, but we should consider this aspect as another type of study.

2.4.4 Attendees

- Study leader (to conduct questions and record all results. Reporting into the summary table is to be done for only those items where an action or comment is required)
- Designer/manufacturer's representative
- Project manager (to represent the business interests and to supply knowledge of the intended usage and staffing plans)
- Production supervisor (to represent the practical working conditions)
- Control/electrical engineer (for control system issues and hazardous areas)
- Pharmaceutical engineer (for process knowledge and application advice).

The team leader will record the names and attendance of each participant into a log sheet kept for study records. This record is essential for validation and quality management of the study process and is useful for all follow-up work.

2.4.5 Study proceedings

Let us now step through the guideword procedure remembering that this study is seeking to identify any significant hazards or essential design changes at the early stage of design. We will have the back up of a full Hazop once the detail P&IDs have been drawn; so it is not necessary to look for detail.

Guideword: External fires ... can it happen

Do we have fuel?	Yes, ETO is a flammable gas on its own and with air. No other fuels involved.
And ... could it be released externally to the process?	Yes, but only under accident conditions. This is only possible where the ETO liquefied gas is held in bottles, i.e. the gas room or during delivery. Each time a cylinder is changed there is the potential for a leak or loose/broken connection. It is likely that a pool of ETO will form on the ground.
And ... can it ignite?	Possible. The liquid will vaporize. There will be lights and instruments in the gas room. People will be in the area.
Consequences (see diagram 2)	Fire within the gas room could cause possible burns. The knock-on effect could be severe if remaining stored bottles are damaged or overheated.
How to prevent or reduce likelihood?	Remove sources of ignition. Define a hazardous atmosphere zone. Ensure maximum ventilation.
How to minimize consequences?	Ensure easy means of escape, provide fire hose station.
Actions required	See above.

Guideword: Internal fires ... can it happen

Do we have a flammable mixture in the process?	Yes, ETO and air but only in the feed pipe and transiently during purging. Note also the presence of combustible material in the chamber. Fire is only possible, once air is available.
And ... can it ignite? ... sparks, static, friction, etc.?	Not possible in the pipeline. Vaporizer heating cannot raise temperatures to ignition point. For chamber gas fire see internal explosion.

Guideword: Internal explosion ... can it happen

Physical overpressure?	Not possible, seals and relief devices will blow
Or ... uncontrolled reaction	Not possible
Or ... flammable mixture	Yes, ETO. Possible causes: ♦ Transiently during purging ♦ If the mixture control goes wrong during injection ♦ Leakage of air into the chamber after ETO injection.
And ... can it ignite?	Hardly possible, but there are electrical instruments in the chamber. So, yes it is possible. It has happened in the past with other sterilizer designs. Also possible if circulation fan strikes casing and causes sparks.
What if anyone left a cigarette live during the manual loading? Consequences	The normal cycle removes most of the air first, so the fire would probably go out before ETO was delivered. Unclear. Energy content of normal gas load is substantial. Possible rupture of chamber, possible fatalities. Investigation needed.

How to prevent or reduce likelihood?	♦ Provide automatic leak tests with interlock before ETO injection.
	♦ Ensure integrity of operating cycle controls to prevent wrong dosage of gas.
	♦ Avoid process using explosive mixture in the chamber.
	♦ Remove sources of ignition. Define a hazardous atmosphere zone in the chamber and hence use explosion proof or intrinsically safe instruments.
How to minimize consequences?	Consider blow-out panels in roof.
Actions required	Development of points in prevention list.

Guideword: Confined explosion in building ... can it happen

Do we have fuel?	Yes, ETO. No other fuels involved.
And ... could it be released externally to the process?	See under 'external fire' for gas room.
Can a gas could escape from the chamber?	No, because it is under negative pressure. If positive pressure did occur the gas would escape slowly and be very dilute.
Actions required	For gas room see as for fire. For chamber – no action.

Guideword: Unconfined explosion ... can it happen

This is the same as above for external fire.

Guideword: Acute/harmful/noxious exposures ... can it happen

Acute/harmful/noxious substance	ETO vapor cloud.
Exposure mechanism	There is risk of high-level exposure from leakages during bottle changes. High levels could also be experienced if the gas scrubber/extractor fan is not working when the de-gassing phase operates.
	High levels of gas exposure would occur if the chamber was opened without having been purged. This could happen if there is a power failure and the sequence controls are not allowed to perform the purge before the doors are opened.
Consequences	Temporary disabling injury. Possible fetal harm.
How to prevent or reduce likelihood?	♦ Use of respirators in gas room when changing bottles
	♦ Standard operating procedures. Good maintenance of fittings
	♦ Control system design to do purge cycles after outages
	♦ Door seals must stay inflated during power outages
	♦ Vacuum pump must not run if scrubber exhaust fan is stopped.

How to minimize consequences?	Medical supervision and health monitoring.
	Gas monitoring hand-held or fixed units to be used for bottle changes and chamber entries.
Actions required	Specify safety-instrumented functions to meet above needs. Write SOPs to attend to above needs. Plan for use of respirators in gas room.

Guideword: Chronic/harmful/noxious exposures ... can it happen

Acute/harmful/noxious substance	ETO vapor cloud.
Exposure mechanism	There is risk of persistent low-level exposure from residual gas in pipe connections during routine bottle changes.
	The maximum low-level exposure is expected as operators enter the chamber after a sterilization cycle to remove a product.
Consequences	Long-term health hazard with life-threatening consequences.
How to prevent or reduce likelihood?	♦ Provide high levels of fresh air ventilation using forced air distribution in chamber when either door is opened. Survey gas levels in chamber and decide optimum number of air purge cycles after ETO removal.
	♦ Ensure best operating practice and control, as for acute exposure, above.
How to minimize consequences?	Medical supervision and health monitoring.
Actions required	Implement design requirements for ventilation systems. Write SOPs to attend to above needs.

Guideword: Pollution ... can it happen

Pollutant	ETO vapor. Ethylene glycol liquid.
Release mechanism	♦ Low-level concentrations of ETO to atmosphere will arise from scrubber as its efficiency falls
	♦ High levels of release will occur if scrubber is not working
	♦ Glycol builds up in scrubber pond. Accidental release to ground during replacement.
Consequences	♦ Atmospheric pollution hazard from ETO with undefined consequences
	♦ Ground water contamination, violation of operating license.
How to prevent or reduce likelihood?	♦ Insist on scrubber working at all times, consider standby unit for availability
	♦ Regular replacement of scrubber fluid (water + dilute acid). Contract disposal of glycol. Avoid spills.

How to minimize consequences?	Medical supervision and health monitoring.
Actions required	Implement design requirements for ventilation systems. Write SOPs to attend to above needs.

Guideword: Violent releases of energy ... can it happen

Energy source	No significant energy source.
Actions required	None.

Guideword: Noise ... can it happen

Sources	Vacuum pump.
How to prevent or reduce likelihood?	Vacuum/scrubber room to be fitted with solid doors.
How to minimize consequences?	Do not plan occupancy of room, except for storage. Keep doors closed.
Actions required	Production manager to note.

Guideword: Visual impact ... is it a problem?

Appearance	Mainly enclosed at side of a low-profile building. Vent stacks will show.
Consequences	Local regulations may limit stack height.
Actions required	Check local regulations.

Guideword: Major financial effects ... can it happen

Factors to consider	Single stream unit. All production must pass through this unit. Availability is critical.
Consequences	Sterilizer outage exceeding 2 days will have major impact on plant throughput.
How to prevent or reduce likelihood?	Consider standby unit.
How to minimize consequences?	♦ Build in overcapacity in sterilizer. Ensure product storage areas on non-sterile side can buffer one week's production. ♦ Keep critical long delivery spares at plant.
Actions required	Evaluate production economics to decode best choices.

The above notes complete a hypothetical run through the flow sheet for the sterilizer with the guidewords as prompts. If the team leader and team members are satisfied that all the essential points have been covered, as far as they can see, the study session can be brought to a conclusion stage.

Remember this is the point where the team leader must get the draft summary sheets in good order and obtain agreement from the team on the outcomes of the study session. The suggested hazard summary table for this session is shown on pages 62 and 63.

Please note that blank versions of this form are to be found in the pages of Practical Exercise No. 3 (p. 268).

Hazard Study 2 Summary Form	Project: IDC workshop … ETO sterilizer			Drawing number: Figure 2.11	Rev no. 1
Team Members: Team Leader, Project Manager, Vendor Expert, Control /Elect Engineer, Pharmacist, QA Manager, Production Supervisor, Staff SHE representative.				**Date: 15/11/2002**	**Sheet no. 1**

Hazardous Event or Situation	Caused by / Sequence of Events	Consequences Immediate / Ultimate	Estimated Likelihood Suggested Measures to Reduce Likelihood	Emergency Measures (to Reduce Consequences)	Action Required
External fire in gas room.	Spillage/breakage in the gas room during bottle changes/handling. Ignition from electrical equipment.	Flash fire, risk of burns to operator. Other stored bottles could explode.	Possible. Design of fittings to reduce spillage.	Forced ventilation. Keep room open to outdoors. Fire fighting hose station.	Advise building designers. Advise piping designer. Add to equipment list.
Internal explosion.	Flammable mixture forms in chamber due to leakage under vacuum, or due to error in control cycle, or during air purge. Ignition from electrical instruments.	Unclear. Possible structural damage to chamber. Possible injuries or fatalities.	Possible. Auto leak test in program cycle. Abort sequence on leak detection or excessive pressure rise. Chamber internals to be Zone 1 hazardous area. Investigate safer operating cycles.	Blow-out panels in chamber and roof of building.	Investigate potential damage levels from chamber explosion. Hazard analysis required to decide risk levels. Review safety of control system.
Explosion in gas room.	Same as for fire, above, but flammable vapor cloud forms in gas room before ignition.	Injuries/fatality.	Unlikely. Ventilation, as for fire.	See as for fire.	See as for fire.

Hazardous Event or Situation	Caused by / Sequence of Events	Consequences Immediate / Ultimate	Estimated Likelihood / Suggested Measures to Reduce Likelihood	Emergency Measures (to Reduce Consequences)	Action Required
Acute exposure to ETO in gas room.	There is risk of high-level exposure from leakages during bottle changes.	Temporary disabling injury. Possible fetal harm.	Possible. Use of respirators in gas room when changing bottles. Standard operating procedures. Good maintenance of fittings.	Medical supervision and health monitoring.	Project team to organize respirators and standard procedures.
Acute exposure to ETO in scrubber room.	The gas scrubber/extractor fan is not working when the de-gassing phase operates.	Temporary disabling injury. Possible fetal harm.	Probable. Vacuum pump must not run if scrubber exhaust fan is stopped.	Gas monitor/alarm in scrubber room. Evacuation of room.	Project team to implement details.
Acute exposure to ETO in chamber.	High levels of gas exposure would occur if the chamber were opened without having been purged. This could happen if there is a power failure and the sequence controls are not allowed to perform the purge before the doors are opened.	Temporary disabling injury. Possible fetal harm.	Probable. Vacuum pump must not run if scrubber exhaust fan is stopped		

2.4.6   Software tools for PHA

It should be clear from all that we have seen in the study that it will be very helpful to have some automation of the study tasks. The potential scope for a software application is typically as follows:

- Study sheets or worksheets to be shown on display screen and updated as the study proceeds
- Deviation and guideword prompts from drop-down lists
- Built-in risk parameter descriptors would assist in consistent usage of qualitative terms
- Built-in risk matrix, customized to suit agreed company risk profiles charts
- Personnel and attendance records, session participants
- Action sheets linked to hazard study item entries
- Action sheets sorted by persons
- Templates for new or existing studies
- Automatic generation of reports, such as Hazard Summary
- Database for commonly found failure modes of plant equipment.

Several products are available worldwide, designed to meet many of the above requirements, with other features not listed here. Many companies make extensive use of such tools or develop their own versions to streamline their hazard study activities. See Appendix A for a short-listing of some established suppliers found by survey.

2.5   Case study

Please refer to Appendix E for directions on how to obtain a download copy of a very interesting advisory case study issued by the US Environmental Protection Agency. This study presents an example of a catastrophic accident and analyzes the causes. It goes on to describe the essential steps that should be taken to try to minimize the risk of such accidents. These steps are very similar to the measures we have been discussing in this chapter. A review of this paper will serve as a useful revision for the work so far.

2.6   Conclusion on hazard studies 1 and 2

We have seen how preliminary hazard studies provide valuable support and guidance to early stages of a plant design project. These studies lay the foundation for an SMS by providing:

Hazard study 1:   Baseline information on overall context and scope of the plant hazards.

Hazard study 2:   Identification of major potential hazards and action plans to deal with them.

We shall see later how these studies also form an essential part of the SLC records and requirements for alarm and trip systems. For this reason, it is strongly recommended that companies adopt a standard, defensible methodology for conducting PHAs to avoid missing the complete benefits of the PHA effort and to ensure compliance. To ensure that it is followed consistently, quality assurance guidelines should also be provided.

3

Risk reduction measures using alarms and trips

3.1   Risk reduction measures

3.1.1   Introduction

In this chapter we shall see how risk reduction measures are recognized and how they are applied to the problem of hazards. The hazard study team seeks to reduce risks by calling for risk reduction measures.

It will be helpful if we know how to identify the limits of the plant and its control system and recognize where the safety measures begin:

- Do we know where the basic plant ends and where the risk reduction measures begin?
- Is the control system part of the problem or is it a solution?
- How can we be sure that the risk reduction measures will be effective?
- How good are the alarms and trips?

We look at the concepts of applying protection layers to provide desired levels of risk reduction. Then we take a closer look at the role of alarm and trip systems as risk reduction measures.

3.1.2   Scope

- The terminologies of safety systems
- Operators and control systems as causes of hazardous events
- Why protection layers are important
- Alarms as risk reduction measures
- Limitations of alarm systems
- SIS as risk reduction measures
- The role of Hazops in defining alarms and trips.

3.1.3   Objectives

The objectives of this work are:

- To introduce the concepts of SIS
- To show the relationship between hazard studies and the specification of risk reduction measures
- To explain the role of alarms and trips as risk reduction measures.

The overall objective here is to strengthen the alignment between Hazops and the new standards for functional safety systems.

3.2   Terminologies and standards for safety systems

Before we go any further we need to be sure of understanding some of the strange terms that crop up in safety systems. The terminologies we are using here are based on those used in the current and upcoming international standards for safety systems, that we are introducing here.

Again, if we are going to align hazard studies with safety system standards, we need to understand the terminologies being used by the practitioners.

Terms explained here are:

- SIS
- Safety function
- Functional safety
- Safety integrity
- SIL
- Equipment under control (EUC).

3.2.1   Definition of SIS

(Origin: UK Health and Safety Executive: 'Out of Control')
'Safety instrumented systems are designed to respond to conditions of a plant that may be hazardous in themselves or if no action were taken could eventually give rise to a hazard. They must generate the correct outputs to prevent the hazard or mitigate the consequences.'

Abbreviation: SIS

We all probably know the subject by other names because of the different ways in which these systems have been applied. Here are some of the other names in use:

- Trip and alarm system
- ESD
- Safety shutdown system
- Safety interlock system
- Safety-related system (more general term for any system that maintains a safe state for EUC)
- Safety-related electrical control system (the term used in machinery safety for any electrical system that performs a safety function).

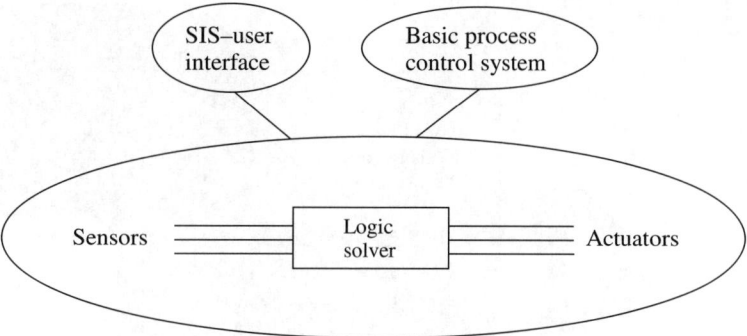

Figure 3.1
Block diagram definition of an SIS

Figure 3.1 defines the SIS as bounded by sensors, logic solver and actuators with associated interfaces to users and the basic process control system. We are talking about automatic control systems or devices that will protect persons, plant equipment or the environment against harm that may arise from specified hazardous conditions.

What do we mean by safety function?

We mean any function that specifically provides safety in any situation, e.g. a seat belt in a car, an air bag, a pressure relief valve on a boiler or an instrumented shutdown system. Thus an air bag has a safety function to prevent injury in the event of collision. The safety system of an air bag comprises the sensor, the release mechanism, the inflator and the bag itself.

3.2.2   Functional safety

'Functional safety' is a concept directed at the safety device or safety system itself. It describes the aspect of safety that is associated with the functioning of any device or system that is intended to provide safety.

'Functional safety is that part of the overall safety of a plant that depends on the correct functioning of its safety related systems.' (modified from IEC 61508 part 4)

'In order to achieve functional safety of a machine or a plant the safety related protective or control system must function correctly and, when a failure occurs, must behave in a defined manner so that the plant or machine remains in safe state or is brought into a safe state.'

From 'Functional safety in the field of industrial automation' by Hartmut von Krosigk. Computing and Control Engineering Journal (UK IEE) February 2000.

Figure 3.2 based on the above paper shows how functional safety compares with other safety aspects such as electrical safety or protection against fire. So we can conclude that functional safety is about the correct functioning of a unit or system designed to protect people and equipment from hazards.

3.3   Equipment under control

The term EUC is widely used in the IEC standard and has become accepted as the basis for describing the process or machinery for which a protection system may be required. Figure 3.3, taken from the HSE book, illustrates what is meant by EUC.

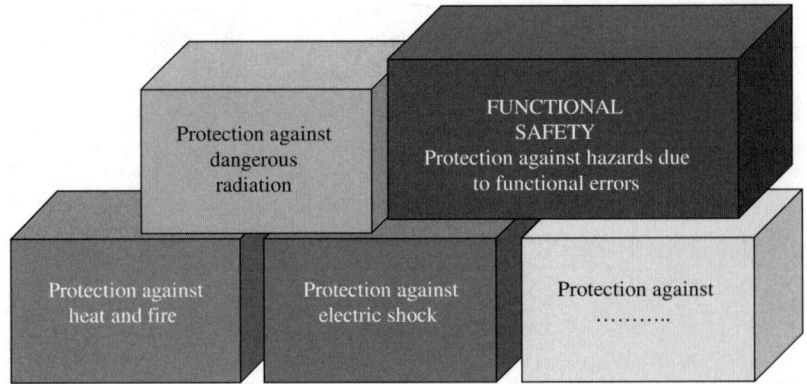

Figure 3.2
Functional safety is part of overall safety

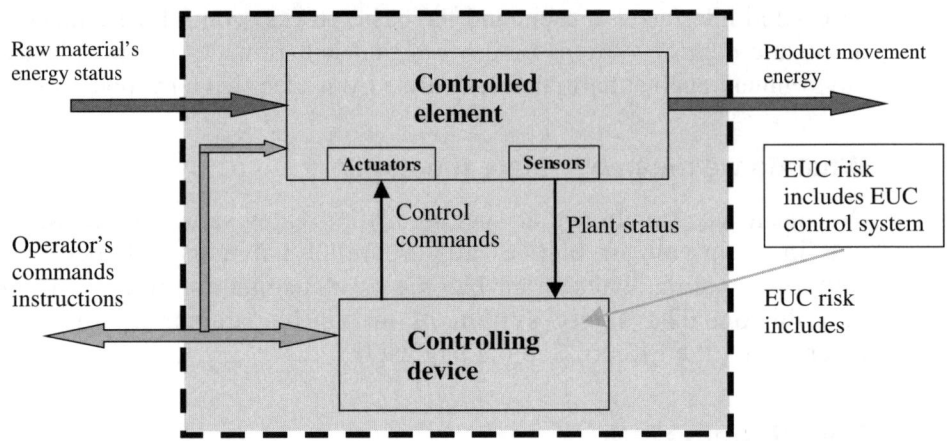

Figure 3.3
Equipment under control

The definition given in the IEC standards is 'Equipment, machinery, apparatus or plant used for manufacturing, process, transportation, medical or other activities'. As shown in Figure 3.3, this includes the plant control system and the human activities associated with operating the plant.

In this diagram, the controlling device is known as the EUC control system and it includes the operator's essential actions, even if these are simple manual operations. Hence, everything in the shaded area represents the EUC, and the risks we have to consider include everything that could cause trouble, including the operator and any malfunction of the control system.

In a way, this is saying that the distributed control system (DCS) on a hazardous process may be part of the problem. This requires that we include for all possible malfunctions of the control system when estimating the risk of a hazardous event, which is where the subject of computer Hazops or CHazops comes into being.

3.3.1 What is the BPCS?

BPCS is the abbreviation used in the process industry standard IEC 61511 for 'Basic Process Control System'. This means the normal control and instrumentation systems used to provide process control for the plant.

Hopefully, in our plant designs, we shall find that the BPCS and SIS are properly separated. This is often not the case, however, and it raises the important question: 'What if the BPCS performs a safety function?' The answer is 'The safety function may not achieve its intended degree of risk reduction'.

We shall look into this point later in the chapter. At this stage, we first need to introduce the engineering standards for the design and application of safety instrumentation.

3.3.2   Standards for safety instrumentation

Up until the 1980s the management of safety in hazardous processes was left to the individual companies within the process industries. Responsible companies evolved sensible guidelines out of the knowledge that if they didn't take care of the problem they would be the nearest people to the explosion when it happened. The chemical industry, for example, was always aware that self-regulation would be better than rules imposed by a worried public through government action.

Nothing much has changed except that the industry guidelines have matured into international standards and government regulators are seeing the potential benefits of asking companies and products to conform to what are becoming generally agreed standards.

3.3.3   Introducing standard IEC 61508

Many years of committee work within the IEC have resulted in the publication, in full, of the generic standard for functional safety systems: IEC 61508. It consists of seven parts as shown in Figure 3.4.

International Electrotechnical Commission

Title:

Functional safety of electrical/electronic/programmable electronic safety-related systems

All sections of IEC 61508 now published

Part 1:   General requirements
Part 2:   Requirements for electrical/electronic/programmable
           electronic systems
Part 3:   Software requirements
Part 4:   Definitions and abbreviations
Part 5:   Examples of methods for the determination of safety
           integrity levels
Part 6:   Guidelines on the application of Parts 2, 3
Part 7:   Overview of techniques and measures
See Appendix 1 for Framework Diagram

Figure 3.4
Titles of IEC 61508

3.3.4   Key elements of IEC 61508

This standard is far reaching and seeks to ensure safety system performance which is addressed in all respects including:

- Management of functional safety using a complete SLC-based quality assurance system
- Technical safety requirements for hardware and software

- Applies to manufacturer of sensors, PLCs and actuators
- Applies to design, installation and upkeep of safety systems
- Defines documentation records and MOC
- Requires persons to be competent for their level of duties.

3.3.5 Features of IEC 61508

Some important points to note about the standard at this stage are:

- Applies to safety systems using Electrical/Electronic/Programmable Electronic Systems (abbreviation: E/E/PES), e.g. Relays, PLCs, Software, Smart Sensors, Actuators, Networks
- Considers all phases of the SLC, including software life cycle
- Designed to cater for rapidly developing technology
- Sets out a 'Generic Approach' for SLC activities for E/E/PES
- Objective to 'facilitate the development of application sector standards'.

The SLC scope includes all stages from initial concepts and hazard studies through to operation, maintenance and modification. The standard covers electrical, electronic and programmable electronic systems and lays down standards of engineering and quality assurance for both hardware and software.

3.3.6 Introducing draft standard IEC 61511

IEC 61511 is a process sector implementation of IEC 61508 and was issued in February 2003. As shown in Figure 3.5 it comprises three parts. Part 1 defines how we should manage a safety instrumentation-based project and how the hardware and software should be designed and installed. Part 2 provides extensive guidance on the technical content of Part 1. Part 3 has very useful guidance on determination of target SILs that are to be set by the process design team at the start of the design phase of a protection system.

IEC 61511 is directed at the end-user in the process industries who has the task of designing and operating a safety control system in a hazardous plant. It follows all the requirements of IEC 61508 but modifies them to suit the practical situation in a process planct. It does not cover design and manufactue of products for use in safety, as these remain covered by IEC 61508. This distinction between the two standards and the link from ISA S84 is illustrated in Figure 3.6.

As IEC 61511 becomes established the process industries will be able to use it for systems and plant applications, whilst devices such as safety-certified PLCs and safety-related instruments will be built in compliance with IEC 61508. IEC 61511 is expected to be accepted in the USA as an approved code of practice for safety instrumentation intended to achieve compliance with OSHA requirements in the process industries. ISA S84, which is currently approved for this role, will then be superseded.

3.3.7 Impact of FS standards on hazard studies

Let us briefly consider here what the new functional safety standards are telling us about hazard studies. The SLC procedures laid down in IEC 61508 and 61511 ask us to carry out a hazard and risk analysis on the plant (EUC inc control system).

International Electrotechnical Commission

Title:

FUNCTIONAL SAFETY: SAFETY INSTRUMENTED SYSTEMS FOR THE PROCESS INDUSTRY SECTOR

Draft version

Part 1: Framework, definitions, system, hardware and software requirements

Part 2: Guidelines in the application of Part 1

Part 3: Guidance for the determination of safety integrity levels

Figure 3.5
Titles of IEC 61511

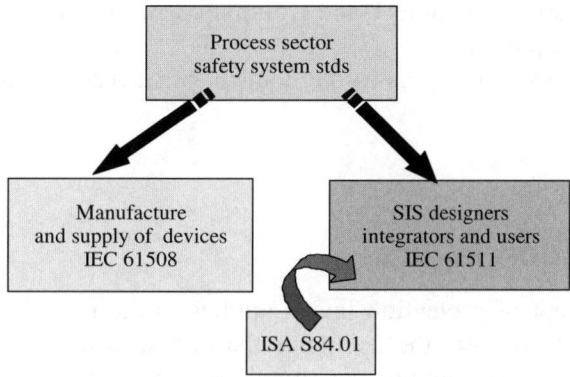

Figure 3.6
Relationships of IEC 61508 and IEC 61511

Here is description of the requirements of hazard and risk analysis contained in IEC 615111 Part 1 paragraph 8.2.1:

- 'A hazard and risk analysis shall be carried out on the process and its associated equipment (e.g., BPCS). It shall result in:
- A description of each identified hazardous event and the factors that contribute to it;
- A description of the consequences and likelihood of the event;
- The determination of requirements for additional risk reduction necessary to achieve the required safety;
- A description of, or references to information on, the measures taken to reduce or remove hazards and risk;
- A description of the assumptions made during the analysis of the risks including probable demand rates and equipment failure rates and any credit taken for operational constraints or human intervention shall be detailed;
- Allocation of the safety functions to layers of protection (9);
- Identification of those safety function(s) applied as safety instrumented function(s).'

The 'shopping list' above gives us a good idea of the steps between the basic hazard identification and the delivery of instructions to the design team for provision of a safety system. This list asks us to go a lot further than we have been so far with the hazard study level 2 results. We have covered the first three steps but the next step requires us to:

- Determine additional requirements for risk reduction
- Describe the measures taken to remove hazards and risks
- Analyze risks to determine probable demand rates and equipment failure rates
- Allow credits for operational constraints or human intervention
- Allocate safety functions to layers of protection
- Define those safety functions allocated to safety-instrumented functions.

Steps 1 and 2 are quite straightforward and often part of the hazard study material. Normally the hazard study will need to call for a risk analysis to be done by an individual to cover steps 3 and 4. This analysis will deliver an estimate for the demand rate, typically expressed as number of hazardous events per year. We saw this in the risk matrix charts in module 1 and it implies work such as fault tree analysis as we shall see later in Chapter 7.

Steps 5 and 6 bring us to the concept of 'layers of protection'. Let us investigate what this means.

3.4   Protection layers

The concept of protection layers applies to the use of a number of safety measures, all designed to prevent the accidents that are seen to be possible. Essentially, this concept identifies all 'belts and braces' involved in providing protection against a hazardous event or in reducing its consequences.

IEC 61511 provides a definition of protection layers for use in risk assessment models. A summary is shown in Figure 3.7. This definition is rigorous because it supports the formalized use of a safety layer in the determination of SIL for SIS designs.

Protection layers can be divided into two main types: Prevention and Mitigation, as seen in Figure 3.8. Each layer must be independent of the other, so that if one layer fails, the next layer can be expected to provide back-up protection.

IEC 61511-3 draft

> A protection layer consists of a grouping of equipment and/or administrative controls that function in concert with other protection layers to control or mitigate process risk.
>
> An independent protection layer (IPL):
> - Reduces the identified risk by at least a factor of 10
> - Has high availability (0.9)
> - Is designed for a specific event
> - Is independent of other protection layers
> - Is dependable and auditable.

Figure 3.7
Definition of a layer of protection

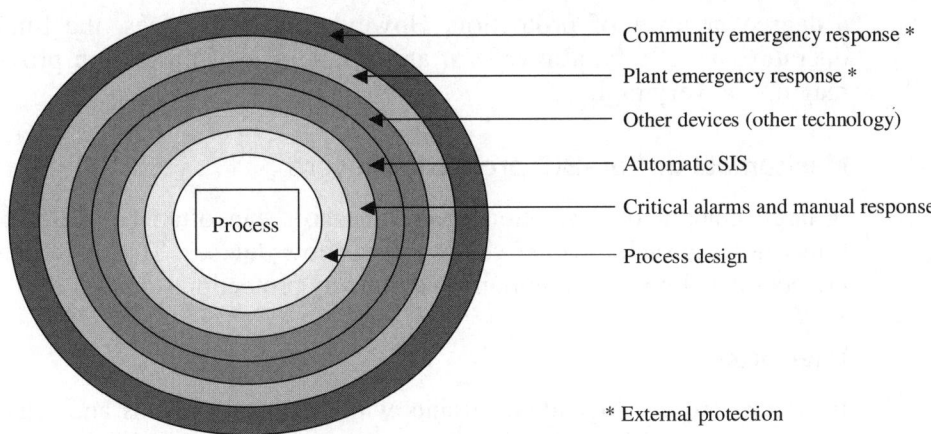

Community emergency response *

Plant emergency response *

Other devices (other technology)

Automatic SIS

Critical alarms and manual response

Process design

Process

* External protection

Figure 3.8
A simple representation of safety layers surrounding and reducing risk

Beware of common cause failures

It is important in any application to check any risk of common cause failures in the safety layers. In other words, we have to be very sure that there is little chance of two or more layers failing at the same time. This is not unusual and is an area of concern to safety specialists.

3.4.1   Prevention layers

Prevention layers try to stop the hazardous event from occurring. Examples of prevention layers:

Plant design

Plants should be designed, as far as possible, to be inherently safe. This is the first step in safety and techniques, such as the use of low-pressure designs, and low inventories are obviously the most desirable route to follow wherever possible.

Process control system

The control system plays an important role in providing a safety function since it tries to keep the machinery or process within safe bounds. As we have noted earlier, the control system cannot provide the ultimate safety function since that requires independence from the control system.

In fact, the accepted practice, as we shall see later, is to evaluate the risks of any operating process in terms of the EUC working with its control system (EUC control system).

Alarm systems

Alarms are provided to draw the attention of operators to a condition that is outside the desired range of conditions for normal operation. Such conditions require some decision or intervention by persons. Where this intervention requires a response that may prevent the situation becoming dangerous, the alarm can be seen to be providing a degree or layer of protection. However, in such cases, the limitations of human operators have to be allowed for and the degree of protection provided by the alarm may not be very high.

Mechanical or non-SIS protection layers

A large amount of protection against hazards can often be performed by mechanical safety devices such as relief valves or overflow devices. These are independent layers of protection and play an important role in many protection schemes.

Interlocks

Interlocks provide logical constraints within control systems and often provide a safety-related function. These functions may be embedded within the basic control system in the form of software, or they may be relay or mechanical interlocks directly linked to the equipment.

However, the contribution of interlocks to safety must be carefully checked during hazard studies by using the following guidelines:

If interlocks are built into the basic control system their performance should be evaluated as one of the failure modes of the control system, rather than as a protection layer.

If interlocks are provided in the safety system, they can be regarded as protection layers.

Therefore, in each application, decide if the interlock is part of the control system, an independent safety device or part of the SIS, and rate the protection accordingly.

Shutdown systems (SIS)

This is a protection layer in which automatic and independent action is taken by the shutdown system to protect the personnel and the plant equipment against potentially serious harm. An automatic shutdown system provides a good level of protection because it is able to take direct action and does not require a response from an operator. If an operator action is the only way to initiate a shutdown system its performance must be de-rated to allow for a correct response from the operator.

3.4.2 Mitigation layers

Mitigation layers reduce the consequences after the hazardous event has taken place. Mitigation layers include fire & gas systems, containments and evacuation procedures. They could also include toxic gas refuges. Anything that contributes to reducing the severity of harm, after the hazardous event has taken place, can be considered to be a mitigation layer.

3.4.3 Diversification and common cause failures

Using more than one method of protection is generally the most successful way of reducing risk. The safety standards rate this approach very highly and it is particularly strong where an SIS is backed up with, say, a mechanical system or another SIS working on a completely different parameter.

Diversification generally reduces the risks of common cause failures, as we mentioned earlier. But beware of residual common cause failures. For example, the pesticides chemical plant in Bhopal, India, released a toxic gas cloud that is believed to have killed more than 16 000 people. 44 tons of deadly methyl-isocyanate (MIC) liquid decomposed into cyanide and phosgene gases and escaped across the city. This happened despite at least three layers of protection:

- Chilled glycol circulation to keep stored MIC stable below ambient temperature.
- Gas scrubber system to absorb vapors arising from the storage tanks.
- Flare stack to burn-off any vapors that escaped the scrubber.

All the protection layers failed at the same time, as a runaway reaction was induced by an incorrect cleaning procedure for pipelines leading to the storage tanks.

Common cause of failure

All protection systems in the Bhopal plant were switched off by instruction from local company management. They wanted to save money whilst the plant was shutdown and did not realize that MIC would decompose whilst under storage if it got warm, or perhaps they thought there would be no reason for it to be heated. The international owners also seem to have neglected to ensure local management was adequately informed.

Whilst the original design engineers for the plant knew the hazards and engineered the appropriate safety systems, the subsequent upkeep of the on-site SLC activities failed when the factory became a loss-making business.

(Ref. D Lapierre and J Moro: Five Past Midnight in Bhopal. Simon & Shuster UK, ISBN 0-7432 3088 4.)

3.4.4   Risk reduction models

The IEC standard explains that the layers of protection are effectively separate risk reduction functions. A general risk reduction model based on the version drafted for IEC 61511 is shown in Figure 3.9.

Risks indicated in Figure 3.9 are:

- EUC risk for the specified hazardous event, including its control system and the human factors.
- Tolerable risk based on what is considered acceptable to society.
- Residual risk being the risk remaining after all external risk reduction facilities, SIS and any other related safety technologies have been taken into account.

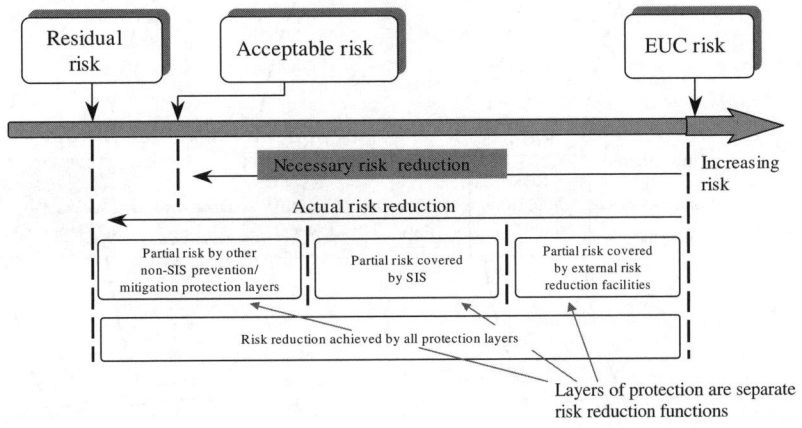

Figure 3.9
Risk reduction: general concepts (from IEC 61511)

3.4.5   Risk classification

The problem is that risk doesn't come in convenient units like volts or kilograms. There is no universal scale of risk. Scales for one industry may not suit those in another industry. Fortunately, the method of calculation is generally consistent and it is possible to arrive at a reasonable scale of values for a given industry.

For example, Suppose we have a critical consequence risk such as one death and there is a risk frequency of one per year. If we decide that the risk frequency is not tolerable unless it is less than once per 500 years, we would need to reduce the frequency of the unprotected hazardous event) from one per year ($F_{np} = 1$) to at least as low as one per 500 years ($F_t = 0.02$). The risk reduction factor is then obtained from RRF = F_{np}/F_t.

In this case, RRF = $500/1 = 500$. This is the overall risk reduction factor.

We can go one step further and allocate a measure of risk reduction to each layer of protection that has been identified for a given hazard. Figure 3.10 illustrates this allocation where it shows two layers of protection, each layer having its own contribution to risk reduction. In this example, the SIS has been allocated a target risk reduction factor of 50.

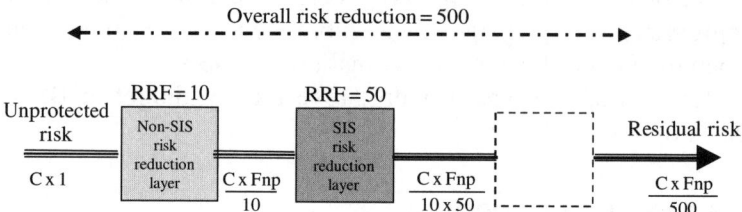

Figure 3.10
Overall risk reduction

This type of diagram (Figure 3.10) is called a risk reduction model and is a powerful method of showing the details of risk reductions for any application. If we return to the concept of layers of protection it should be clear now how each layer of protection provides a separate and independent risk reduction function.

The safety layer diagram we saw in Figure 3.10 can be enlarged as shown in Figure 3.11 to show how each layer reduces risk by a certain amount.

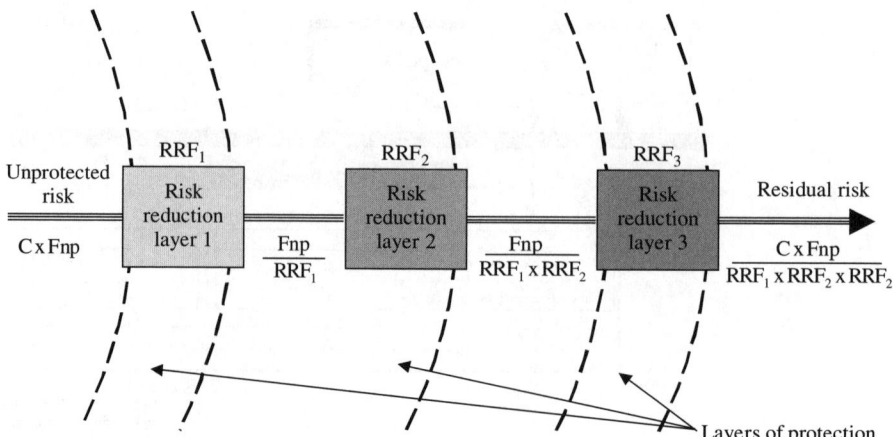

Figure 3.11
The unprotected risk is reduced by each layer of protection

3.4.6   Risk reduction terms and equations

There is a choice of the terms we can use to define the risk reduction. The following notes and Figure 3.12 should help clarify the use of these terms.

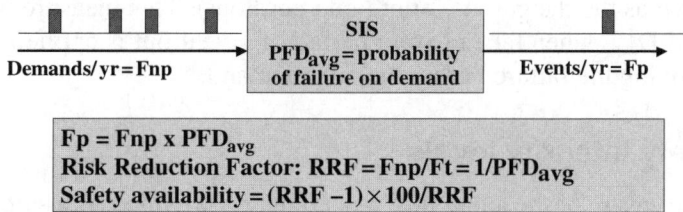

Figure 3.12
Introducing PFD$_{avg}$

Risk reduction factor

It is often convenient to talk in terms of risk reduction factor. In our example, above, when the tolerable frequency of a hazardous event is Ft = 0.02 per year and the unprotected hazard frequency is Fnp = 1 per year, the overall RRF = 500.

The RRF of the SIS in our example is 50 and the non-SIS layer contributes an RRF of 10.

Safety availability

This is simply the percentage of time that the protection system has to be available to meet the risk reduction target. As you would expect, this figure is typically a high percentage.

$$\text{Required safety availability} = (\text{RRF} - 1) \times \frac{100}{\text{RRF}}$$

For our example of 500, the SA will be 99.8% for the overall scheme. The SIS will require an availability of 98%. This is the average probability of failure to perform the design function on demand (for a low demand mode of operation).

Average probability of failure on demand (PFD$_{avg}$) is described in standards as the average probability of failure to perform the design function on demand (for a low demand mode of operation). It is important to be clear on the concept of probability of failure on demand. Figure 3.12 illustrates the principle.

The diagram in Figure 3.12 illustrates how a series of demands on the safety trip system will not lead to a hazardous event, unless the system has failed at the time of the demand. The rate at which events occurs is then Demand rate × average PFD of the safety system.

The alternative name for PFD is the fractional dead time, FDT. The meaning is clear – the fraction of time for which the safety system is dead!

For example, if the SIS must reduce the rate of accidents by a factor of 100 it must not fail more than once in 100 demands. In this case, the SIS requires a PFD average of 0.01.

Basically,

$$\text{PFD}_{avg} = \Delta R = \frac{1}{\text{RRF}}$$

For our RRF example of 500, the PFD$_{avg}$ will be 0.02 or 10^{-2}

Low demand vs high demand rate

If the demand on the safety system is less than one per year, the PFD_{avg} serves as the measure of risk reduction that can be expected from the safety system.

If the demand rate is higher than one per year or is continuous, the IEC standard requires that the term to be used is the probability of a dangerous failure per hour (also known as the dangerous failure rate per hour). This measure is approximately the same as the PFD_{avg}, when taken over a period of 1 year but is applied when there is a poor chance of finding the failure before the next demand.

3.4.7   Safety integrity levels

The measure of the amount of risk reduction provided by a safety system is called the safety integrity. In order to get a scale of performance safety, practitioners have adopted the concept of SILs. The SILs are derived from earlier concepts of grading or classification of safety systems. We have seen that the risk reduction factor provided by the SIS can be converted to a PFD_{avg}, Safety integrity levels are simply broad bands of PFD_{avg} values or RRF values. Here is a table relating the SIL values to RRF and safety availability.

SIL	PFD_{avg}	Risk Reduction Factor	Safety Availability (%)
1	10^{-1}–10^{-2}	10–100	90–99
2	10^{-2}–10^{-3}	100–1000	99–99.9
3	10^{-3}–10^{-4}	1000–10 000	99.9–99.99
4	10^{-4}–10^{-5}	10 000–100 000	>99.99

So, for example, if we have an SIL-2 rated safety system, it means the risk reduction it provides lies in the range of 100–1000. Using this logarithmic scale means that a broad classification of safety systems can be achieved in four classes ranging from 10:1 to over 10 000:1 risk reduction capability. In practice, the process industries will use SIL-1, -2 and -3, whilst SIL-4 is only used for very severe problems. SIL-3 represents the upper limit of normal engineering practices achievable in most process environments.

Hence, the performance level of safety instrumentation needed to meet the SIL is divided into a small number of categories or grades.

3.4.8   Implications of SILs for the Hazop study

An SIL-1 system is not as reliable in the role of providing risk reduction as SIL-2; an SIL-3 is even more reliable. Once we have the SIL, we know what quality, complexity and cost we are going to have to consider. It should be clear from this that if we can calculate the risk reduction factor (and convert it to a PFD_{avg} requirement or as safety availability), we can look up the tables and define the SIL we are going to need from our safety system.

It should also be clear that if we call for more risk reduction than is really needed, the SIL rating will become high and the cost is likely to be drastically higher. As you would expect, hazard study teams are always conscious of the cost of measures they prescribe.

3.5   The role of alarms in safety

The motivation for many of the alarms in a control room comes from the hazard studies. It follows that the hazard study team should have a clear understanding of the ground rules for the alarm system in their subject plant. This is often called the 'Alarm

Philosophy' and is written up by the design team or operating team for the plant. Often, the alarm philosophy is inherited from practices established elsewhere in the same complex and which has formed the basis of training practices for control room operators.

Here lies a potential source of trouble because the existing alarm philosophy may be designed around a particular control system equipment and long-established practices but it may not provide much guidance for the Hazop team trying to specify the alarm functions and their priorities. In other words, we may be accepting a 'bottom up' approach to the specification of alarms. This carries the risk of an unstructured growth of alarms, which can have disastrous consequences when a major plant upset occurs. What we really should have is a top-down design.

Top-down design:

- The process design team proposes alarms for process management and for safety-related functions.
- The Hazop team reviews the alarms in terms of the hazard and operability requirements and may suggest additional alarms for protection against hazards. The Hazop team must ensure there is clarity on the purpose and risk reduction expectations of all alarms that are safety-related. The alarm scheme is built around these needs.

Bottom-up design:

- The Hazop team generates a list of alarms in terms of pre-existing categories e.g. first level process alarms, critical process alarms, safety-related alarms.
- Designers add these alarms to the list provided by the process designers and the list is simply transferred into the control systems.

It is not the job of the Hazop team to design the alarm management system; they have enough to do just identifying the problems and suggesting corrective measures. But the engineer who has to follow up on the Hazop actions list will be expected to set up the alarm system to deliver on all the requirements as reliably as possible. To do this, the engineer needs to have fundamental information on each alarm including:

- Its expected frequency
- The desired response and the consequences of failed response
- Is the alarm classed as safety-related?

Hazops have a notorious record of generating large quantities of 'add-in' alarms as a 'quick fix' solution to numerous operability problems found as the studies progress. The bottom-up approach tends to encourage this practice as it accepts all alarms into the list without questioning their performance intent. This carries the risk of alarm overload at times of plant upset, thus diminishing the value of all the alarms that arise.

Let us take a look at the key issues in the management and the use of alarm systems so that we have a better understanding of what the Hazop team should expect from alarms.

3.5.1   Principles of alarm systems

Alarms support the operation of industrial plants by alerting operators to a variety of conditions. The majority of alarms found in a process control system will be there to help the operators to keep the manufacturing process running in the intended manner and to help achieve the best possible production performance. Other alarms will warn of deviations that are linked to possible hazards.

3.5.2   Alarm indicators and annunciation

Alarm displays are an integral part of the human interface to the production process. Any device that is able to bring the operator's attention directly to the alarm condition will be classed as an alarm indicator but typically it will consist of a display such as an alarm annunciator window in a hardwired panel or a color-coded faceplate or message line on a control room VDU. Audible alarm tones or sirens frequently support the display to grab attention. Similarly, if an operator is likely to be outside of the control room or when operating with a large distributed process such as gas distribution or water schemes, alarms can be transmitted to mobile receivers.

3.5.3   Alarm equipment options

A complete alarm system consists of both hardware and software including field signal sensors, transmitters, alarm generators and handlers, alarm processors, alarm displays, annunciator window panels, alarm recorders and printers. Which of these features is used and how they are deployed depends on the type of alarm being handled and the scope of the application. Let us consider the most common arrangements.

3.5.4   Integrated control and alarms

In most chemical processing plants we would expect to see much of the alarm system embedded in the basic control system, typically an SCADA system or a DCS. This brings the benefits of integration with the main controls for the process and makes it simple to provide alarms related to the process conditions. Similarly, the displays of such alarms can be achieved through the normal VDU workstations in the control system network. Software tools in the SCADA and DCS packages allow the alarms to be configured easily and offer the facilities to group and prioritize the alarms.

One of the long-standing attractions of SCADA and DCS systems has been their ability to record the occurrence of each and every alarm in the system with high degree of time resolution, hence supporting problem analysis and the reconstruction of a sequence of events during a plant outage.

3.5.5   Discrete alarms

Before the widespread use of computers in process control, the normal arrangement was to organize alarms through discrete sensing and switching devices and perform the interface functions of light indication and audible warning through an alarm annunciator module. This method is still widely used for many plants and is commonly used in equipment packages where a small number of equipment alarms, such as bearing overheats or low oil pressure warning alarms, are provided before the safety tripping level is reached.

3.5.6   Alarm management systems

The most significant development in recent years has been the realization that the quantity of plant alarms can get out of hand and become a serious problem when abnormal situations develop in the production plant. Most alarms have been installed to deal with detailed and localized problems of process or equipment control. Very often, this leads to a backlog of standing alarms or frequent minor alarms from process conditions running in and out of limits at short intervals. This is a nuisance in times of steady operation, but when a major upset occurs, the large number of alarms becomes a real problem as they all switch into alarm as the disturbance takes hold.

Alarm management systems are software-driven resources within the DCS or SCADA that serve to group, rationalize and prioritize the alarms to provide a more effective response to the most important alarms.

They assist operators to pack away less important alarms whilst ensuring they do not get lost completely. The system may also provide:

- Tools to assist in the tuning and adjustment of alarms
- Alarm log and selected history files
- Sequence of event recording.

All the facilities of an alarm management system are of little help if the plant does not have a clear and well-supported 'plant alarm management strategy'. This involves process designers and the end-users working to a common set of rules for defining and maintaining the alarm system. For guidelines on how alarms systems should be managed we would suggest that users make a careful study of the EEMUA guide.

The most important lesson is to realize that:

- Hazard studies teams must be aware of the benefits of well-managed alarm systems to plant operability.
- Well-defined pre-trip alarms can save a great deal of money for the business by helping operators to avoid tripping of plants.
- A badly managed alarm system can be a serious danger to the plant by overloading the operator with useless information at the time of crisis.
- All alarms must be defined within the framework of the alarm philosophy.

3.6 Alarm types and do they qualify as safeguards?

From the outset of any alarm study or design task, it is important to recognize characteristic types of alarm. From the designer's viewpoint and for hazard studies, it is very important to make a clear distinction between alarms that provide production support and those that are there as safeguards against hazards. Here are some fundamental types of alarm.

3.6.1 Process alarms

These alarms may be to do with efficiency of the process or indicate deviations from intent. Process alarms are normally incorporated into the plant control system (typically a DCS or PLC SCADA system) and often share the same sensors as the control system as shown in Figure 3.13. Where sensor reliability is an issue, the alarms may need to be driven from an independent sensor.

3.6.2 Machinery or equipment alarms

These alarms assist with detection of problems with equipment and indirectly affect the operation of the process. An equipment alarm may simply be an important status relevant to the operation of a process (e.g. 'Compressor stopped') or it may be critical to the equipment (e.g. 'bearing temperature high').

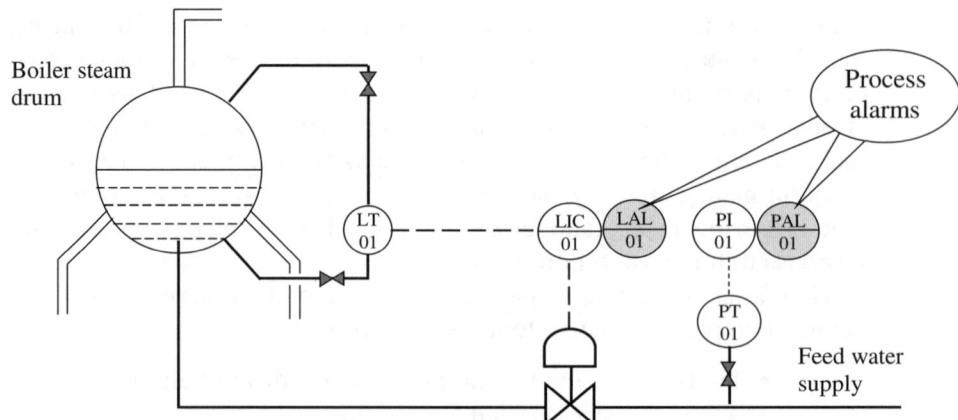

Figure 3.13
Typical process alarms

3.6.3   Safety-related alarms

These alarms are used to alert operators to a condition that may be potentially dangerous
or damaging for the plant. They will always demand a defined and prompt response from
the operator if they are to be of any real value in reducing risk. Safety-related alarms
might occur with or without an associated automatic trip. We shall look at these more
closely in a moment.

3.6.4   Shutdown (or ESD) alarm

This type of alarm tells the operator that the SIS has initiated an automatic
shutdown event. It is basically a monitoring function for the SIS. It contributes as a
layer of protection because it informs on the status of an emergency shutdown,
supporting corrective and subsequent actions by the operator. It may also be the
trigger for a mitigation action such as closure of other parts of plant or isolation of
fuel supplies.

3.6.5   Fire and gas alarms

These types of alarms belong to the safety-related types but are usually built within
dedicated and entirely independent fire and gas detection systems. This approach is
essential for protection of personnel since it must be assumed that all other control
systems may be shutdown or damaged due to fire or gas conditions.

3.7   Identification and design of safety-related alarms

Here we examine factors that help to identify and specify a safety-related alarm. We
begin by taking a simple example as might present itself at a hazard study.

In Figure 3.14 we see a safety-related alarm scheme for the boiler drum.

The first alarm is LAL-01. It is a process alarm but it may have a safety role as well. It
is integral to the process control scheme and warns of the need to take action as a first
threshold level is crossed. The operator should respond by correcting the control settings
(maybe they were left on 'manual') or perhaps reduce the steam load on the boiler. If this
alarm fails or the operator actions fail to solve the problem, the level may fall lower and a

'final warning' alarm LAL-02 sounds and the operator is required to shutdown the burners or furnace by his own actions within, say, the next 10 minutes or else the boiler will be damaged and there may be up to two injuries. In the example shown here, the safety-related alarm is LAL-02 and it operates from a level-sensing instrument that is independent of the level transmitter used in the control system.

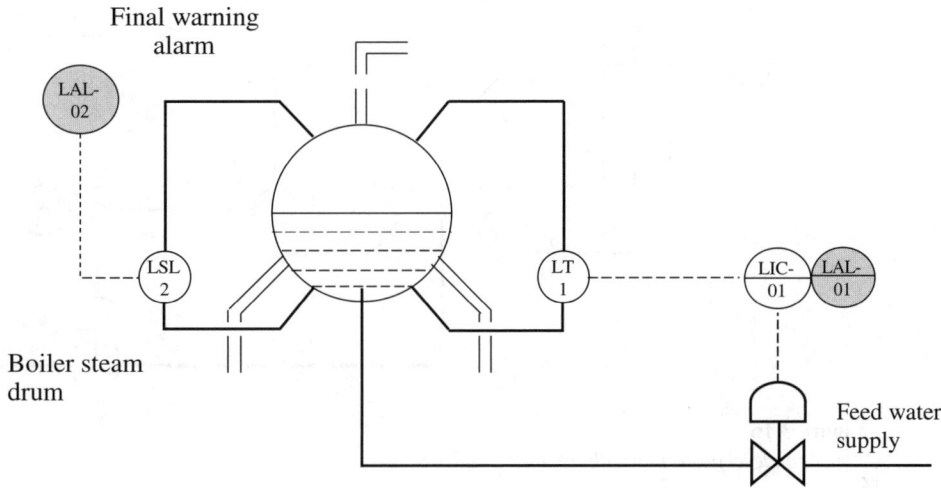

Figure 3.14
Safety-related alarm with independent sensor and display

With no automatic trip, in this example, the dependency on the alarms and operator are high, and both alarms carry the common failure mode that the operator may not be in the control room at the time of the event. The (approximate) risk reduction diagram for this situation looks like that in Figure 3.15.

The predicted risk of boiler damage and injuries at one per 20 years may well be unacceptable; hence the normal practice here is to provide an automatic trip of the boiler furnace. In Figure 3.16 the trip has been added and the final warning alarm has been set up as a pre-trip alarm operating at a level higher than the trip point.

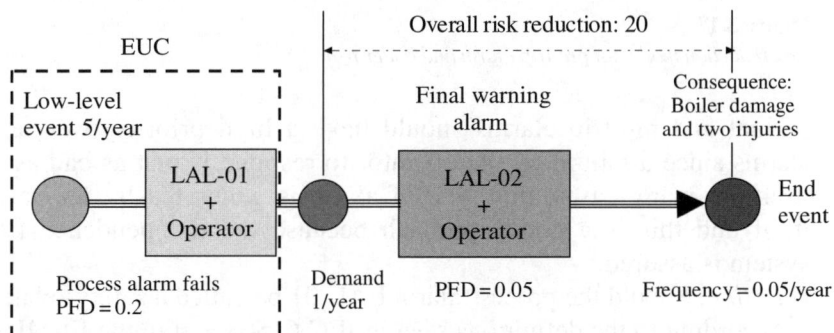

Figure 3.15
Risk reduction diagram for 'final warning' alarm

The next risk reduction model in Figure 3.17 shows how this arrangement increases the risk reduction factor to 1000 and hence drops the predicted event rate to one in 1000 years. In this version, the alarm LAL-02 becomes a 'pre-trip alarm' but it still performs a safety function when combined with the operator's intended response. Pre-trip alarms can provide a significant protection against hazards because they act to reduce the demand rate on the SIS. Even the process alarm has a role to play here.

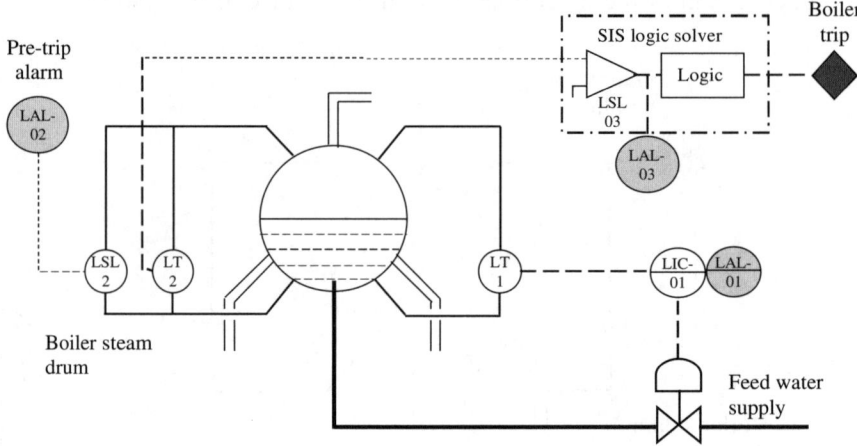

Figure 3.16
Low drum level protection with pre-alarm and SIS trip

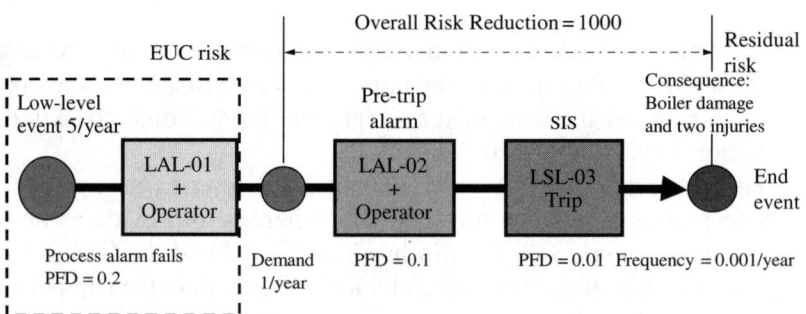

Figure 3.17
Risk reduction model for *pre-trip alarm and boiler trip*

Final and pre-trip alarms should have a high priority over process and equipment alarms since a failure by the operator to respond is just as bad as a failure of the alarm itself. In many cases, this type of alarm is generated by the safety shutdown system itself and this is a good approach because the independence from the main control system is assured.

Question: Should the process alarm LAL-01 be called a safety-related alarm?

According to the definitions seen in IEC 61508 and in the EEMUA guide the answer is NO. The alarm LAL-02, on the other hand, is definitely safety-related. Let us see why.

IEC 61508 says that an alarm can be considered to be safety-related if:

- It is a claimed part of the facilities for reducing the risks from hazards to people to a tolerable level.
- The claimed reduction in risk provided by the alarm system is significant. 'Significant' means a claimed PFD_{avg} of less than 0.1.

In our example, the combined PFD of the pre-trip alarm and operator response has been set at 0.1. If we set the operator response at an optimistic 0.05, this means the alarm will have to be at least as good as PFD = 0.05.

3.7.1 Design implications of safety-related alarms

This definition also implies that the alarm system equipment falls within the category of at least SIL-1. Hence it follows (and as stated in the EEMUA guide and in IEC 61508) that:

- It should be designed, operated and maintained in accordance with the requirements set out in the standard (IEC 61508).
- It should be independent and separate from the process control system (unless the control system itself has been designated as a safety system).

This requirement implies that an alarm sourced or delivered via the main plant DCS cannot qualify as a safety-related alarm. Such alarms should be provided by an independent alarm system or via the SIS.

3.7.2 Reliability considerations for the safety-related alarm

We have seen that the overall reliability (expressed as PFD_{avg}) of a safety-related alarm depends on the combined response of the equipment and the operator. Hence human factors have be taken into account when designing the alarm function. These factors also limit the risk reduction claims that can be made for any alarm function. What are the factors that affect the operator's response?

Principally, they are:

- Loss of concentration
- Lack of experience, knowledge of required action or inadequate training
- Errors due to time pressure
- Stress due to overload situations at the time of an event.

The design of the alarm system can go a long way to overcome some of these factors by using measures such as:

- Clear presentation and annunciation of alarms
- Designing and setting alarms to allow adequate time for response
- Structured and prioritized alarm presentation
- Alarm management methods, including rationalizing the installed alarms and automatic suppression of redundant alarms, reduce the quantity of alarms.

With these features in place, the probability of failure of the operator's response can be expected to be lower than 1 in 10 ($PFD_{avg} < 0.1$) but should never be expected to be better than 1 in 100 ($PFD_{avg} > 0.01$). Under conditions of stress and alarm overload, the failure rate may rise well above 1 in 10 and in reported accident events, operators have been forced by alarm floods and other concerns to ignore all alarms.

The instrumentation of the alarm system can usually achieve a PFD_{avg} range from 0.1 to 0.01 (i.e. SIL-1 performance) without too much difficulty, provided the equipment is designed as safety-related, preferably to IEC 61508. There seems little point in specifying the instrumentation higher than SIL-1 if the operator portion can never achieve the same.

3.7.3   Summary

From the above descriptions, we can now summarize the points to bear in mind as alarm requirements emerge from the hazard studies:

- All safety-related alarms should be clearly identified by the hazard study reports.
- A listing of these alarms should always be maintained and should subject to MOC control. Other alarms will be treated as non-safety-related.
- Separation of safety-related alarms from basic control systems should be ensured and checked by the hazard studies.
- Consider the possibility of common mode failures due to any factors, but, including:

    - Sharing of the sensors used by the control system, either due to shared instruments or due to factors in the process (e.g. plugging of instruments due to sludge)
    - Any shared equipment such as power supplies, DCS or SCADA displays
    - The workload of operators at the time of alarms, including that due to other non-safety-related alarms
    - Any maintenance activities.

- Safety-related alarms to be given the highest priority in display and annunciation schemes. Other alarms to be of lower priority.
- Reliability of each safety-related alarm function should be specified in terms of the required or claimed risk reduction or PFD_{avg}. This statement will then link the hazard analysis and final risk assessment to the design of the alarm function. See EEMUA guide page 15 for a summary of the reliability requirements for alarms vs risk reduction claims.

3.7.4   Problems of alarm overload

In recent years, it has become clear that 'alarm floods' and poorly managed alarm systems present a considerable threat to plant safety. Let us take a few minutes to look at some of the well-documented case histories.

- From a paper by Michael S Carey: Human factors in the design of safety-related systems, UK Computing and Control Engineering Journal February 2000: An HSE Report surveyed 96 operators at 13 different chemical plants and power stations. Averaged results included:

    - Normal operations experienced approximately one alarm per 2 minutes throughout the shifts.
    - Around 50% were repeat alarms of that which appeared and reset again in after a few seconds and repeated again within 5 minutes. These were of little value.
    - Of the many standing alarms, about 6% were related to actual operational problems.
    - The report found that when large plant disturbances occurred, the alarm systems became even less usable.

- Typically, 90 alarms in the first minute and another 70 in the next 10 minutes.
- About half the operators sometimes felt forced to accept the alarms during upsets, without reading or understanding them.
- Sometimes, during plant upsets, operators would not make a full check of alarms for up to half an hour.

From Health and Safety Executive: The explosions and fires at Texaco refinery, Milford haven 4 July 1994: Causes of an accident that cost an estimated £48 million to rebuild were attributed to several items including a modification of a protection system from automatic to manual without any formal 'safety assessment', poor instrument maintenance, badly designed VDU presentations on the DCS and alarm overloads in the control room.

- Several hundred alarms occurred in the control room with a peak rate estimated at one every 2 or 3 s.
- Prioritization of alarms was poor with 87% of alarms classified as high priority.

These events, amongst others, have led to a strong drive in the process and control world to get to trips with the problem of uncontrolled growth of alarms in control rooms. Many companies have embarked on alarm improvement projects and have been able to show good results.

The EEMUA guide provides substantial guidance on how to set about an alarm improvement exercise, and if you suspect that your plant is a candidate for such an exercise we would strongly recommend that you study the EEMUA guide. See also the HSE free publication: 'Better Alarm handling' (via www.hse.gov.org).

How can the alarm situation be improved?

- Ensure the risk reduction expectations of each alarm are clearly understood by all parties.
- Make sure the safety-related alarms are separated from standard process and equipment alarms in the equipment and in the method of display.
- Follow key design principles in planning an alarm.
- Use prioritization techniques such as color coding and sound to ensure the safety-related alarms get the highest priority.
- Weed out all badly defined and poorly performing alarms and present for an alarm reduction review. Use Hazop methods where safety-related alarms are involved.
- Develop alarm suppression techniques to reduce the number of unnecessary alarms according to the status of the plant.
- If alarm suppression is used, a hazard study session may be needed to ensure critical information has not been accidentally hidden.

The following notes are a brief outline of some key points that may be relevant to hazard study participants to consider in the development of an alarm system.

3.8 Key design principles for alarms

Not all design principles for alarms can be included here but those that have relevance to the hazard study stages have been identified.

3.8.1 Present only relevant and useful alarms to the operator

An effective alarm system presents only the alarms that help an operator in monitoring and controlling the plant/equipment rather than being a nuisance or hindrance. The operator's time and attention should not be diverted by the alarms, which do not require any response or an intervention from the operator and can be ignored. Otherwise, there is a possibility of the 'Cry Wolf' effect and the operators may lapse into a frame of mind that the alarms can be ignored.

3.8.2 Each alarm should have a defined response from the operator

For an alarm system to be effective, every alarm should have a defined response from an operator. The response should be in the form of a preventive and/or corrective action or an acknowledgement. If a response to an alarm cannot be defined or identified, such signal should not be configured or presented as an alarm. Operational steps, for example, during startup should be classed as message events presented as separate 'Events List' displays.

3.8.3 Allow adequate time for an operator to respond to an alarm

There are two things to consider here:

- It is essential that the alarm should occur early enough to allow the operator to correct the fault.
- The time used in handling the fault should be estimated as this adds to the workload of the operator. The total workload must not be allowed to reach the point where critical alarms cannot be handled.

3.8.4 Establish and enforce an alarm priority structure

Alarms are usually prioritized according to:

- The severity of consequence (in safety environmental and economic terms) that they could prevent if the operator responds correctly. We have seen that safety-related alarms will automatically get to be high priority but maybe one or two will be classified as 'critical'.
- The time available compared with the time required for the corrective action to take effect.

In any one alarm system, experience has shown that three levels or bands of priority are most effective. If there is an independent SIS or alarm system as well, this may justify an additional band for 'critical' alarms. The structure shown in Figure 3.14 would then apply.

The EMUA guide suggests the allocation of alarms across the bands as shown in Figure 3.18.

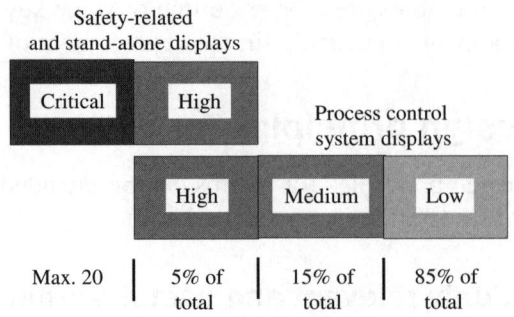

Figure 3.18
Alarm priority bands in a mixed alarm system

It will be helpful to the hazard study if the prioritization rules are clearly defined at the time of studies. The rules will be similar to those defined for the risk matrix, and if these can be aligned at the time of hazard studies, consistent definition of alarm priority band for each alarm may be achieved.

3.8.5 Consider the need for an alarm management system

An alarm management system has the potential to reduce stress and workload imposed on the operator by suppression of temporarily redundant alarms and by dealing with nuisance alarms in a secure manner. Techniques include:

Suppression

Alarm suppression involves the conditional blocking of an alarm signal that is:

- Invalid at the current state of the plant
- Redundant because the same condition is already signaled through another alarm
- Part of another and more serious alarm condition that is already present.

As alarm requirements differ under different plant conditions, it is important that the alarms configured and presented are context-sensitive or plant condition-sensitive. Some signals may be required as alarms during normal plant running, but the same may not be relevant during plant startups and other operational conditions.

Shelving is a term used for the techniques that allow an operator to prevent or 'hide' alarms that are causing a nuisance without any benefit to the operator at that time. Variations include 'auto shelving' that hides away endlessly repeating alarms until their behavior pattern changes.

3.8.6 Alarm suppression hazard study

Whilst we have may the software available in our control system to carry out some sophisticated alarm suppression strategies, there will always be the risk of accidentally hiding an important safety-related condition. Alarm improvement exercises therefore require the safeguards of a hazard study type of review to ensure that safety functions are not compromised in the enthusiasm to rationalize alarms.

The EEMUA guide has a very useful example of an Alarm Suppression Hazard in its Appendix 17.

3.8.7 Conclusion for alarms in the risk reduction role

This completes an overview of those factors in alarm system applications that form a background to hazard study activities. What should we know in the hazard study team?

- Safety-related alarms are identified by consequence of their failure. These include major financial loss events.
- Pre-trip alarms can save a lot of money.
- Risk reduction values define the required alarm reliability.
- Alarm reliability depends to a large extent on operator reliability.
- Operator reliability degrades during plant upsets unless alarm system is well-managed and workload is minimized.

The key to the application appears to lie in understanding the risk reduction role of each alarm being considered by the study team.

3.9 SIS, principles of separation

3.9.1 Historical

Safety functions were originally performed in different hardware from the process control functions. This was a natural feature because all control systems were discrete single function devices. It was not really inconvenient for instrument designs to achieve the separation and extra features need for the safety shutdown devices.

Only with the advent of DCS and PLC controllers did we have to pay attention to the question of combining safety and control in the same systems. There is a temptation nowadays to combine safety and control functions in the same equipment and it can be acceptable under certain conditions but it should be resisted wherever possible. It also depends on what is meant by 'combine'.

3.9.2 Separation

All standards and guidelines clearly recommend the separation of the control and safety functions. Here are some examples from the various standards and guidelines to back up this statement.

Figure 3.19 is taken from the UK health and Safety Executive's publication 'Programmable Electronic Systems in Safety Related Applications'. Basic Reasons for separation are also clear from the next two quotations.

Quotation from IEEE 'The safety system design shall be such that credible failures in and consequential actions by other systems shall not prevent the safety system from meeting the requirements.'

Quotation from ANSI/ISA S84.01 'Separation between BPCS and SIS functions reduces the probability that both control and safety functions become unavailable at the same time, or that inadvertent changes affect the safety functionality of the SIS.'

Process Control vs Safety control

Separation of safety controls from process controls
(from UK Health and Safety Executive recommendations)

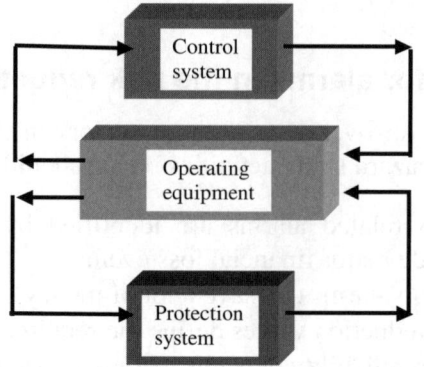

Figure 3.19
Separation of protection system from control system

3.9.3 Functional differences

Some basic differences between the two types of control system support the idea that control and safety should be separated.

Process controls act positively to maintain or change process conditions. They are there to help obtain best performance from the process and often are used to push the performance to the limits that can safely be achieved. They are not built with safety in mind and are not dedicated to the task. Because they are operating at all times they are not expected to have diagnostic routines searching for faults. Generally a fault in process control is not catastrophic although it can be hazardous, hence the need to include the EUC control system in the risk analysis.

Most significantly, one of the most valuable assets of a process control system is its flexibility and ease of access for making changes. Operators may need to disable or bypass some portions of the control system at different times. This is exactly what we want to avoid in a safety control system.

Safety controls are passive/dormant

Safety controls are there as policemen and security guards. They need to be kept to a fixed set of rules and their access for changes must be carefully restricted, i.e. they should be incorruptible! They must be highly reliable and be able to respond instantly when there is trouble!

3.9.4 Specials: integrated safety and control systems

Having said so much in favor of separation of control and safety functions there are always going to be situations where the control system also performs a safety role or where an integrated package is needed. An example would be a large turbo compressor set where anti-surge control and load control are integrated into one scheme. Burner management systems are another example where the fuel/air ratio controls provide a safety role, and where the startup ignition and shutdown purge sequences protect against possible explosions of unburned fuel in the furnace.

The philosophy in the new standards is clear on this issue. If the control system performs a safety-related function it should be considered to be an SIS. This leaves only two options as shown in Figures 3.20 and 3.21.

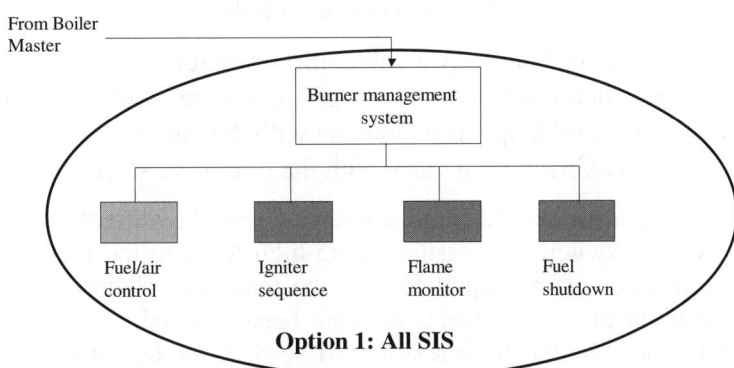

Figure 3.20
Integrated control and safety system all to SIS standards

What if a process controller has a safety function???

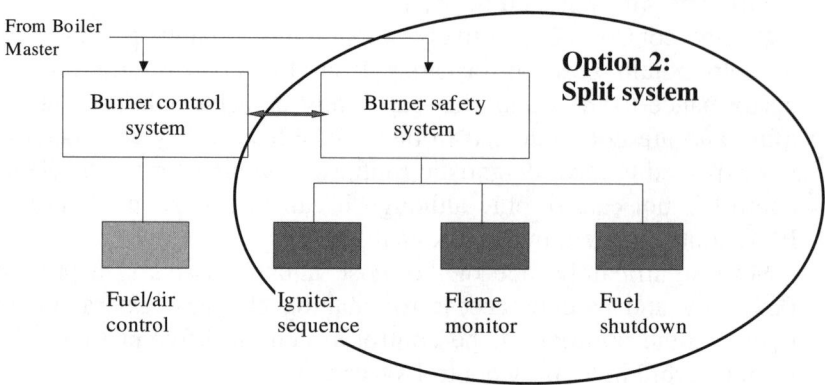

Figure 3.21
Separated control system with only the safety system to SIS standards

In cases where the process control functions are well-defined, the solution is often to use option 1 and build a single high-integrity control system. Internal architectures of the SIS support subdividing of high- and low-integrity safety functions for economy.

3.10   Simple and complex shutdown sequences, examples

At this point it may be helpful to illustrate some typical features of SIS before we move on to risk reduction concepts.

3.10.1   Simple shutdown sequence

A simple shutdown device is shown in Figure 3.22. It shows a flammable liquid being drawn from a process source into a buffer tank from which it is to be pumped onwards to a treatment stage. A typical level control loop is provided in the process control system to maintain level at, say, 50% full. A hazard will occur if the level control fails for any reason and the tank becomes full. The tank must have a pressure relief valve by law and if the liquid has to escape via this valve it will form a dangerous vapor cloud.

The reasons for loss of level control include:

- Jammed open level valve (dirt or seizure)
- Failed level transmitter – giving a false low-level signal
- Control loop left on manual with the valve open
- Leaking control valve with the pump-out stage shutdown.

A simple instrumented shutdown device would require the features shown in Figure 3.23, i.e. A level switch set to detect extra-high level in the tank causes an automatic shutoff valve to close-off all liquid feed to the tank. The shutoff valve remains closed until the defect in the process control system has been rectified.

Once the level has been restored to normal, the operator is allowed to reset the shutoff device and continue operations.

Note how even the simplest of protection systems requires a degree of logic functionality. Hence the logic solver role in the SIS quickly becomes complex.

Simple Shutdown System: Example 1

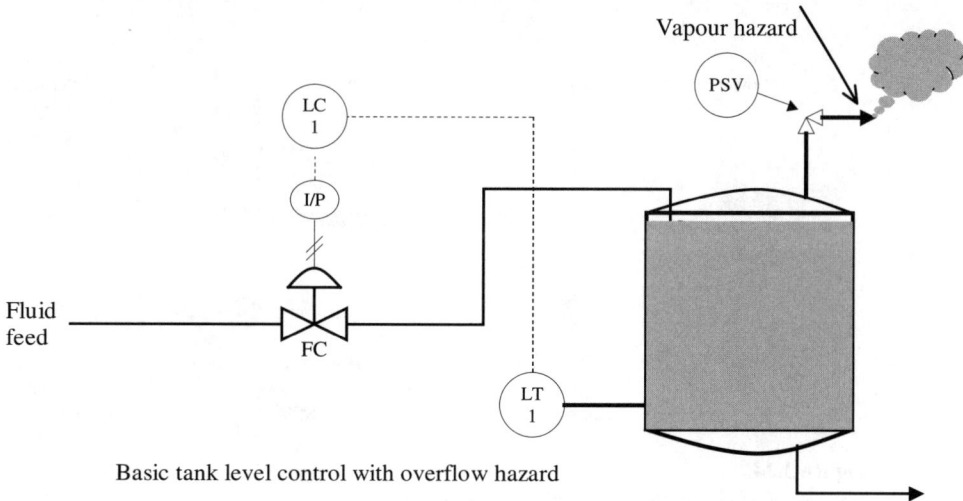

Basic tank level control with overflow hazard

Figure 3.22
Simple process hazard example

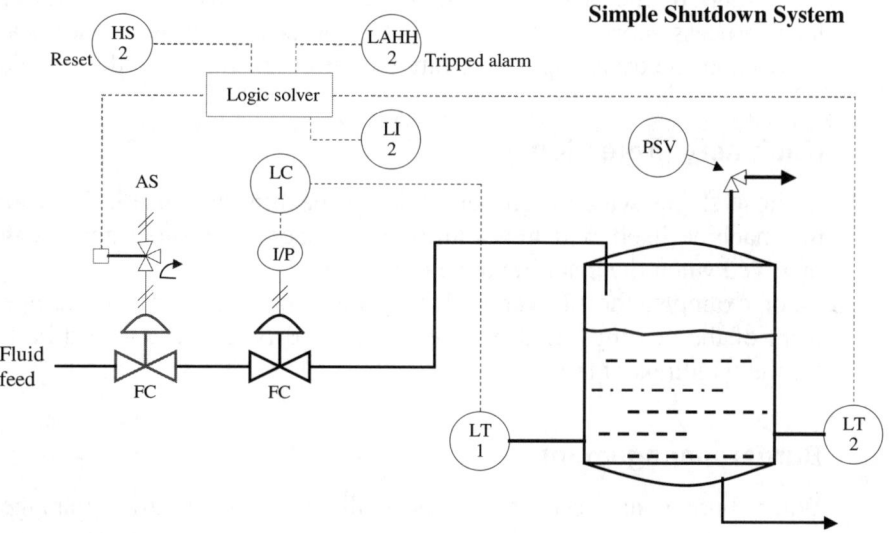

Figure 3.23
Simple shutdown system example

3.10.2 Complex shutdown sequences

This example is based on a section of a large chemical process plant. Figure 3.24 shows a process gas blower (steam turbine driven) delivering gas via a pre-heater stage into a set of burners in a processing furnace. Natural gas is used for preheating and for base heating the furnace. Oxygen is injected into the gas burner stages, also an acid stream is sprayed into the furnace to neutralize the product. The hot gases are then passed via a quench tower before being delivered into the next stage of the process.

The safety systems for this plant will be typical for a large processing plant. There are some common characteristics of the safety systems for such plants, and this example illustrates them.

Example of process requiring complex trip sequence

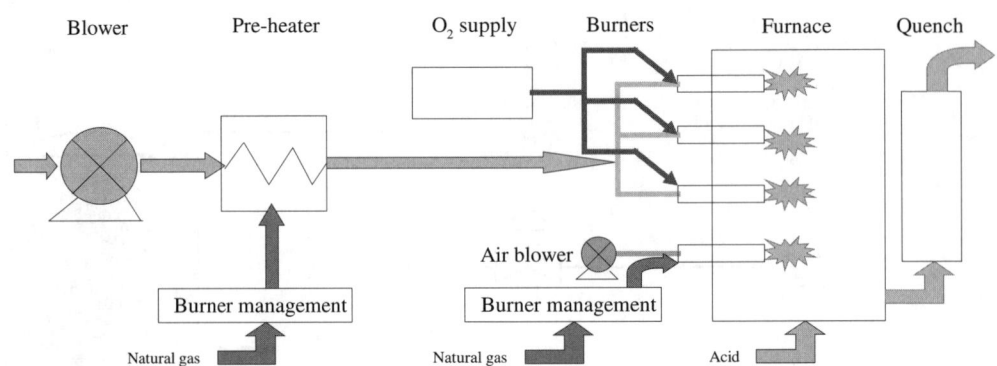

Figure 3.24
Example of process requiring complex trip sequence

Local process conditions

Each stage of the process has its own inherent hazards and limiting conditions. Hence, local process sensors are used to detect process hazards in each stage. Actuated control valves and electrical trip signals are then used to shutdown the stage that has the problem.

Machinery protection

In those stages where high-performance machinery is used, for example the gas blower, the machine itself will have numerous sensors to protect against the hazards and costs involved when the machine develops a fault.

For example, the blower will have detectors for vibration, temperature and displacement of the bearings. If it were electrically driven, there would be temperature detectors for the windings of the motor.

Burner management

Where burners are used, there is normally a complete burner management control system with its own set of protective instruments and a safety control system.

Stages are interdependent

Each stage in this process is dependent on the correct working of at least one other stage. Hence, the SIS for any one stage is likely to have additional shutdown commands from the status of the other critical stages. The trip logic diagram shown in Figure 3.25 shows how the shutdown or tripping of one stage leads to the tripping of another. The entire plant will shutdown if the Quench tower process trips out. However, if one burner trips out in the furnace, the logic diagram shows that the plant will continue operating. All the trip logic is performed in the safety system.

It is important to note that all the actions required by the trip logic diagram have been evaluated through hazard studies as being essential for the safety of personnel and plant. Hence, the entire system shown is a safety system and remains independent of the BPCS for the plant. The trip system plays no role in the startup of the plant except that it will prevent the startup of any one stage if it is not safe to do so by virtue of the trip logic. We shall be returning to this example later, in the course, when we look at design features. What we are trying to show by the examples given is the typical nature of SIS, as applied to process operations.

Complex trip sequence for process example

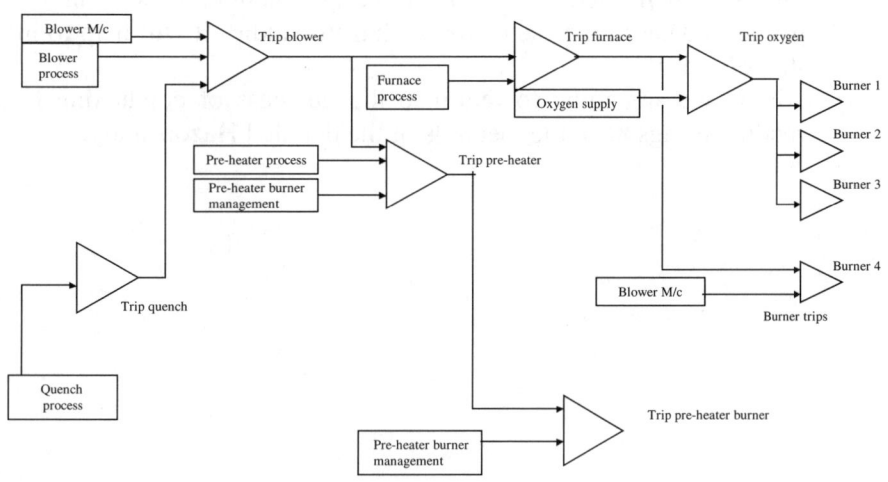

Figure 3.25
Complex trip sequence for process example

3.11   Conclusions: the role of Hazops in defining alarms and trips

We should now have some good indicators of what we would like to see coming out of hazard studies when a situation has been found that requires risk reduction. We would like to see:

- Risk reduction measures of all types identified as discrete protection layers.
- Trips and alarms are regarded as protection layers in the risk reduction models.
- Protection layers are not considered to be truly effective unless they are able to operate independently of the rest of the plant or EUC, as it is formally known.
- The safety functions to be performed must be clearly defined and the required degree of risk reduction should be specified. We have seen that this leads directly to the SIL requirements of the SIS and alarms.
- The safety functions cannot be defined in detail unless the basic causes of the hazardous condition are stated. (This may require various types of cause and effect diagram or tables.)
- For those trip actions involving shutdown or stabilizing of the process, the safe state of the process must be defined.

- The actions of the safety function to bring the plant into a safe condition must be defined. We have seen, for example, how the response to an alarm must not be left undefined.
- It is not realistic to expect that all the above information can be supplied at the time of the hazard study session. It is however reasonable to expect that at least some of the team members are aware that this information must be agreed before the protection measures can be claimed to have been put in place.

We shall see in Chapter 6 how the 'Safety requirements specification' describes all the information needed to call up a safety function. If we can find ways to capture the essential data during the course of hazard studies it will help to integrate the studies with the solutions.

Now we are going to return to the business of conducting hazard studies. Our next module brings us to the methods of the detailed Hazop study.

4

Hazop method

Objectives

The scope of this chapter covers the essential basics of Hazop study procedures and includes:

- Introduction to Hazop methods
- Guideword examination procedure
- Design representation and defining of parts for study
- Selection of elements and deviations
- Study procedure and examples for continuous processes
- Practical Exercise No. 3: Trial Hazop study and recording of results
- Guidewords and procedures for batch processes
- Examples of wider applications of Hazop.

The objectives of this chapter are:

- To introduce the basic method of guideword-based Hazop
- To provide entry level skills sufficient for a person to attend and actively participate in Hazops.

The overall objective here is to provide some basic training in Hazop method and to place the work in the context of the drive for effective safety systems.

4.1 Introduction

In this chapter, we shall learn about the basic principles and techniques of the detailed level hazard study, best known as Hazop. These methods have been well-documented in guide manuals over the past 20 years and there is general agreement of the technique itself. What has changed in recent years is the application of the guideword methodology to a range of applications beyond their original use in chemical process plants.

An indication of the success of the Hazop method can be seen in the recent publication of an IEC standard (61882 Ref. 7) providing guidelines on the application of Hazop. We have used this standard along with another well-established publication (EPCS Hazop Guide Ref. 10) as the basis for the method described here. We would recommend that if you wish to continue to acquire skills in this field you should obtain these or similar specialist publications and consider attending further detailed training from specialist companies.

4.2   Introduction to Hazop

The Hazop study method is the term applied to a detailed method of for systematic examination of a well-defined process or operation, either planned or existing. Hazop is the abbreviation for Hazard and Operability Study.

We have seen from the work in Chapter 2 that this type of study is to be used at the detailed examination stage when sufficient well-defined information is available for the process. In the sequence of studies we have been looking at this, the 'level 3 hazard study'.

4.2.1   Objectives of the Hazop study

The main purposes of a Hazop are:

- To identify and evaluate hazards within a planned process or operation
- To identify significant operating or quality problems
- To identify practical problems associated with maintenance operations.

Operating or quality problems may be an optional purpose depending on the company and its application. However, the EPSC Guide (Ref. 10 Section 3.2) reports that a survey of European Process Safety Centre members found that 'over 90% of the respondents included significant operability problems in the scope of the search'. Issues of quality of product and sources of downtime may also be defined as objectives of the study.

Maintenance operations including isolation, preparation and removal for maintenance are important areas to study because they often create hazards as well as an operability problem.

Perhaps the most common example of this last point is the question of standby pumps and their removal for maintenance. An experienced maintenance person in the Hazop study will very quickly notice if there is no isolation valve for removal of an in-line duty/standby pump. Where hazardous chemicals are involved, this isolation must extend to suitable drain and flush facilities to make it safe to remove the pump.

Then there are the questions:

- How can we remove it safely?
- Is there enough headroom?
- Do we need a crane?
- Can we get the crane in?
- Can we get the crane out?

Sometimes these types of questions are best dealt with by having a separate 'Maintenance Hazop'. So it is important to decide the scope and objectives of each Hazop. The objectives and scope must be defined and agreed to avoid loss of focus or confusion in the approach to the study.

4.2.2   Terminology and sources

Several textbooks and handbooks describe the Hazop method and it is useful to note that there is a general consistency in the descriptions and advice contained in the sources we have referenced.

In the following description of the Hazop method the terminologies used are based on those published in the IEC standard (61882 Ref. 7). Descriptions of the Hazop method in this standard align closely with those given in EPSC Guide to Hazops (Ref. 10). We have referenced both of these publications in order to ensure that the most consistent descriptions of the method are used in this book.

4.3 Overview of Hazop method

The overall Hazop procedure comprises four sequential steps as shown in Figure 4.1.

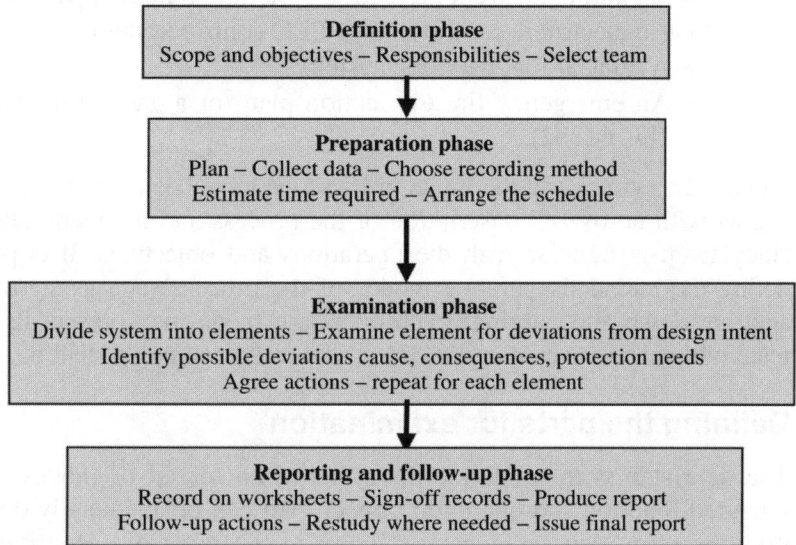

Figure 4.1
Overall Hazop study procedure

The basics of each phase are clear from Figure 4.1 but there are some important points to be learnt about each phase. The best approach is, probably, to learn about the examination phase first. This takes us to the core of the study procedure. All other activities support the preparation for the study and then try to ensure adequate follow-up and long-term record-keeping.

4.3.1 The examination phase

The basis of Hazop is the 'guideword examination' of parts of a plant or system to find credible deviations from the design intent. This is an inductive method because we are 'inducing' changes to the intended working conditions and testing in our minds to see if these changes produce any problematic effects.

4.3.2 Define the system

The first step is to divide an overall plant into operations, units or systems that are to be studied. In a process plant this will be similar to a unit operation, a processing stage or a pipeline operation. It may be that an entire Hazop will be dedicated to one such unit or system.

Process examples would include:

- A batch reactor for PVC polymerization (Autoclave)
- A distillation column
- An evaporator or concentrator
- An offshore wellhead gas platform linked to a central process facility (example from EPSC Hazop Guide)
- A chlorine electrolyzer cell.

Elsewhere in industry the system might be:

- A robotic assembly or welding unit
- A warehouse conveyer system
- An automatic train protection system (example from IEC 61882)
- An electronic control unit or PLC control system
- A nuclear fuel handling system
- An emergency fire evacuation plan for a gas platform (another example from IEC 61882).

Once the system and the limits of its scope have been agreed, the examination procedure begins with an overall description of the process and its operational aspects to ensure the study team is familiar with the operations and objectives. It is particularly important to define the intended operating modes of the process such as starting-up, recycle, on-line, shutting down and purging. Each operating mode may present its own problems and the team will need to decide on the extent to which each mode shall be studied.

4.3.3   Defining the parts for examination

The next step is to subdivide the system into 'parts' or 'nodes' in such a way that the intended function and operation of each part can be adequately defined. The EPSC guide uses the term 'section' when referring to the parts of a continuous process whilst the equivalent in batch process is 'step' or 'stage'.

4.3.4   How do we define the parts?

The diagrams in Figures 4.2 and 4.3 provide an example. Figure 4.2 represents a very simplified version of a P&ID for an oxonation reactor commonly used in the first stage of a butanols production plant.

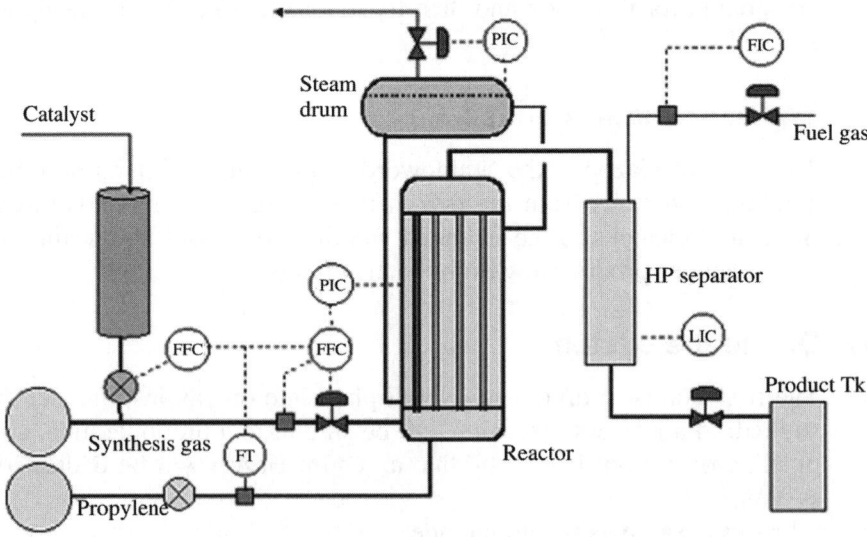

Figure 4.2
First steps in the Hazop examination: example of a system (simplified P&ID)

Synthesis gas with catalyst added is fed to a plug flow reactor. Propylene gas is fed to the reactor and continuous reaction occurs as the mixture flows through the tubes. The reaction is exothermic and cooling is provided by closed circulation of water to a steam drum. The resulting product is fed to a separation stage where the heavier products are condensed and the lighter gases are taken off for purification and used as fuel gas.The process details are unimportant for our illustration but the reactor would normally be presented in the form of a nearly completed P&ID with all piping details and instrument functions clearly shown.

Figure 4.3 shows how some of the parts might be chosen so that the examination procedure can be applied efficiently to each part.

- Choose small parts where the system is complex or the hazards are likely to be high.
- Choose larger parts where the system is simple or if the hazards are low.

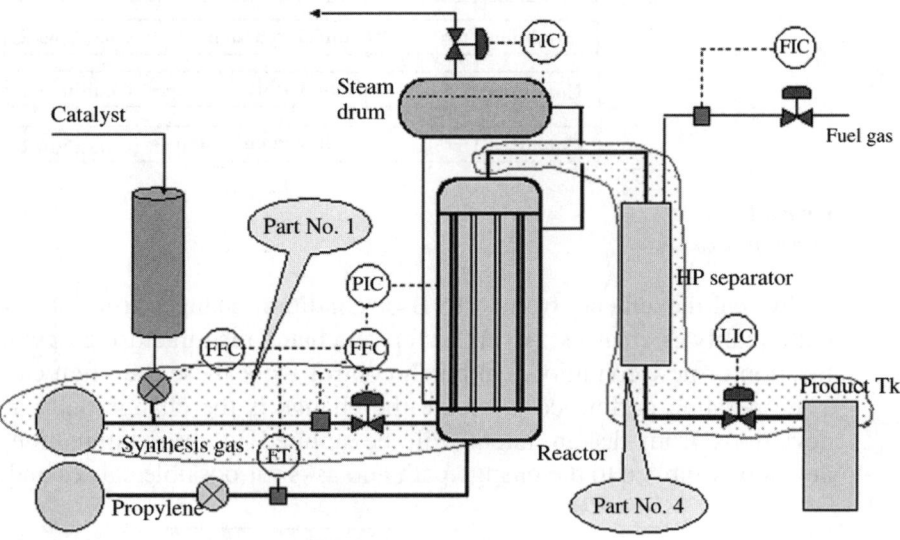

Figure 4.3
First steps in the Hazop examination: defining the parts for study

Getting the choice right is important because each part will take time to examine and hence this influences the length and depth of the study work:

- *Small parts*: With many small parts, the study will become slow and tedious. It is also a weakness of the Hazop method that the interaction between parts is easy to miss. So if parts have a very strong coupling they may be better treated as one part.
- *Larger parts*: With large parts, the study will be faster and it may help to find interactions. But larger parts carry the risk that important details within the part will be missed.

Defining the way parts are selected for study is best done by an experienced Hazop study team leader, supported by the person with best knowledge of the process. However the selection of parts can be assisted by using the concept of 'change paths'.

4.3.5 Concept of change paths

Typically, a part for study will include the transfer of material from source to destination. The function of a part can then be seen as:

- Input material from a source
- Perform an activity on the material
- Product delivered to the destination.

Study manuals from ICI and AECI describe this typical scope as a 'change path'. It is easy to recognize a suitable change path because it becomes practical to apply 'deviations' to the operation. Figure 4.4 shows a generalized model for the change path concepts.

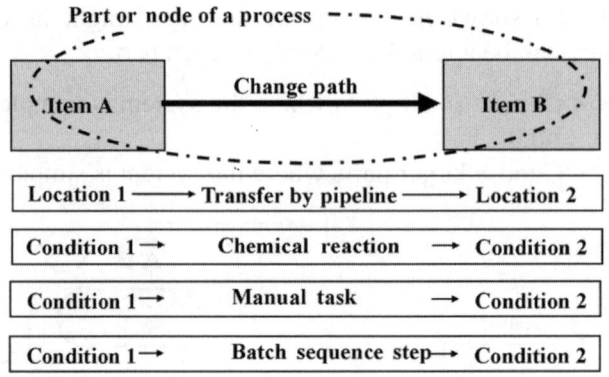

Figure 4.4
Change path concepts

Physical movements from A to B or condition changes from one substance to another, both qualify as changes. A single step of a batch manufacturing operation produces a new location or new condition. (In batch control work these are often called 'state changes'.) The operation performed to create the change is the change path, and this is where the deviations from design intent will be applied in the examination. Figure 4.5 shows deviations applied to the change path and asks for possible causes and consequences.

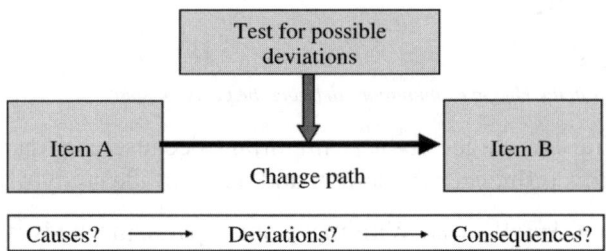

Figure 4.5
Deviations applied to the change path

If this seems rather detached from reality try creating change path applications for real-life examples. Here we have tried an operational step in assembling a rotary lawn mower (Figure 4.6). The change path is to fasten the blade to the shaft on the mower.

In Figure 4.6, a change of state is intended in the mower. The task of fastening the blade is the change path and having the wrong size of nut causes the deviation. The elements involved in this operation are the worker, the blade and the nut. It is the elements of the task that are capable of deviating.

Let us recall where we have got to so far:

- The overall system has been explained to the team.
- The system has been divided into parts such that each part can be described as a meaningful operation.
- Parts are chosen by agreement in the team such that it can test a realistic set of deviations.

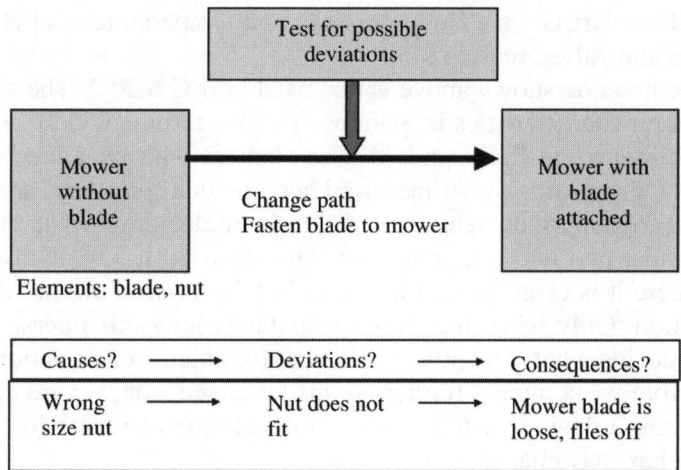

Figure 4.6
Cange path example for assembly task

- The parts have been identified on a large diagram.
- The parts have been arranged and labeled in a logical order so that the study can begin with input stages to the process and work towards the output stages.

Now, before we can apply deviations we have to ask 'what is going to deviate?'. We have to identify the elements or parameters.

4.3.6   Identify elements

An element is defined in IEC 61882 as 'constituent of a part which serves to identify the part's essential features'. The standard goes on to note that 'elements can include features such as the material involved, the activity being carried out, the equipment employed etc. Material should be considered in a general sense and includes data, software etc'.

Any element in a part may be able to change in some way that will affect the operation or safety of that part. It is easiest to see this if we take a process example as shown in Figure 4.7.

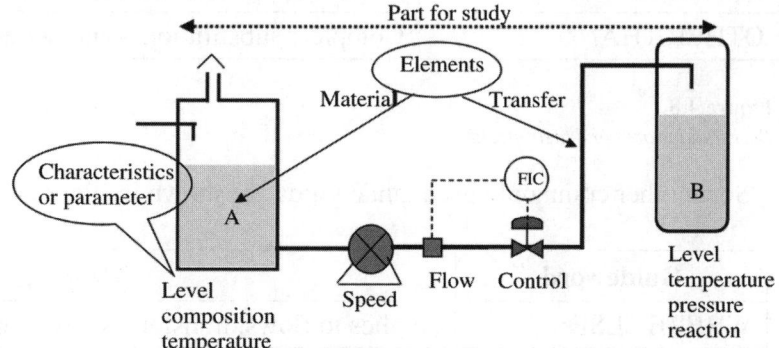

Figure 4.7
Examples of elements in a part

Figure 4.7 indicates that the 'elements' of the part are: material in A, the transfer of material from A to B and the material in tank B. Process variables such as flow and pressure are characteristics of particular elements. For example, in the transfer operation, the characteristics are flow, pressure and temperature, and abrasion. The tanks, pipeline, pump and valve are 'items'.

The notation shown above is that used in IEC 61882. The alternative name for elements and their characteristics is 'parameter'. This term is widely used in process industry work but it seems the IEC standard uses element with characteristics to give a more generic term. Other words sometimes used here include 'property' and 'keyword'.

The Hazop study team has the task of deciding what elements are applicable to a particular part under examination. This decision has to be taken for each part, in turn this is where it is essential that the team is fully briefed on the 'design intent' for this part in question. Only when the team understands how this operation is supposed to work can they decide what elements are relevant for change or deviation.

Before we comment further on this let us see what we are going to do to the elements to generate a deviation. It is easier to decide the elements for study when you understand how they may change.

4.3.7   Generating deviations

We know that a deviation is to be considered for an element or parameter. The most common types of deviation can be listed as a set of guidewords. A common guideword list is used for Hazop work in any particular industry sector with additional guidewords being available to help stimulate other possibilities in the minds of the team.

The guideword system begins with a set of basic guidewords that will always apply to any element. The basic guidewords and their generic meanings are shown in Figure 4.8.

Guideword	Meaning
NO or NOT (or none)	None of the design intent is achieved
MORE (more of, higher)	Quantitative increase
LESS	Quantitative decrease
AS WELL AS (more than)	Qualitative modification or additional activity occurs
PART OF	Only some of the design intent is achieved
REVERSE	Logical opposite of design intent
OTHER THAN	Complete substitution – another activity takes place

Figure 4.8
Basic guidewords and their meanings

Some other commonly used guidewords are shown in Figure 4.9.

Guideword	Meaning
WHERE ELSE	Applies to flows, transfers, sources and destinations
BEFORE/AFTER	Relates to order of sequence
EARLY/LATE	The timing is different from intention
FASTER/SLOWER	The step is done faster or slower than the intended timing

Figure 4.9
Guidewords relating to location, order or timing

The basic guidewords lack any real meaning until they are combined with elements or characteristics within elements.

Combining guidewords with elements generates a matrix of deviations, some of which are credible and some not credible. It therefore falls to the study team to decide which deviations from the matrix they are going to consider, and the deviations list may be different for each part under examination. Here is a situation that requires a graphical tool or computer chart for assistance. Here, in Figure 4.10, we can try out a matrix example for the simple fluid transfer we saw in Figure 4.5.

Blanks indicate the deviation is invalid or not feasible. The ticked boxes in the matrix clearly indicate deviations that will have to be considered for this part. Other specific deviations such as those due to maintenance work or flushing out will normally be listed against 'Other'.

In the process industry it is common practice to have a prompting list to help with the interpretation of the basic guidewords as applied to typical process conditions. We have placed a typical guideword expansion at the end of this chapter.

At this stage of the examination the study team has basically established its raw list of possible deviations for this particular part of the process. This raw list also serves the study team leader for the purpose of prompting and recording the progress of the examination. We should note that at no point should this list be considered closed. If any deviations occur to the team as they move along they should always be allowed to add it to the list.

Element (parameter)	Guidewords								
	NO	MORE	LESS	REVERSE	PART OF	AS WELL AS	WHERE ELSE	EARLY/ LATE	OTHER
Tank A level	X	X	X						
Tank A composition		X	X			X			
Flow in pipe	X	X	X	X		X			
Temperature in pipe		X	X						
Pressure in pipe	X	X	X						X
Speed of pump	X	X	X	X					
Control valve opening	X	X	X						
Tank B level	X	X	X						
Tank B composition		X	X		X	X			X
Tank B pressure	X	X	X	X					
Tank B reaction	X	X	X		X	X			

Figure 4.10
Example of guideword/element matrix for process example

We are now ready to proceed with the examination procedure by considering each element and its characteristic in turn to see if the deviation is possible and to consider its possible consequences.

4.3.8   Guideword examination procedure

Here our starting point can assume that the stages of selecting the part and describing its design intent have been done. We have a draft table of parameters to consider, with each element of the activity also identified.

There is one more choice to be made before proceeding. The team has to decide if the study will work best by using 'guideword first' or 'element first method'.

Element (parameter) first method

The flowchart in Figure 4.11 illustrates that this method allows the study to begin with an element (such as material in tank A) and test its deviations against the guidewords. For example, in the part shown in Figure 4.8 the result will be that we consider level deviations, then temperature, then composition all in tank A before we move along to consider the transfer operation.

The sequence shown in Figures 4.11 and 4.12 allows the team to concentrate on all the possible deviations of one element or parameter before moving on to the next. This method is probably most favored in chemical process Hazops but the EPSC Guide suggests that it may not be as good as guideword first method when it comes to encouraging creative thinking in the study team. This is particularly directed at the later guidewords such as 'part of' and 'other than'. The reason given is that 'If the parameter first approach is used, there is a tendency to list the parameters at the start of the analysis and when the original list is exhausted to move on to the next section without a final search for more parameters. This is not good practice'.

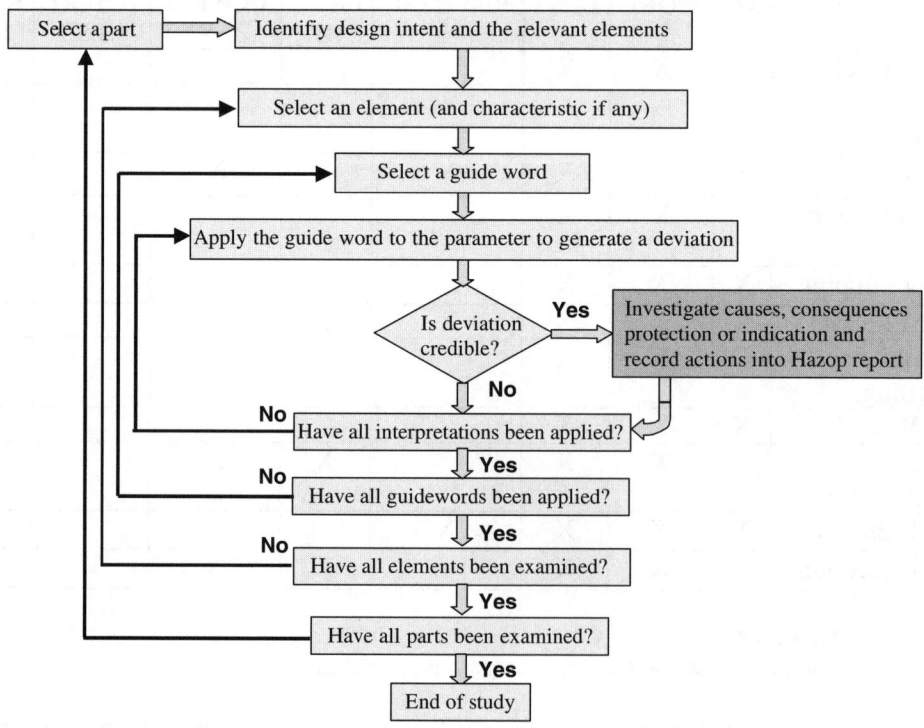

Figure 4.11
Element-first examination procedure

Element first examination procedure

Element (parameter)	Guidewords								
	NO	MORE	LESS	REVERSE	PART OF	AS WELL AS	WHERE ELSE	EARY/ LATE	OTHER
Tank A level	X	X	X						
Tank A composition		X	X			X			
Flow in pipe	X	X	X	X		X			
Temperature in pipe		X	X						
Pressure in pipe	X	X	X						X

Figure 4.12
Element-first applied to guideword matrix

Guideword first method

This method puts the guideword first, testing the same type of deviation on all possible elements or parameters before moving to the next. The procedure is otherwise identical to the element-first approach. Figure 4.13 illustrates the effect of guideword first method.

Guideword first examination procedure

Element (parameter)	Guidewords								
	NO	MORE	LESS	REVERSE	PART OF	AS WELL AS	WHERE ELSE	EARY/ LATE	OTHER
Tank A level	X	X	X						
Tank A composition		X	X			X			
Flow in pipe		X	X	X		X			
Temperature in pipe		X	X						
Pressure in pipe	X	X	X						X

Figure 4.13
Guideword first applied to guideword matrix

As noted in the EPSC Guide, 'the element-first approach may provide convenience, but it demands a greater understanding and application by the team leader and team members if the best results are to be obtained.'

This completes our run through the procedures for selecting parts and elements and for generating deviations. We are almost ready to try out the method on some simple examples but lastly we need to re-enforce the ability to translate the basic guideword/parameter deviations into some more tangible meanings for practical work.

4.3.9   Derived guidewords

Here is a table based on the EPSC guide for typical derived guidewords generated by the parameter and guideword combinations for typical process parameters. This table makes it easier to visualize the possible deviations.

Parameter	Guidewords That Can Give a Meaningful Combination
Flow	None, more of, less of, reverse, elsewhere, as well as
Temperature	Higher, lower
Pressure	Higher, lower, reverse
Level	None, higher, lower
Mixing	Less, more, none
Reaction	Higher (rate of), lower (rate of), none, reverse, as well as
Phase	Other, reverse, as well as
Composition	Part of, as well as
Communication	None, part of, more of, less of, other, as well as

Generating the derived guidewords such as those shown above is part of the team leader's responsibility in each study session. In summary, we see the relationship between guidewords and the parts to be studied in Figure 4.14 is based on diagrams used in the AECI Hazop manual.

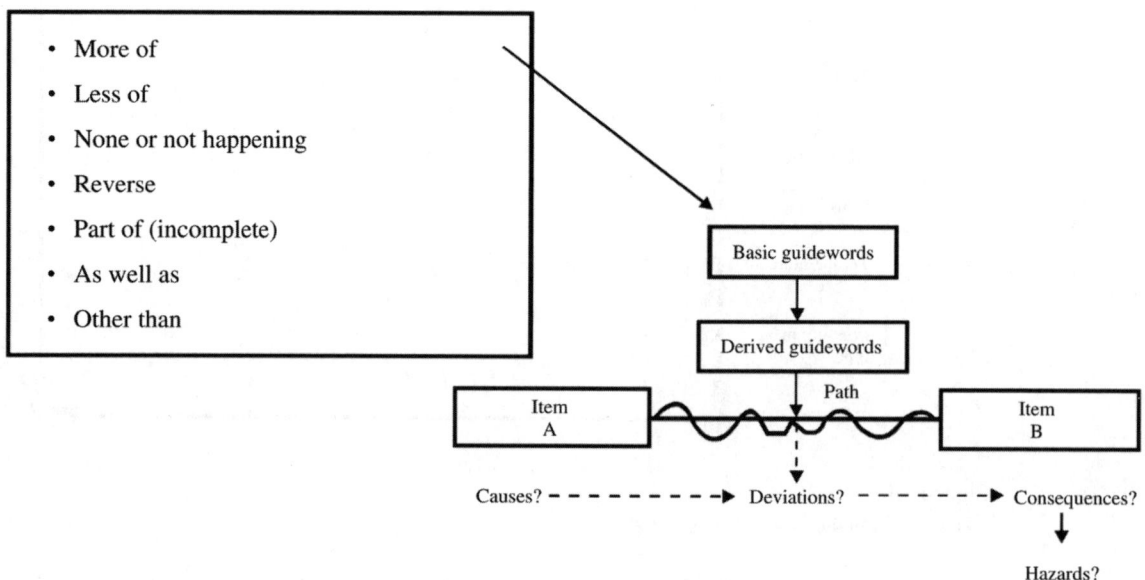

Figure 4.14

Application of guidewords to the change paths

4.3.10 Study procedure

The hazard study team leader or facilitator has the task of taking the team members methodically through a sequence of questions for each recognized deviation. Figure 4.15 shows the sequence of questions that should be asked of the team.

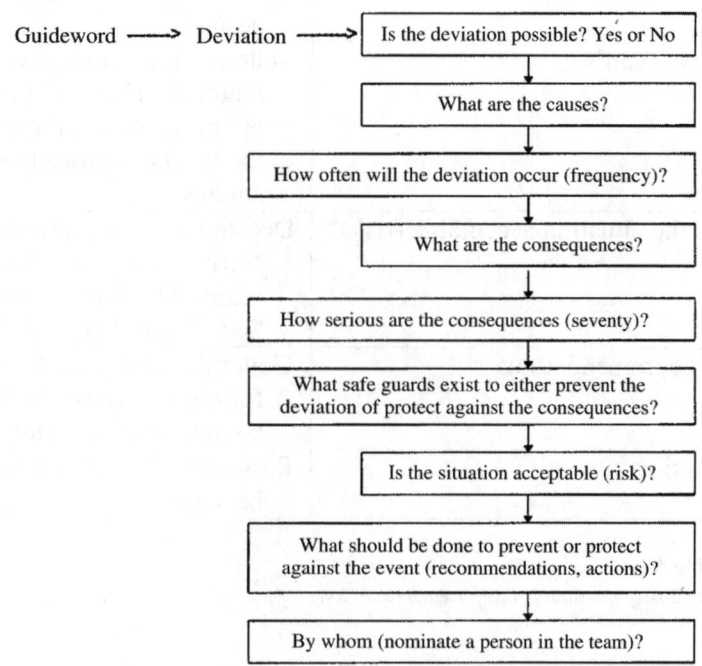

Figure 4.15
Logical steps in the processing of each deviation

The sequence shown in Figure 4.15 will naturally change in nature according to the problem but it represents the core of the procedure that has to be followed for each legitimate deviation. Before we try a practical exercise, here is a summary of the initial points (Table 4.1) to remember about each stage of the Hazop examination steps:

Is the deviation possible?	Decide Yes/No based on simple physical rules. Decision is made jointly with the next item, 'causes'
What are the causes?	Team discussion will establish possible causes. Using a prompting list or checklist assists the search for causes. See 'checklists'
	If consequences of the deviation are trivial, detail causes are not needed. The sequence can be terminated at this point
How often will the deviation occur (alternative)? What is the likelihood?	Optional at the initial study.
	The expected frequency or likelihood should be stated for the event without any of the operational safeguards described below. See 'discussion points'

What are the consequences?	Consider under categories including: Harm to persons, harm to environment, damage to equipment, loss of quality, loss of production. Follow-up study may be called for Identify consequences both inside and outside the part under review. Team leader decides on when to study 'knock-on'consequences. See 'discussion points'
Safeguards	State existing safeguards in the design or operating methods. These will include design features, operating procedures, alarms and trips. These are the layers of protection described in hazard 2 studies
Is the situation acceptable (risk)?	Decision does not have to be taken in the study. Normally any uncertainty in this decision is to be referred for hazard analysis and risk assessment. See 'discussion points'
Recommendations	Generally problems are referred back to designers for corrective study. Some obvious solutions can be agreed in the study
Actions	Persons or department responsible for action must be stated

Table 4.1
Summary of question points for each deviation

After we have tried the following example there are some more points for consideration that are easier to follow after some practice.

4.3.11   Worked example of continuous operation

Here we can try testing the procedure on a simple example (Figure 4.16) taken from the transfer operation shown earlier.

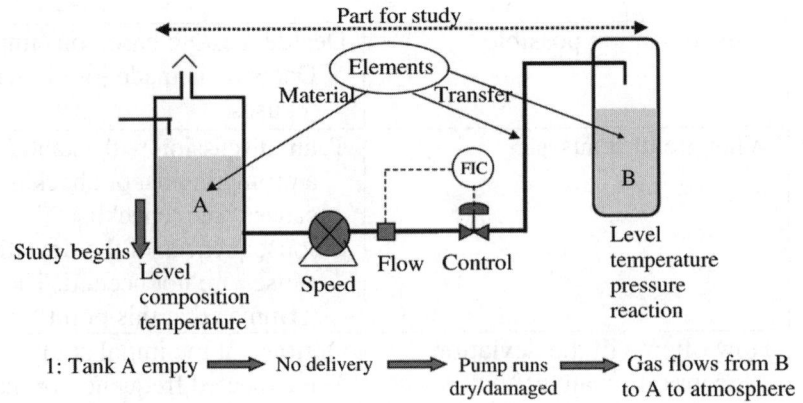

Figure 4.16
Simplified diagram of transfer of acid into pressurized reactor

Some added information here: The fluid in A is highly acidic and has high viscosity when cold. The fluid mix in tank B is at high pressure and has a saturated vapor in the space at the top of the tank.

The elements for study in this part were defined as: contents of tank A, transfer operation from A to B, contents of tank B.

Assume an element-first procedure, choose the first element – tank A contents and then choose the parameters:

Part: Transfer of Acid from A to B	Element: Contents of Tank A	Parameter: Level	
Deviation	**NONE**	**Meaning/Effect**	**Tank is Empty**
Is it possible	YES		
Causes	1:No supply	2: Extraction exceeds inflow	3:
How often?	Monthly	Monthly	
Consequences	1: No transfer	2: Pump damage	
Severity	Nil	Moderate + Loss of production	
Safeguards	Operational	None	
Acceptable risk	N/A	NO	
What should be done		Low-level detection and interlock on pump	
Action	Specify safety trip	Process and Instrument engineers	

Figure 4.17
Sequential worksheet. Page 1

Part: Transfer of Acid from A to B	Element: Contents of Tank A	Parameter: Level	
Deviation	**MORE**	**Meaning/Effect**	**Tank Overfills**
Is it possible	YES		
Causes	1: Uncontrolled input	2:	3:
How often?	Probable		
Consequences	1: Overflow to drain to effluent drains	2: Acid spills	
Severity	Minor losses of material	Moderate risk to persons	
Safeguards	Operational + High-level alarm		
Acceptable risk	Yes	NO	
What should be done		Enclose tank overflow pipe outlet to drain	
Action	Ensure correct response to alarm is in ops. manual	Piping designer	

Figure 4.18
Sequential worksheet. Page 2

Part: Transfer of Acid from A to B	Element: Contents of Tank A	Parameter: Level	
Deviation	LESS	Meaning/Effect	Tank Runs Low
Is it possible	YES Same as for No Level: see page 1		

Figure 4.19
Sequential worksheet. Page 3

Part: Transfer of Acid from A to B	Element: Contents of Tank A	Parameter: Temperature	
Deviation	MORE	Meaning/Effect	Material Hotter than Intended
Is it possible	YES		
Causes	1: Temperature control fault on steam jacket controls	2:	3:
How often?	Possible		
Consequences	1: Excessive vapor from the acid. Enviro. pollution	2: Acute toxic risk to persons	3: High rate of corrosion
Severity	Moderate	Serious	Minor
Safeguards	High temperature alarm	Extended vent stack	Rubber-lined tanks and vent stack
Acceptable risk	No	No	Yes
What should be done	1: Consider design change to hot water jacket instead of steam or see 2	2: Provide high temperature trip on heating	
Action	Process to review design and decide on cost of design change vs cost of trips. Instruments to estimate cost of trips		

Figure 4.20
Sequential worksheet. Page 4

Part: Transfer of Acid from A to B	Element: Contents of Tank A		Parameter: Temperature	
Deviation	LESS		Meaning/Effect	Tank Contents Cooler than Required
Is it possible	YES			
Causes	1: Loss of steam supply		2: Instrument fault	3:
How often?	1 per year		0.2 per year	
Consequences	1: High viscosity, pump cannot deliver. Risk of clogging pipes			
Severity	Minor production losses			
Safeguards	Operational + low temperature alarm			
Acceptable risk	Yes			
What should be done				
Action	Ensure correct response to alarm is in ops. manual			

Figure 4.21
Sequential worksheet. Page 5

Part: Transfer of Acid from A to B	Element: Contents of Tank A	Parameter: Composition	
Deviation	MORE	Meaning/Effect	Too Viscous
Is it possible	YES		
Causes	1: Upstream mixture error	2: Low temperature, see above	3:
How often?	Probable		
Consequences	1: Pump drive overload	2: Pump overheats	
Severity	Minor	Moderate risk of fire in pump	
Safeguards	Electrical overload switch fitted	None	
Acceptable risk	Yes	No	
What should be done	Ensure correct response to alarm is in ops. manual	High temperature at discharge to trip pump	
Action	Process	Instruments	

Figure 4.22
Sequential worksheet. Page 6

Part: Transfer of Acid from A to B	Element: Contents of Tank A		Parameter: Composition	
Deviation	**LESS**		**Meaning/Effect**	**Weak Mix**
Is it possible	YES			
Causes	1: Uncontrolled input		2:	3:
How often?	Probable			
Consequences	1: Reduced yield in tank B		2:	
Severity	Minor losses			
Safeguards	Operational			
Acceptable risk	Yes			
What should be done	Nil			
Action				

Figure 4.23
Sequential worksheet. Page 7

Only the flow, temperature and composition were considered as relevant parameters for the material tank A. The examination then repeats for the next element – the transfer operation in the pipeline.

Part: Transfer of Acid from A to B	Element: Transfer Pumping Operation along Pipeline to Tank B	Parameter: Flow	
Deviation	**None**	**Meaning/Effect**	**No Flow**
Is it possible	YES		
Causes	1: Pump stopped 4: Tank empty see sheet 1	2: Line blocked	3: CV closed by operator
How often?	Frequent	Possible due to sticky liquid	Likely
Consequences	Loss of production	Maintenance call out will be needed	
Severity	Minor	Minor	
Safeguards	Low flow alarm	Flushing, see under low flow	Low flow alarm
Acceptable risk	Yes	Yes	Yes
What should be done	Nil		
Action	Nil		

Figure 4.24
Sequential worksheet. Page 8

Part: Transfer of Acid from A to B	Element: Transfer Pump and Pipeline to Tank B	Parameter: Flow	
Deviation	**Less**	**Meaning/Effect**	**Low Flow**
Is it possible	YES		
Causes	Line clogged		
How often?	Frequent		
Consequences	Loss of production		
Severity	Minor		
Safeguards	Regular flushing		
Acceptable risk	Yes		
What should be done	Scheduled flushing in standard operating procedure		
Action:	Writer of SOPs		

Figure 4.25
Sequential worksheet. Page 9

Part: Transfer of Acid from A to B	Element: Transfer Pump and Pipeline to Tank B	Parameter: Flow	
Deviation	**More**	**Meaning/Effect**	**Higher Flow than Intended**
Is it possible	YES		
Causes	CV open too far		
How often?	Possible		
Consequences	Loss of quality in tank		
Severity	Minor		
Safeguards	Operational procedures		
Acceptable risk	Yes		
What should be done	Nil		
Action	None		

Figure 4.26
Sequential worksheet. Page 10

Part: Transfer of Acid from A to B	Element: Transfer Pump and Pipeline to Tank B		Parameter: Flow	
Deviation	**Reverse**		**Meaning/Effect**	**Flow from Tank B to Tank A**
Is it possible	YES			
Causes	1. Pump stops, CV open, pressure in B >A		2 CV leakage when pump stopped	
How often?	Frequent			
Consequences	Gases forced into tank A and escape to atmosphere			
Severity	Moderate			
Safeguards	CV closed by operator before pump stop			
Acceptable risk	No			
What should be done	Consider non-return valve or auto shutoff valve to trip on negative differential pressure			
Action	Process engineer/instrument engineer			

Figure 4.27
Sequential worksheet. Page 11

The worksheets shown in Figures 4.17–4.27 would be continued in the Hazop study until all credible parameter/guideword combinations have been completed. It should be clear from this exercise how the study will systematically reveal most operating problems, provided all the right information is available at the start of the study. It still depends on the skill of the team leader and the dedication of the team members for the best possible depth of examination to be achieved.

Before we try another practical exercise here is a further discussion of some aspects of the study steps we have been using.

4.4   Points to note on the examination procedure

4.4.1   Worksheet formats

There is much scope for refinement of the worksheet formats for use in the Hazop sessions. Most study groups will find ways of improving the efficiency of the descriptive terms and the recording methods.

Some key points:

- Worksheet formats should be flexible to suit the study.
- Worksheets may need to include for risk matrix values to be entered during the initial study. The team leader decides on the degree of risk estimation to be done in or out of session.

- The team leader needs a worksheet to assist in tracking the progress through the deviations and to provide a simple record of each guideword/element combination that has been tested.
- Worksheets must identify information collected from the team during the sessions. Subsequent work such as risk estimates done outside the session can be added to the same worksheet as long as it is identified as an 'add in'.
- Companies may develop their own spreadsheet designs, or software products can be bought in.

4.4.2   Identifying causes

If the plant design is basically the sound cause of a deviation parameters will nearly always be due to a failure of some kind. The following categories are common:

- *Hardware:*   Equipment, piping, instrumentation, design, construction, materials
- *Software:*   Procedures, instructions, specifications
- *Human:*   Management, operators, maintenance
- *External:*   Services (steam, power), natural (rain, freezing), sabotage.

Checklists

It is helpful to have a checklist for your particular industry of typical causes of faults or failures to achieve design intent. A problem to avoid in checklists is that they should not encourage exclusive thinking about the possible causes.

EUC risks

Remember that, as in hazard study 2, we need to consider 'EUC risk'. Hence, the potential causes of the deviation must always include the controls and instruments, as well as the operator. The hazard 3 study is very likely to see that safeguards provided against a hazard are all in place and include an alarm or SIS protection. These safeguards are to be taken into account only after the possible causes if the deviation event have been identified.

So the approach is:

- Pretend there are no safeguards
- Identify deviations and causes
- Identify consequences, again without protection
- Recognize the protection measures provided (describe the safeguards)
- Decide if the protection measures are good enough. This decision may not be sensible to decide in the session, rather refer it for analysis and add-in later.

This procedure ensures that the hazard study either verifies or disputes the safety measures that have been provided. It follows from the needs of the last step that the design team is obliged to provide sufficient evidence of the design basis for the protection measures to allow the study team to be satisfied.

Free discussion of faults

It is also important to create an atmosphere of free discussion in the Hazop with regard to possible defects in design or equipment. The design team should not be made to feel they are being criticized or forced into a defensive attitude. For example, the instrument engineer should not be offended if someone suggest that an instrument system fault is the most likely cause of trouble.

Do not waste time on minor consequences

Acceptable risk involves a trade-off between frequency and severity. Hence low frequency events with minor consequences do not merit great effort from the study and they may often be declared as acceptable. If the consequence is minor the team can often dismiss an event without wasting time on searching for causes.

Do not try to do risk estimates in the session

Beyond this level of judgement the team may find itself drawn into making 'on-line risk assessments' and some caution is needed here. When a consequence is seen to be serious the team would need to be very careful about estimating frequencies during the Hazop session. The EPSC guide suggests engineers can estimate high frequency events quite well but says that frequencies below 1 event per 10 years are not something you can rely on by judgement alone. When in any doubt the best approach is to identify all causes of the deviation event and then refer the event for separate risk study.

Operability

Experienced team members will question the detailed practicalities associated with maintenance or startup activities. These are part of the operability function of the study and are of great value. The questions raised in these areas are very often prompted by the deviations found when studying different modes of the operation prompted by the 'other than' guideword.

Consequences of deviations

As we can see, the consequences of a deviation may be an operational problem or it may be a hazard of minor or major concern. For safety systems we are going to be interested in those hazards that justify a SIS, i.e. a safety-related alarm, a trip device or an ESD system. The duty of the Hazop team is to accurately state what is known about the consequences, including the possible hazards that may arise, e.g. tank overflow may be trivial if it is water, but disastrous if it is a strong acid or a flammable liquid that could create a fire.

Many of the items found by the study will be of minor consequence requiring good operating practice. These will be recorded for actions such as 'item for standard operating procedures'.

Operator responses

When considering consequences, the team must take into account the dependency on operators to take corrective actions and/or the possibility that the response may lead to a bigger consequence. As noted under EUC risk, this is part of questioning the validity of the claimed safeguards.

Safeguards

The best approach is to ignore existing safeguards in the design until the event consequences and frequencies have first been declared by assuming there are no safeguards. By this method the performance or safety integrity requirements of the safeguards can be recognized. As noted under 'EUC risk', this is particularly important for the validation of the alarm and trip functions.

The study team can then decide if the existing safeguards are adequate in the light of the expected frequency of demand and severity of consequence. This approach aligns with the risk reduction models we saw in Chapter 3 and allows us to properly recognize the layers of protection.

Recommendations/actions

It is essential to avoid the study becoming a design meeting. Beware of digressions causing lost momentum in the study. The EPSC guide recognizes that there is a range of approaches to this problem:

- Refer all potential problems to outside the meeting.
- Propose and record the suggested solution to each problem as it is found.
- Intermediate approach – Recommend a solution only if there is a breach of standards, or if it is agreed within the team, and the team is authorized to do so.

The third solution appears to be the most practicable. The first item has the problem that a second review meeting would be needed to cover all the feedback points. The second solution may be acceptable in a small organization.

Adding protection layers

The application of protection layers by the Hazop study team should have already been done in the hazard 2 stages. If any remaining problems are found at the detail level, then more detailed protection layers may be proposed at this stage.

Some typical solutions in the chemical industry can be found from the following list:

- Provide a relief system in the piping (overflow, return path, vacuum breakers)
- Provide manual isolation valves or non-return valves. However these solutions sometimes introduce their own problems and need to be treated with caution
- Standby pumps, auto starting on alarm or stop of the first pump
- Mechanical restrictors against high flows
- Reduce exposure of operators hazard (reduces consequence rather than frequency)
- Additional actions sequenced in the process control system. Limited risk reduction
- Alarms on process controller excursions to sound in control room
- Back up/independent alarms/louder alarms
- Simple interlocks (e.g. pump to trip on low level before it runs dry)
- Safety trip devices
- Chemical bombs to kill reactions
- Emergency cooling
- Gas detectors
- Fire deluge systems
- Blast protection walls.

Where the Hazop team suggests new protection measures, the overall risk reduction requirements and the existing risk reduction measures will need to be stated. We shall look into these requirements more closely in Chapter 6.

4.4.3   Recording of Hazop results and safety functions

We have seen how the basic steps of the Hazop study are recorded on worksheets. These sheets are essential to have hour-to-hour working and progress control of the Hazop. Further to this short-term need, there is requirement for a permanent record of the conclusions of the hazard study, very much the same as the preliminary hazard study report we saw in Chapter 2.

The EPSC guide reminds us 'The Hazop report typically represents the only comprehensive record of the study and of the operating strategy intended by the designers of the plant. The report should be regarded as one of the suite of key documentation handed forward to the operators of the plant.'

We shall look at the overall documentation needs of Hazops in the next chapter. At this stage it is important to be aware that the worksheet records form part of the final Hazop report package.

It follows that it will be most efficient if the Hazop worksheet format is either already in the reporting format ultimately required or that it is created on a computer package as a reliable database with a report generator facility that allows the final version to be produced accurately and easily.

4.5   Practical exercise: continuous process example

Now is a good time to try Exercise No. 3 for some basic exercise in using the guideword system.

See Exercise No. 4 in the 'Practicals' section of this book.

Our suggestions for the next steps in the workshop:

- Work through Exercise No. 4: Hazop study of a continuous process
- Identify options and points in the worksheet details
- Study the guideword methods for batch processes
- Consider the duties of the team leader
- Work through Exercise No.5: Composite batch and continuous process.

4.6   Hazop for batch processes and sequential operations

Sequential operations are involved in most manufacturing activities. For example:

- Assembly of a machine
- Startup of a continuous process
- Flushing or cleaning of a pipeline
- Batch production processes such as paint manufacture, polymers, explosives.

The Hazop method of all of these and many other sequential operations is basically the same as for continuous processes but with the following essential differences:

- The process must first be carefully analyzed or resolved into clearly defined operations or phases.
- The parts or sections of plant involved in the operations must be identified.
- The Hazop examination must be performed on each part for each operation.
- The Hazop guidewords must include prompts for deviations in the order of execution and in the timing of execution.

In effect, a mini Hazop study has to be performed on each operation that can be defined for the process. Here we can use an example of a typical batch process. Figure 4.28 shows a very simplified version of the hydrogenation process for vegetable oils.

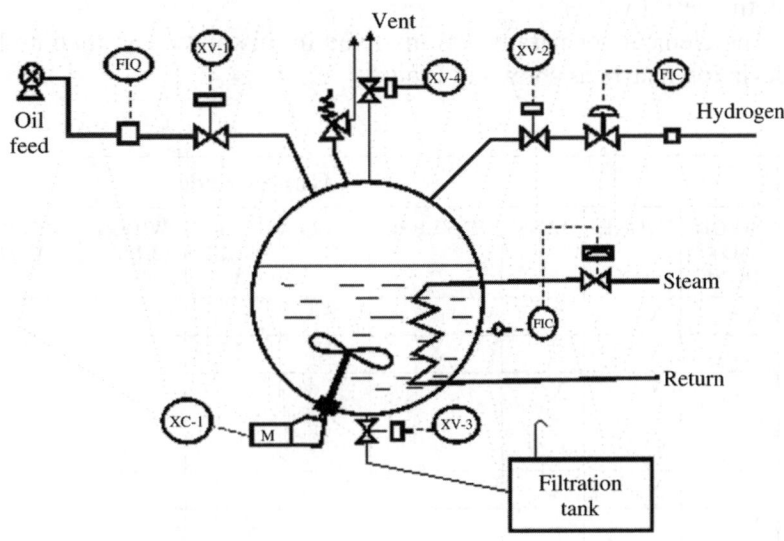

Figure 4.28
Simplified diagram for hydrogenation stage

4.6.1   Example of batch study method

A PLC controller will perform all sequential controlled operations. Manual control overrides will only be permitted under authorization from the supervisor. In batch control terminology a controlled operation is usually described as a 'Phase' and the phase comprises one or more 'Steps'.

The phases of the batch operation are shown here:

Phase No.	Description
1	Add prescribed volume of oil
2	Heat up to controlled temperature
3	Remove air from headspace by vacuum system and test for leaks
4	Add hydrogen at controlled rate to a set pressure and allow reaction to proceed for prescribed time
5	Remove hydrogen residual and purge with air
6	Drain oil to filtration system

The steps of Phase 1 are:

Step No.	Operation
1	Close all vessel valves and confirm shut
2	Open vent valve XV-4 and confirm open
3	Zero oil flow meter totalizer FIQ-1, set target value
4	Open oil valve XV-1 and deliver oil to target value, close oil valve
5	Close vent valve XV-4 and confirm shut

The remaining phases will have similar step details, as required to perform all the operations.

The Hazop examination for Phase 1 will begin with the definition of the parts for study. These would be the oil transfer line and the reaction vessel with its heater and stirrer and all the vessel valves.

The elements or parameters involved in this phase can then be listed and placed in the deviation matrix as shown in Figure 4.29.

Element (para-meter)	Guidewords									
	NO OR NOT DONE	MORE	LESS	REVERSE	PART OF	AS WELL AS	WHERE ELSE	BEFORE/ AFTER	EARLY/ LATE	OTHER
Oil quantity	X	X	X							
Oil flow	X	X	X	X	X	X		X	X	
Oil temperature		X	X			X				
Oil pressure		X	X							
Step 1	X				X	X		X	X	
Step 2	X				X	X		X	X	X
Step 3	X	X	X		X	X		X	X	
Step 4	X				X	X	X	X	X	X
Step 5	X				X	X		X	X	
Operator action						X				X

Figure 4.29
Example of guideword/element matrix for Phase 1 of hydrogenation batch

The guideword/element matrix now contains the sequence-dependent deviations 'Before/after' and the timing deviations 'Early/late'. There is also a greater likelihood that the destination guidewords 'Where else?' will be applicable since many batch processes involve multiple paths for the materials.

Using the guidewords in the normal manner the team can now address the issues in this operation in terms of the possible deviations from the intended steps of this phase as well as the possible deviations in the oil transfer operation. When considering the steps of this phase the individual valve operations will be checked for deviations. Hence:

Step 1: close all vessel valves

'No step' means the step is missed out . . . valve positions uncertain.

'Part of step' means one or more valves failed to respond or are not confirmed. It would also serve as a derived guideword for 'valve does not seal'.

'Before/after' will prompt for 'out of sequence'.

In an application such as this, where the sequence is automated, the likelihood of out of sequence or unconfirmed operations causing a hazard or operability problem is low. However, the Hazop must check for serious consequences of a failed step operation but will probably dismiss most low consequence items.

4.6.2 Elements subject to repeated study

As the study moves along to the next phase of the operation, it is very likely that the same elements will be visited again and again. This will happen because each phase of the operation will present new working conditions that may be hazardous. For example, if the vent valve XV-4 has a leaking seal this will not be important in Phase 1 but it will present a hazard at Phase 4 when hydrogen is added because it will first allow air to leak back into the vessel whilst it is under vacuum, then it will allow hydrogen to leak out when it is under pressure.

4.6.3 Control system deviations

In the above example, it becomes clear that the leak test is a very important phase. The 'before/after' guideword provides an obvious prompt to consider the hazards if the vacuum stage is missed out or incorrectly executed. Control failures will be critical at this phase. The study team will look for safeguards against control system failures as well as for loss of power at critical stages.

This point highlights a difficult problem for automated batch processing plants. The control system provides an initial level of protection from hazards due to mal-operation of the sequences needed for safe working of the plant. As noted in Chapter 3, we need to evaluate the 'EUC risk' including risk due to failures of the control system. Whenever the Hazop encounters a serious consequence it is important that the risk is highlighted for checking of safeguards. As we shall see in Chapter 6 the basic batch control system should not be expected to reduce the chances of a hazardous event by a factor of more than 10. Alternatively it should not be expected to have a dangerous failure rate of less than 1 failure per 10 years.

These requirements mean that the Hazop team should recommend additional safeguards where the dependency on the batch control system exceeds the levels stated above.

4.6.4 Further practice with batch Hazop

You can work through the phases of this study in to get some more practice.

There is also an excellent example of a batch Hazop in the EPSC guide that you can look at.

4.6.5 Conclusions on Hazops for batch processes

- Techniques for batch plants are similar to those for continuous processes but carry the addition of sequence and time dependent deviations.
- Each phase of a batch operation must be considered as a separate part for study.
- Preparation work for a batch study is usually more demanding than for a continuous process. The preparation work must identify the phases of the operation and plan for each phase to be explained with its design intent. Detailed flowcharts and functional specifications of the sequences could be made available to a study team.
- Hazop studies are carried out on each phase of a unit operation and this may involve repeated examination of the same elements under different conditions. Careful planning of the scope of each part for study can help to reduce the workload.

- Critical conditions and trip settings may change according the phase. Interlock protection may be removed at certain phases and applied in others.
- Interactions with other sections of plant must be checked. If these generate different conditions, according to the relative phases of two sections the combinations must be checked.

Complex batch processes are normally designed for control and operability by using structured design principles, typically, as seen in the ISA S 88 design standard. Operability issues are then dealt with by designing and controlling the plant as a set of units and subsystems. This approach simplifies the Hazop studies as random interactions between units are minimized.

4.7   Hazops for other disciplines

There is limited space in this book to study Hazop methods as they are applied to other fields but we have summarized some of the key points here.

4.7.1   Hazops for mechanical plant and machines

In general, Hazop methods can be extended to machines and automated plant in very much the same way as for processes. There are three basic steps to doing this.

1. The machine actions or operations must first be identified into subsystems or defined tasks. This is very much the same as a phase or step in a batch plant. For example, in an injection molding machine, the first step is to close the die.
2. The activities in the operation must then be defined. For example, In closing a die the activities might be: Clear the mold of residue ... Open hydraulic valve ... Advance the die along the slide ... Meet the opposite face ... Apply pressure to set point ... Check closure limit switch:
3. Each of the activities can then be considered for deviation using the standard guidewords.

For example:

- *Clear the mold:*   Deviations are: No clearance (leaving residues of past components), More clearance (damage to mold by removing inserts), Other than (introducing foreign bodies by action of clearing the mold).
- *Open hydraulic valve:*   Deviations are: Too soon, too late, incomplete, etc.

This type of activity analysis is also used in risk assessment methods such as the Job Safety Analysis (JSA) we saw in the introduction to this book.

4.7.2   Hazops for electrical systems

Hazops have been applied to electrical power systems where the systems are divided into distinct operational sections. Derived guidewords can then be applied for deviations in voltage, current, power, power factors, alternative paths, frequencies and surges, etc.

4.7.3   Hazops for electronic control systems

Hazops can be applied to detailed electronic circuits to examine the effects of deviations from design intent on the performance of particular operations. These are relevant if the electronic circuit is linked to a safety function or a hazardous operation.

IEC 61882 shows an example of an electronic control circuit for a piezo-electric valve device.

4.7.4 Hazops for control systems and PLCs (CHazops)

The purpose of a CHazop is to examine how control systems deviate from designed function (e.g. produce incorrect output for a given input) and the effect that this would have. As for Hazops, CHazops examine whether the safeguards currently in place or proposed will adequately prevent the deviation or mitigate its consequences.

Most requirements for a hazard study on a control system will focus on its main functions to identify ways in which it could create a hazard. We have already seen that individual loop or sequence control functions are likely to be tested for deviations by the process Hazop using standard guidewords. In such cases we always assume a loop function failure is possible and in critical cases a separate safety system is installed (independent trip or alarm).

What the process Hazop may not show is the effect of failures of subsystems, controller units, communications or power failures within the control system impacting the overall plant or affecting the ability of an operator to take corrective action. Hazops on plants are not very effective at detecting this type of problem as they concentrate on individual parts of plant and may not look at the interactions between sections. In such cases the typical DCS may have failure modes that present a unique set of deviations not previously anticipated.

Here are some suggested applications where a Control Hazop would be considered:

- Control system for multistage process where co-ordination and balancing of the process is handled by a DCS or multiple PLCs interconnected by communication links
- Alarm management systems (Loss of system will create an alarm flood or no alarms)
- Multistage batch processes where sequencing of vessel feeds and material transfers are controlled by a PLC network
- The PLC networks on large automation plants, car assembly lines, and robot lines
- Offshore platform fire and gas detection systems
- Safety-related control systems applications … emergency shutdown systems (ESDs or Trip systems)
- Automated crane controls
- Military defense systems (see Ref. 15).

The general, approach for a CHazop is to begin by defining the scope of the control system in block flow diagram format depicting the main functional components with their data transfer paths identified. These will include the interfaces to the plant sensors and actuators and the operators. The operational diagrams then represent the 'design representation' as an equivalent to the P&I diagram used in process Hazop. In some applications there will be a hardware system diagram and a software system functional diagram.

For each diagram, the study team will identify the parts for study as in Hazop and will apply the deviations from intent based on guidewords in the usual way. The problem is to find a method that quickly identifies the critical components or tasks within the control system. Then these parts must be examined in more detail for potential deviations.

Specialist groups are still developing CHazop methods in various forms and a variety of papers have been presented on this subject. See Ref. 16 for a very helpful and straightforward presentation by Raman and Sylvester from Halliburton (Australia) Pty Limited. Their approach begins by using FMEA to identify which main components will have a hazardous impact on plant operations. Then, for each component, Hazop guidewords are used to identify unplanned deviations around that operation.

Suggested CHazop Guidewords applied to control system components (from Ref. 16).

Hardware

- Device error
- Bad measurement

  - Loss of signal (zero read)
  - Loss of signal(full-scale read)
  - Signal erratic
  - I/O failure
  - Control card failure
  - Processor rack failure
  - Abnormal temperatures.

Software

- Program corruption
- Memory management error
- Addressing errors/data failure
- Timing overload
- Timeout failure (sequence control problems)
- Endless loops
- Message failures
- Failure of counters
- Sequence interpretation error
- Data validation problems
- Operator override.

When all components have been checked for their impact on the process under the above fault conditions the identified parts are then subjected to an overview set of guidewords to test for general failures and their impact on the controlled process.

Suggested CHazop Overview Guidewords applied to control system (from Ref. 16):

- Power failure
- Power surges
- Field 24 V supply failure
- Rack 9 V supply failure
- Static electricity
- Physical contamination
- Abnormal environment
- Redundancy in system architecture
- Error diagnostics adequacy

- Operator training
- Code security
- System crash recovery.

4.7.5   When is a control Hazop needed?

Here is a suggested approach to deciding the need for a more detailed study of a computer-based control system.

Use a first level review to carry out the following checks:

- Establish the limits of the computer system and its network.
- Identify what plant's units depend on or interact with the computer system for their operations.
- Develop a block flow diagram of the functions of the computer in controlling the plant units.
- Identify hazards of the units as defined by a hazard study 2 or process Hazop and include for any hazards associated with interactions between units.
- List those computer functions associated with the hazards in any way.
- Are any of these functions NOT already provided with independent safeguards such as a separate safety-related trip or alarm system?

If this procedure indicates the computer system is involved in the control or supervision of hazardous functions, that have not been provided with safeguards, the project should consider conducting a Hazop of the control system. Even if the independent safeguards are provided, the project team should check that the safety functions have been specified to respond to deviations that could arise from the control system actions.

4.7.6   Studies for safety-related computers

The hardware and software requirements for safety-related programmable systems are defined by IEC 61508. Approved testing bodies such as TUV normally undertake the evaluation of these systems using FMEA and criticality analysis. All embedded software is subjected to rigorous examination of its structure, its security and error control features, its testing methods and the quality assurance history. This means that if you buy an approved product the security of the system is assured and its failure mode behavior is predictable. Thus, a detailed Hazop would be confined to a check of the safety applications as configured by the end-user.

Where non-certified PES devices are used in safety controls there is considerable uncertainty about their failure mode performance, and safety studies should be carried out. These installations are generally considered to be unsuitable for safety duties and will have difficulties in reaching compliance with IEC 61508.

4.7.7   Hazops for alarm systems

Many companies have carried out 'Alarm reviews' to try reducing the quantity and frequency of control room alarms. The results of these reviews may lead to the use of an alarm management system as part of the software facilities of the DCS or it may be installed in a SCADA system. Safety-related alarms may or may not be involved in this exercise. Reducing alarm loads often involves automatic prioritization of alarms or automatic suppression of alarms found to be inappropriate under certain running conditions.

Referring to the EEMUA guide (Ref. 11), it suggests these studies be supported by an 'Alarm suppression hazard study'. It states 'Suppression of alarms involves a potential risk of depriving the operator of important, possibly safety-related information.' The guide goes on to describe an example of a method for the study.

In summary, there are a number of associations between alarms and Hazops:

- Hazops often request alarms. These need to have a clear requirement and a clear response defined at the time of the Hazop.
- Safety-related alarms might not achieve the desired amount of risk reduction if the alarms are not adequately notified or if they are frequently in a false state or operate spuriously. Hence the overall alarm situation in the control room affects reliability of alarms.
- Modifications to safety-related alarms must always be referenced back to the original Hazop and the change may require the Hazop item to be studied again.
- An alarm management system may introduce new common cause failure modes (such as might occur if a plant status signal incorrectly blanked out a whole group of alarms) and this should be subject to Hazops.
- Alarm suppression plans should be subjected to an 'Alarm Suppression Hazop'. For example, each suppression mode would be tested for deviations from design intent to check for unsafe conditions that would be missed.

4.7.8   Control and operability studies

For complex processes or highly integrated plants the methods of Hazop are sometimes not well suited to studying the operability of the process. Often these plants will have complex control systems with a hierarchical control strategy. In such cases, some companies, in particular ICI Ltd in the UK, advocated the use of a separate methodology called 'Control and operability study' or Co-op Study. This approach allows the Hazop to concentrate on basic Hazards arising in the process and leaves more time and scope for practical control and operability problems to be studied separately under a different set of guidelines. This approach is helpful for control system design but the detailed physical operations involved on plants should still be examined by the Hazop team due to the close links between these operations and the possibilities of hazards.

The Co-op approach is valuable for the development of good automation practices and falls in line with quality assurance methods for control system design. The Co-op studies employ an incremental approach to the development stages of the project in much the same way as we have seen for the hazard study life cycle. Studies begin with a statement of business and operational objectives and lead to the plant being structured into operational stages. The relationships between stages are then identified, as well as the operating objectives of each stage. Detailed control strategies for each stage are then developed and examined to ensure each stage can be managed to meet the overall performance objectives of the plant. These studies allow process designers to define the best control system actions to meet the design intent.

4.7.9   Hazops for maintenance tasks

These are similar to job safety assessments, as noted in USA practice. The task to be performed is divided into operational steps, very much, as is done for batch plants. The deviations are proposed in the usual way for steps omitted or wrongly performed. For each type of task the study team will develop a set of derived guidewords for typical errors or deviations in the type of work being considered.

4.8 Conclusions

We have seen that Hazops are based on a very simple set of principles:

- Structure the plant or system into component operations.
- Explain each operation and then test it for tolerance to all possible deviations.
- Recommend safeguards and corrective actions where needed.

These principles supported by guideword prompts for deviations make it possible to construct Hazop methods for a wide range of plants and processes. Where no existing examples can be found for a new application, it appears feasible to adapt the principles we have seen to a new application.

4.8.1 Limitations of Hazop

It is also clear that the method is painstaking and lengthy and may not be justified if parties are sure that no hazard potential exists. The Hazop method can be tedious for participants and when loss of concentration sets in the belief that a design has been checked for hazards may become false.

We have also noted that Hazops are not well suited to finding hazards of co-ordination or those hazards that are caused by disturbances that pass from one section of plant another. IEC 61882 notes 'Many accidents have occurred because small local modifications had unforeseen knock-on effects elsewhere. Whilst this problem can be overcome by carrying forward the implications of deviations from one part to another, in practice this is frequently not done'.

Another weakness of Hazop is that the study only considers the parts or activities that appear on the design representation. If they are not on the drawing, they are not considered. This places much responsibility on the planning and presentation of the design.

5

Planning and leadership of Hazops

Objectives

The objectives of this chapter are to provide a brief introduction to the tasks of organizing, planning and reporting on a Hazop study. The depth of training in this chapter will be limited, as many of the skills required for this task must be acquired by experience on the basis of simple principles.

5.1   Introduction

This chapter steps back from the detailed procedures of the Hazop study and considers the tasks involved in organizing a Hazop study. Much of this task falls to the Hazop study leader. It helps to take a look at the leadership role in the studies and the outputs and results that the study is expected to deliver. After this chapter there is another Hazop exercise that may help to consolidate the points learned so far.

5.2   Organizing the Hazop

The detailed procedure we have looked at in Chapter 4 takes place within a planned framework of preparation, good knowledge of the scope and boundaries of the study and an environment supportive of creative thinking. At the same time Hazops require a disciplined and methodical approach. Balancing the need for thoroughness against the need for good ideas is the task of the team leader. Getting the plan right contributes to the quality of the study.

5.2.1   Initiation of the study, issues of planning

The project manager usually initiates the Hazop study on the basis of plans previously laid down for the project. There must be a serious intention to do the job properly:

- What is the motivation for a Hazop study? We saw in Chapter 1 that there are legal requirements in many cases to carry out risk management plans for hazardous plants.
- The first step in a Hazop study is to assign someone with an appropriate level of authority to carry out the actions arising from the study.
- There must be an intention to see that the actions/recommendations of the study are going to be carried out.

The project manager's responsibilities are:

- Determine when a study is to be done
- Appoint a study leader
- Provide the resources to carry it out.

These last three points raise some questions.

Question 1:   When should the study be done?

We saw in Chapter 3 that there is progression from hazard 2 at the flow sheet stage to hazard 3 or Hazop at the P&ID stage.

The problem is that Hazops take up a lot of time and generate interference with the progression of a design task (especially for process design projects):

- If Hazops are too early, there will be incomplete information and more studies will be needed later.
- If Hazops are too late, the cost of rectifications could become high.

Another complication for Hazops is sometimes the linking of design contracts and deadlines to the Hazop completion. In some projects the best study team members cannot be available until later in the progress of the project.

Hazops for any system cannot be started until the line diagrams (P&IDs or detailed system diagrams or equivalents) are complete. But they need to be done as soon as possible once these diagrams are ready (see Figure 5.1). For Hazops on existing plants under review or for modification it is essential that the diagrams are fully up to date before the study begins.

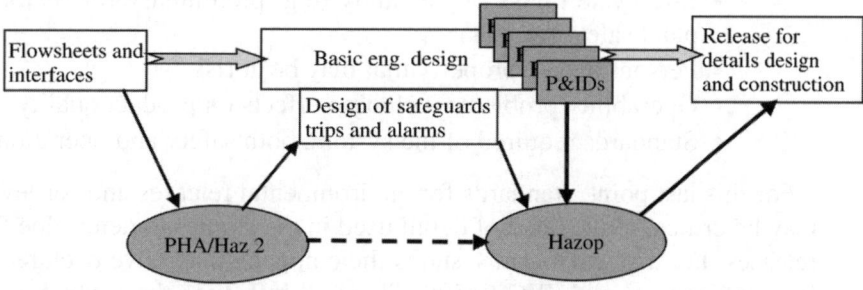

Figure 5.1
Hazops should be done when the design is ready for detail engineering

Question 2:   Where do you find a study leader?

Fortunately the study leader can be someone from outside the immediate project or even from outside the company. The project manager will want to find a leader who will help to get the balance right between safety and overdesign of safeguards.

Question 3:   Where are the resources?

The primary resource used by a Hazop is the time of skilled team members and the team leader. The time resource is not only expensive but it uses the skilled people who are needed at that time for project design work. Fortunately one of the best ways to do a Hazop seems to be part time by doing perhaps 3 or 4 half days per week. This approach avoids fatigue in the study team and leaves the members some working time to continue their other duties or to follow queries raised in the study sessions.

Here is a commentary to help with estimating the time frames, from Trevor Kletz, one of the pioneers of hazard studies. 'A Hazop takes 1½ to 3 hours per main plant item (still, furnace, reactor, heater, etc). If the plant is similar to an existing one it will take 1½ hours but if the process is new it may take 3 hours per item.'

Even this estimate can be optimistic if the team is inexperienced or if the plant is highly interconnected.

5.2.2   Definition of scope and objectives of the study

The scope of the study refers to both the physical boundaries of the plant and to the operations to be studied. A large chemical process may occupy 10 or 20 P&IDs and the complete study may extend over several months. The study may however be limited to certain areas of the plant or may even be limited to a specific operation or modification in one part of the plant.

Essentials for the definition of scope are:

- Physical boundaries of the system and its interfaces with other systems
- The specific drawings and diagrams used to define the system
- The scope of any previous studies including other hazard studies (e.g. Hazard 2) and supporting analyses
- Any regulatory requirements applicable to the site or the process
- The range of operational modes to be covered, e.g. start ups, shutdowns, storage modes
- Identification of any packaged units that are excluded or studied separately.

Essentials for the objectives to declare are:

- Purpose for which results of the study will be used
- Life cycle phase of the study (e.g. preliminary for major hazards or final for plant safety records)
- Persons and/or property that may be at risk
- Operability problems including effects on product quality
- Standards required of the system, both safety and operational.

For this last point, standards for environmental releases and for levels of toxic exposure may be critical to the level of detail used in the studies to determine the possible causes of releases. For trip and alarm systems the company may have declared its intention to work for conformance with IEC 61508. This will influence the evaluation of risks and the way in which risk reduction measures are specified but should not add to the workload, as the procedures are clear.

5.3   The team leader and the team

The first requirement of the team leader is to see that the Hazop methodology is used effectively and productively. The EPSC Guide says, 'The leader needs to have a deep understanding and considerable experience of HAZOP studies. HAZOP study teams work best when there is clear leadership from an experienced leader.'

The IEC 61882 standard describes the study leader's skills and duties as follows:

- Not closely associated with the design team and the project
- Trained and experienced in leading Hazop studies
- Responsible for communications between project management and the Hazop team
- Plans the study

- Agrees study team composition
- Ensures the study team is supplied with a design representation package
- Suggests guidewords and guideword-element/characteristic interpretations to be used in the study
- Conducts the study
- Ensures documentation of the results.

Which pretty well specifies the team leader's job. It is clear that this job requires considerable experience and attention. Let us take a look at who should make up the Hazop team itself.

5.3.1   Who is in the team?

In addition to the team leader the essential players are those given in the following table.

Team Member	Role and Duties
A recorder or 'scribe'	Documents the proceedings including recording attendance. Prepares and completes the worksheets as the study progresses. Reads back conclusions for agreement as each item is covered The study leader may sometimes do this job but it can distract from the promotion of thinking and control of the meeting. Sometimes one of the team members less involved can do this A good software package makes this job much easier and more efficient
Designer (process engineer, control engineer, mechanical engineer, etc. according to project)	Explains the design and its representation on the diagrams and drawings under review. Explains how the system may respond to suggested deviations. This person's knowledge of the system is essential but his/her assumptions can be challenged
Project engineer (may also be designer)	The person who represents the project interests in terms of costs and progress. This person will also know the implications of the recommended actions
User (commissioning manager or production manager)	Explains the operational context for the parts under study. In process plants the commissioning manager is the essential person here. He/she will have to start up and operate the plant and train others to do the same. This person is sure to be keen on making changes that make for more practical operating. Their practical experience is essential to balance the plans of the designers
Maintenance representative	This person may be needed where maintenance of the plant is complex or hazardous. Many operability problems are associated with maintenance and many accidents occur during maintenance

Team Member	Role and Duties
Instrument/control engineer	The instrument engineer represents the technical and functional aspects of the control system as part of the process equipment (EUC). This person can advise on control system responses to deviations and as causes of deviations
	The second role is to advise on the performance of safeguards employing alarms and trips. The instrument engineer will be required to implement any new safety instrumentation measures called for during the studies. He/she will want to collect the best possible information on safety system requirements at the time of the study
SHE expert (mandatory in some countries)	This person will represent the interest of occupational safety and health and may be required to serve as an independent observer to see that the study proceeds in a satisfactory manner
Other specialists	They provide expertise relative to the system and the study as needed. This may only require limited participation but the team leader will have to decide on the times when such persons are needed. Likely candidates include: • Research chemist for new processes • Electrical engineer • Environmental pollution specialist • Effluent treatment specialists • Safety specialist • Control system software engineer
Contractor and client representatives	If the plant is being designed by a contractor, the Hazop team should contain representatives from both contractor and client. This may result in some duplication of the above the roles but is generally necessary do to the alternative perspectives of the parties

The team leader who is often a specialist in the field of safety management may cover the role of safety specialist for the study.

Software specialist

The control system software engineer will be needed for those systems where the operations are largely conducted through automatic control sequences. One of the well-documented problems of batch control lies in transferring the design intent to the control software. The risk is that the functional requirements will be incorrectly defined by the process engineers or will be misunderstood by the software engineer. Having the engineer present during the Hazop will help to avoid this problem.

Trevor Kletz in *An engineer's view of human error* provides an example of this problem: 'Specification error causes discharge to atmosphere':

- In a batch reactor plant the controller specification called for a 'freeze of all output controlled variables' if a fault was detected in the plant.
- The sequence control specification required increased flow of cooling water to the reflux condenser immediately after catalyst addition.
- When a fault occurred at the time of catalyst addition the computer fixed the flow rate of water and the reactor overheated. This resulted in a relief valve opening and a discharge of toxic contents to atmosphere.

This type of error in design is called a 'systematic error' and it is one of the most difficult types of error to eliminate. We shall see in the next chapter how designers of safety instrumented systems have to go to great lengths to reduce the chances of such errors.

There is also the case of the automatic train protection system in service for the London underground railways (see Figure 5.2).

Train safety...............?!!!

A London Underground train rolled backwards half a mile when the driver fell asleep, highlighting a serious flaw in signalling systems: they only work with trains going forward.

The Engineer 14 July 2000

Figure 5.2
Illustration

5.3.2 Revision and continuation of the study

Now that we have the team members and roles in place it just remains to review the overall Hazop study sequence we saw at the beginning of Chapter 4 and to look at how the study meeting is to be conducted (see Figure 5.3).

Planning

We have looked at the planning phase. We have also noted the need to plan the studies with enough time allowed for each plant unit or system, allowing up to 3 h per major equipment item. The planning must allow for study sessions to be uninterrupted but the sessions must be short enough to prevent fatigue.

'No one goes home until this P&ID is finished'

This is not the way to make sure a dangerous hazard is detected!

Recording methods

We saw the recording method requires worksheets to be filled as the discussions progress and should be easy to convert to finished reports. We have discussed methods of recording and would prefer computer-based tools if these are flexible and adaptable.

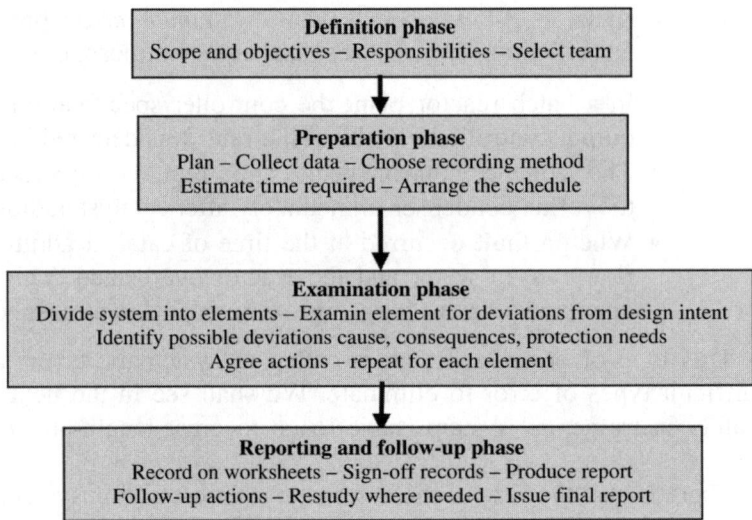

Figure 5.3
Methodology

The study meetings

Let us work through the progress of a study meeting to identify the key points.
  Initial study meeting:

- Introduce the team to the plant, the systems to be studied and the overall context of the plant.
- Make sure the team members are aware of their expected roles in the study and that they are aware their names will be recorded in the history of the studies.
- Ensure that members have all the required background information including scope and purpose of the study.
- Review the intended Hazop procedure and issue a preliminary list of guidewords and typical derived guidewords.
- Explain the recording method to be used and its intended final layout.
- Clarify if the study is to cover only hazards or if it is to include operability.
- Agree the methodc to be used for the prompting of guidewords.

Visual aids

There are two essential components to keeping the team focused.

1. What is the part and change path under discussion?
2. What is the deviation we are looking for?

For example:
  The part under discussion must be clear for all to see. Preferably a large P&ID mounted on a board with the change path highlighted (see Figure 5.4). The parameter and the deviation under review should be displayed on a large board or panel. A popular technique in AECI and ICI was to use 'flash cards', preprinted cards or flip charts displayed in a sequence, each card containing the next deviation prompt.

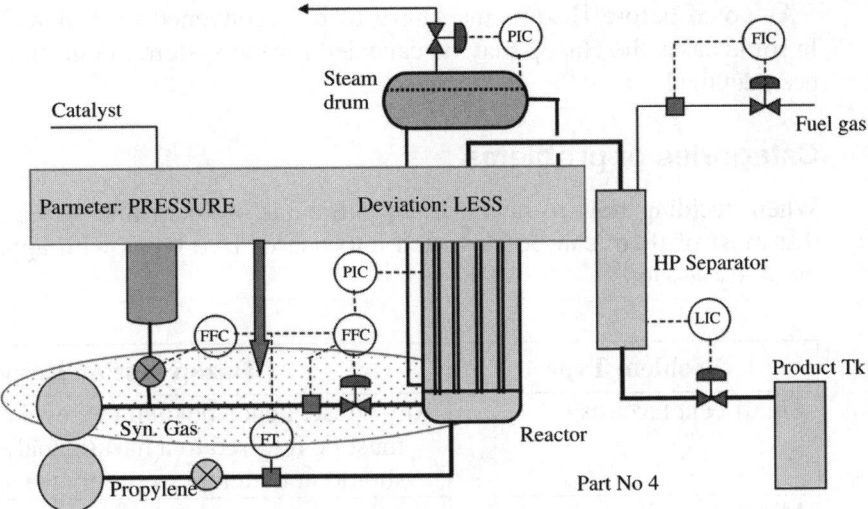

Figure 5.4
P&ID with visual aids to identification

In the example shown in Figure 5.5 a data projector and PC can also be used to create the desired effect. A data projector can also be used to good effect when replaying the results of a PC-based worksheet method.

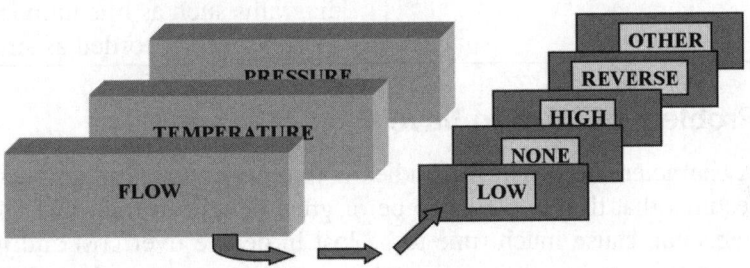

Figure 5.5
Visual aids to prompting for deviations

Reviewing responses

At periodic intervals in the study the team leader should call for a review of responses so far recorded. This is important to do whilst the team is still focused on a particular system or part. Then all the recommended actions can be seen together and discussed to find if all agree they should go forward into the final report.

Key points:

- It is essential to see that the solutions or actions proposed are adequate and do not cause further problems.
- Unresolved issues can only be referred for design review or hazard analysis.
- The leader needs to balance the need to avoid the Hazop becoming a design meeting with the need to avoid unresolved design problems.

As noted before Hazops may have to be reconvened to deal with the revised design. In some cases the Hazop may be canceled for the system in question until the change has been detailed.

5.3.3   Categories of problems

When deciding how to handle each particular problem that arises it helps to recognize that most of them can be classified into one of five types with appropriate responses in the study session.

Problem Type	Hazop Session Response
Significant hazards	Any consequence beyond the level of 'minor'. This must be referred to a hazard analyst. The Hazop team should not attempt to supply the solution
Minor hazards	Consequence does not lead to any serious harm to person or environment. The Hazop team may be able to define a simple safeguard during the study. Otherwise it should be deferred
Standards/operability	Non-conformance to standards or company practices. Experience of team members may allow the remedy to be stated and actioned in the records
Deferred problem	Problem that cannot be resolved in the study due to lack of information or differences of opinion in the team
Line diagram errors	Team members often notice technical errors in the diagrams such as line numbers or wrong symbols. These can be recorded as simple actions

5.3.4   Problem of 'nice to have'

A characteristic of Hazop studies is that experienced plant personnel will call for design features that they consider to be of great benefit to their task of running the plant. These items can cause much time to be lost in debate over cost and justification. They may be just a 'nice to have' feature or they may be of great value. The study leader will need to deal with this situation by capturing the essential details of the debate and closing it off quickly. This may mean recording it as a 'deferred problem'.

5.3.5   Completion and sign-off

When is the Hazop finished? EPSC Guide says: 'when all of the selected plant operations have been examined and all of the problems identified during the examination have been completed'.

The problem is that some items may have been referred for additional study or information and the responses have to be collected by the team or a representative as correct or satisfactory. So the completion becomes staggered with certain exceptions being noted and pursued for closure. This requires a secure follow-up mechanism (see Figure 5.6).

The Hazop report must include details of the items resolved during the follow-up studies. So the team leader will normally collect the responses and arrange for the team members to sign-off on both the study session reports and the follow-up reports. A limited number of exceptions may then be allowed in the sign-off provided the outstanding actions are listed for completion by agreed deadlines.

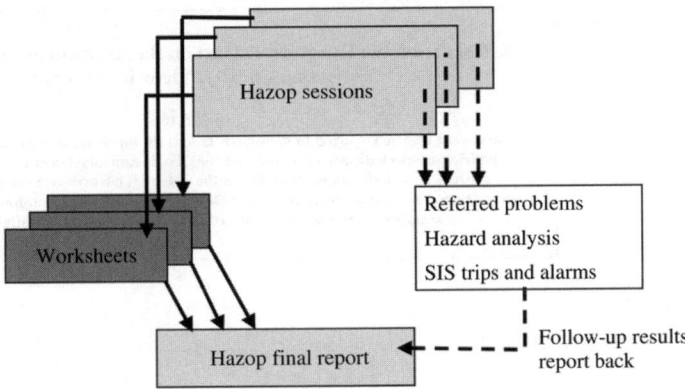

Figure 5.6
Workflow and reporting in Hazops

5.3.6 Follow-up actions and control of change

The follow-up phase of the Hazop is considered to be part of the Hazop study since some items may remain unresolved until the special investigations have been concluded.

For the follow-up actions IEC 61882 suggests the report back contains:

- Recommended action
- Priority/risk ranking
- Responsibility for action
- Status
- Comments.

These items are likely to be associated with additional risk reduction measures and hence we shall see that they will have strong linkage to the provision of safety instrumented systems.

5.3.7 Recording and reporting

In Chapter 4 we discussed the need for session worksheets and looked at their contents. The subject of reporting software packages was also discussed.

Here we need to consider the objectives and functions that the final report will serve.

The final report of a Hazop is all that exists to show the work done by the team. But the Hazop report remains a living document for the life cycle of the plant. What are the uses that this final report may serve?

- The implementation of the actions listed in the report.
- It is needed for linking to any follow-up studies including the Hazop for any modifications to the plant. See below under management of change.
- The parts of the report covering hazards and safeguards against them will be held as part of the plant safety dossier.
- It may be needed to satisfy regulatory obligations. For example: COMAH regulations require evidence of the systematic study of hazards for the plant.
- Regulations often require a regular hazard review. Where regulations do not compel this to be done the need for such action may still be clear (see extract from Longford report in Figure 5.7).

**Extract from the Longford Report on the Accident at the ESSO Gas
Production Facility, Victoria, Australia**

Clause 5.16

'Esso should also be required to demonstrate that its operating standards, practises and
policies are periodically reviewed and that the documentation of each identified
procedure includes an explanation of the potential hazards associated with the
procedure. The critical procedures include start up, controlled shutdown, emergency
shutdown and any deviation from normal operating conditions.'

Figure 5.7
Extract from Longford report

- It forms an essential reference report for the safety life cycle records of any safety instrumented systems built to comply with IEC 61508.
- If a significant incident does occur at the plant the company and the authorities will most likely wish to examine the hazard study record.

5.3.8   Contents of the report

Given the wide-ranging usage of the report what should go into the final documents?

The contents list, based on descriptions in IEC 61882 and EPSC guide, should contain the following:

- Summary, conclusions, scope and objectives
- Statement of the Hazop procedure used in the sessions
- Details of identified hazards and operability problems with details of the provisions for detection and/or mitigation (see recording format)
- Recommendations for any further studies
- Actions required for addressing any uncertainties discovered during the study
- Recommendations for mitigation of problems where available
- Notes for attention in the operating and maintenance procedures
- List of team members for each session
- Hazop worksheets referenced to each study session with numbered actions list
- List of drawings and documents used in the study
- References to previous studies and sources of information.

5.3.9   Level of recording

Three levels of recording are recognized:

1. *Record by exception only:*   Lists only those items resulting in an action.
2. *Intermediate record:*   Records all hazards with or without action and all other operability items where significant discussion was held.
3. *Full record:*   Includes an entry for every deviation that was considered by the team and may include listing the guidewords that were not accepted for deviations.

The choice depends on ultimate uses for the study based on the following considerations.

Exception only   Used to initiate actions in a simple and direct form. Not suitable for safety case records as it cannot be audited to see what items were considered and why they were accepted.

Intermediate This has the advantage of identifying all recognized hazards and operability issues, thus providing a permanent record in the numbered list. This maintains awareness throughout the life of the plant of the existence of all known hazards even if existing safeguards are stated to be adequate. Particularly useful for management of change where there is a chance of accidentally removing safeguards that were not known to be important.

Full recording Used in highly hazardous processes or to meet full audit requirements of regulations.

5.3.10 Recording format and content

The Hazop study sessions are generally recorded and presented in the final report in tabular form. Very often these tabular records are the same as the worksheet records used during the actual study. There is very little point in creating another version other than to tidy up the draft phrases. The format is similar to the one described in Chapter 2 for the level 2 hazard studies.

Each table must have a header section defining the system and the part or section being examined. The dates and times of the section are required as well as the reference documents such as P&ID number and its revision number.

A typical set of columns is

Ref. No.	Parameter	Deviation	Cause	Consequence	Safeguard	Action

Some studies will add in the fields for estimating the risk ranking before and after safeguards in the same way that have seen for the level 2 studies in Chapter 2, para. 2.3.7. See Appendix G for examples of a report format from IEC 61882.

5.3.11 Conclusions

This completes our run through all that the Hazop study procedure is expected to do in its basic functional form. What we see here is a disciplined and structured procedure intended to find the problems in the detail of the design and ensure that they are permanently recognized and dealt with in the subsequent usage of the plant.

What we have not covered in the procedural material are the human factors involved in running a successful Hazop study. These factors involve the skills of a Hazop study leader in making sure that the sessions run smoothly whilst allowing all participants to make their best contribution. At this stage, it will be helpful to try out some more Hazop methodology on another practical exercise.

5.4 Practical exercise: hybrid batch process example

In this practical exercise, we would like to try out some leadership roles as well as practising the skills needed for a Hazop sequence. We are going to try planning and executing a Hazop on the Ethylene Oxide sterilizer that we examined previously in the level 2 study.

This practical has the following stages:

- Plan the study in groups of four or five by appointing a leader first
- Leader checks the available information and proposes the study documents
- Leader appoints a 'scribe'
- Team decides the division of the unit into parts
- Team leader organizes the parameter deviations matrix
- Study proceeds
- Summary report is prepared.

See Exercise No. 5 in the 'Practicals' section of this manual.

6

Specifying safety instrumented systems

Objectives

- This chapter begins with an illustration of how risk reduction measures can be quantified and adjusted to meet desired low target rates for an accident.
- We then look at how IEC 61508 sets a standard for the safety life cycle that lines up well with hazard study results. This leads to an example of how SIL targets can be determined using risk graphs.
- Then we consider some of the features of an instrumented safety system that are essential for meeting the SIL targets and discuss how these can influence cost and availability.
- The chapter concludes with a set of notes on how to prepare the detailed safety requirements specification for the trip system.

(In a workshop session where time is limited it is likely that this material will be summarized for further study by participants in their own time and for further learning.)

The objective of this chapter is for hazard study participants to be able to carry risk reduction requirements through to practical specifications for trip systems and to encourage use of the methods laid down in IEC functional safety standards.

6.1   Introduction

A hazard study on its own does not reduce risks or eliminate hazards. It is how we use the knowledge gained from the study to provide safeguards that matters. This chapter briefly scans through several aspects of the task of developing a safety system to assist participants in understanding some of the well-established ground rules.

The basis of good safety system design is in the measures taken to ensure accuracy in the specification of the requirements. This is supported by the safety life cycle procedures we are going to look at here. These procedures turn out to be a logical extension of the hazard study sequence we have been following so far.

Safety integrity is a measure of confidence we can place in the safety system. In this chapter we see how this relates to risk reduction needs and look at ways of deciding what SIL ratings we require for risk reduction measures.

It is also important to understand how instrument design features such as redundancy can be adjusted to obtain the appropriate level of performance from the SIS. We shall see how online proof testing and diagnostics contribute to safety integrity once the system has been installed. Getting both the SIL target and the SIL performance right for the job should ensure we pay the right price for it.

6.2 Risk reduction by instrumented protection

We saw in Chapter 3 that protection systems must essentially be able to act independently of the equipment or systems that they are to protect. This is a fundamental principle behind the concept of layers of protection.

When it comes to a formal assessment of a hazardous plant and its safety measures the auditor will look carefully at any claims concerning the performance of protection systems. For the SIS this means achieving functional safety with an assured level of safety integrity. We have seen that this comes in four grades according to need and these are the safety integrity levels (SILs).

Recall from Chapter 3 that the SIL defines a range of risk reduction factors that can be assumed or claimed for the SIS function (see Figure 6.1). Safety integrity level is really a measure of the degree of confidence we have that the safety system will do the job intended. This gives us the basic 'before and after safeguards' model we were looking for in the Hazop.

SIL	RRF Range
1	10–00
2	100–1000
3	1000–10 000
4	10 000–100 000

Figure 6.1
Risk reduction factor (RRF) vs SILs

6.2.1 Risk reduction example

Let us take an elementary design example based on the overpressure hazard case we looked at in Chapter 3 (see Figure 6.2).

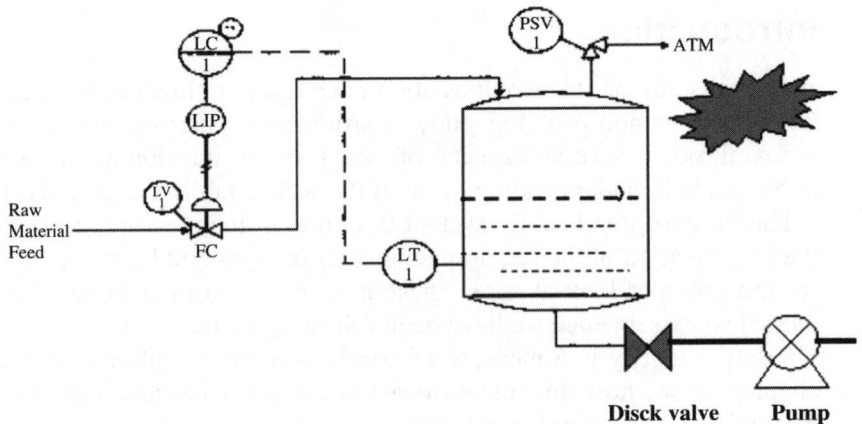

Figure 6.2
Basic tank level control with overpressure hazard

The Hazop study might have a worksheet entry as shown in Figure 6.3.

Part: Transfer of Acid from A to B	Element: Contents of Tank A	Parameter: Level	
Deviation	More	Meaning/effect:	Tank is full and still feeding in
Item ref. no.	A1-1		
Is it possible	YES		
Causes	1: Failed level control and exit shut	2: Operator error opens level valve and exit shut	3:
Likelihood	Probable	Probable	
Consequences	RV opens, possible explosion		
Severity	1 in 5 chance of fatality		
Safeguards	None		
Acceptable risk	No		
What should be done	Consider: Shut off feed on high level and Reduce exposure of persons to the area		
Action	Do hazard analysis. Recommend protection to provide tolerable level of risk	Process and instrument engineers	

Figure 6.3
Sequential worksheet. Page 1

This type of problem should ideally be referred first to the process engineer to improve the basic design. But let us assume we still need to have protection added. The hazard analysis will look at the factors leading to this event and the possible consequences. In this example as shown in Figure 6.4 we are going to assume that a fault tree diagram has been drawn up and the frequencies of possible causes have been estimated.

Figure 6.4 also shows the estimated probabilities of a vapor cloud failing to disperse and its chances of being ignited by a random spark or ignition source. Finally the probability of a person in the immediate vicinity of the tank is combined with the probability of an explosion to arrive at a predicted fatality rate. There is also an FAR calculation based on five persons working at the plant (2000 h per person per year = 10 000 h exposure per year). The FAR without protection is estimated to be 3 deaths per 10^8 exposed hours. Notice that in the figure the target for fatal accident rate has been set at 0.2 which corresponds to a tolerable rate for fatality due to this event of 0.0003 per year or 1 death per 30 000 years.

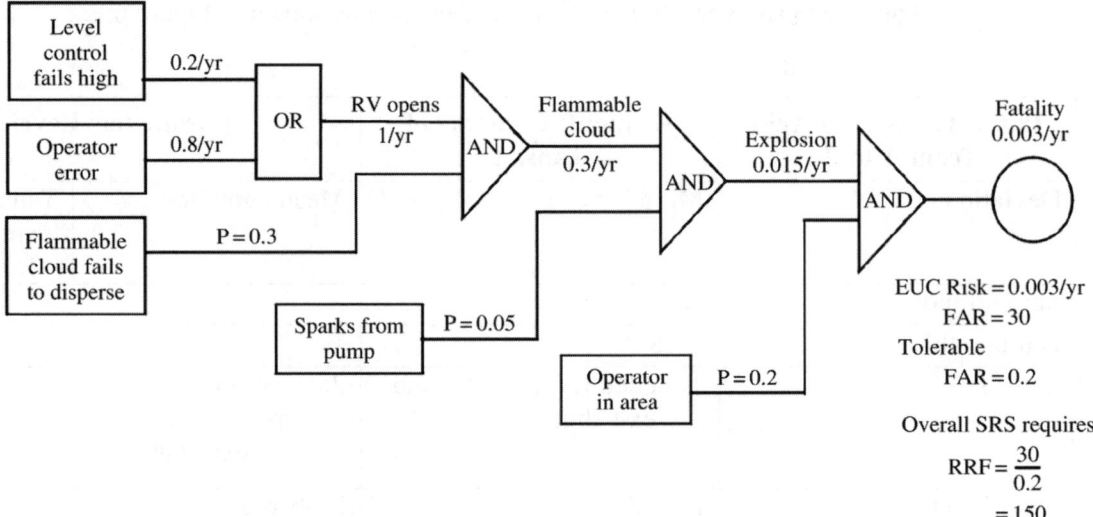

Figure 6.4
Fault tree diagram before safeguards

External or mitigation layer of protection

Now we can add an external means of risk reduction in the form of a fence around the offending vessel so that the probability of an operator being nearby when the explosion occurs is substantially reduced. In fact the analysis suggests at least a 10 to 1 reduction in the risk of a person being close enough to be injured (see Figure 6.5).

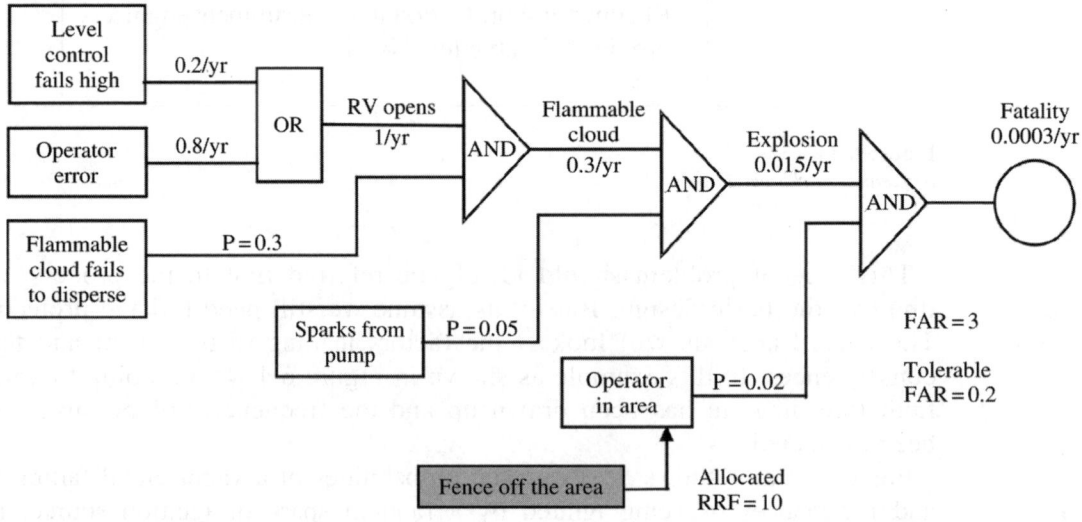

Figure 6.5
Fault tree diagram with mitigation layer of protection

Note that this action does not change the original hazardous event rate but it does reduce the risk frequency of a fatality. In this case the FAR is 10 times lower than before but remains well above target, and this leads us to suggest an SIS protection layer.

6.2.2 Adding the SIS

The suggested SIS is the high-level trip system designed to reduce the risk of overfilling and overpressurizing in the tank.

We can calculate the desired RRF for the SIS using this fault tree as shown in Figure 6.6. Alternatively, we can draw a risk reduction model or diagram as shown in Figure 6.7.

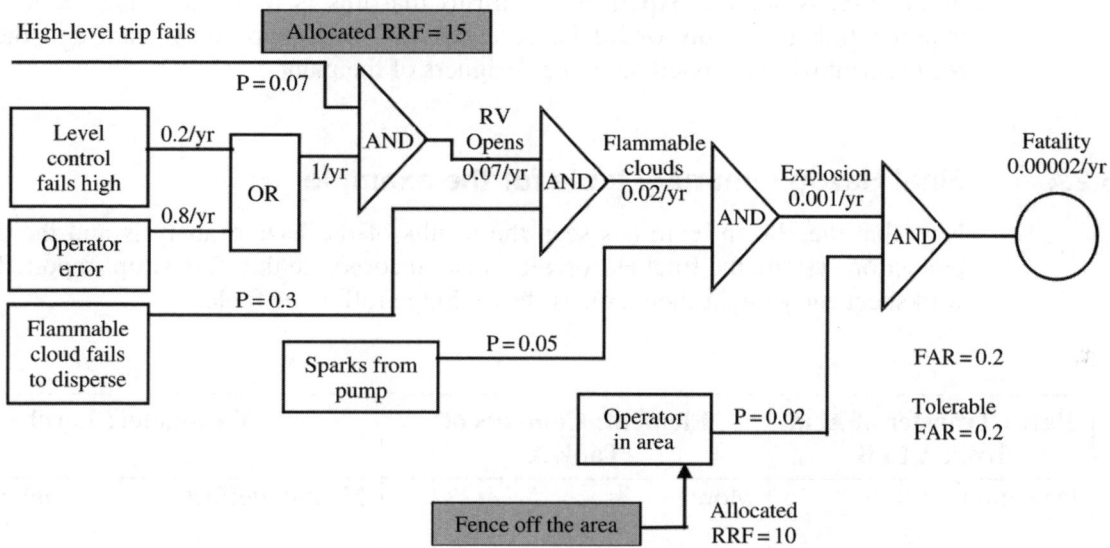

Figure 6.6

Fault tree diagram with mitigation + SIS layer of protection

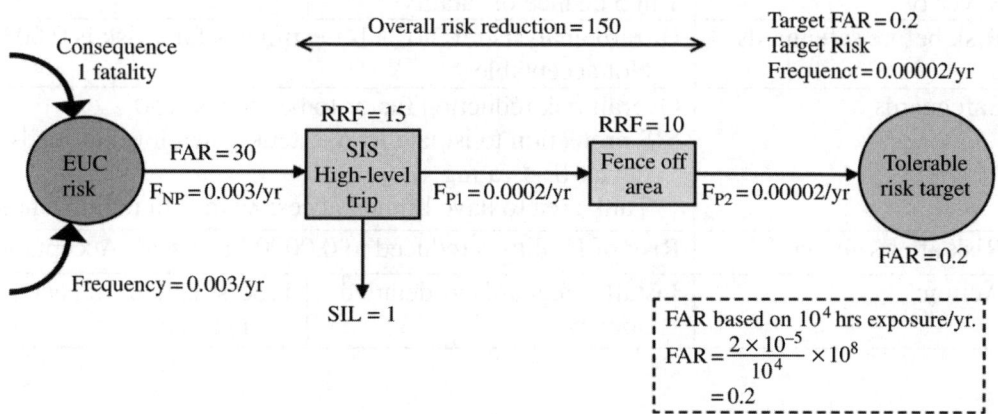

Figure 6.7

Risk reduction diagram for tank protection

Note how the risk reduction factor can be calculated for each component of the safety system. The failure rate target figures can be adjusted within a feasible range until a realistic model of the protection system is achieved.

6.2.3   Obtaining the SIL

Now that we have isolated the risk reduction factors we can see how SIL can simply be obtained from the tables in the standards.

This exercise results in a requirement for a safety integrity level of 1 (SIL-1) to be met by the SIS. Assuming experience confirms that this is reasonable and feasible for the required task the result would be accepted by the Hazop study team and the design requirement will be passed on to the designers of the plant.

6.2.4   Final Hazop summary table for the example

Now that the Hazop team has seen the results of the hazard analysis and the proposed protection system, the final Hazop report can incorporate the 'follow up' report. The final worksheet entry might then look as shown in the following table.

Part : Transfer of Acid from A to B	Element: Contents of Tank A	Parameter: Level	
Deviation	More	Meaning/effect:	Tank is full and still feeding in
Item ref. no.	A1-1		
Causes	1. Failed level control and exit shut	2. Operator error opens level valve and exit shut	3.
Likelihood	Probable	Probable	
Consequences	RV opens, possible explosion		
Severity	1 in 5 chance of fatality		
Risk before safeguards	Hazard analysis ref. no. A1-1 estimates fatal risk is 0.003 per year. Not acceptable		
Safeguards	Overall risk reduction factor to be at least 150. SIS protection to isolate feed to tank when level exceeds 90%. SIS to have SIL-1 rating. Tank area to have limited access with 5-m radius fence		
Risk after safeguards	Risk of fatality is reduced to 0.00002 per year. Acceptable		
Action:	Install safeguards as detailed above	Process and Instrument engineers	

The final Hazop report now contains the details that can be used for managing this particular risk throughout the life of the plant. The Hazop report also contains much of the detail required for the design of the SIS.

6.2.5   Specifying the safety requirements

Now we need to look at what has to be done to achieve the SIL ratings that are needed. And we also need to look at how to define or 'determine' those SIL requirements from the time we first find the need for a safety system.

This is the area where the linkage between hazard studies and the development of alarm and trip systems becomes very important. To do a good job of linking up with the safety instrumentation designer we first need to be aware of the design pitfalls. We will return to the specification stage later.

6.3   What affects the safety integrity of an instrument trip?

The reasons why an instrumented protection system may not be as good as we hope it is can be classed into two basic failure types (see Figure 6.8).

1. Random failures of hardware
2. Systematic errors in design or in software.

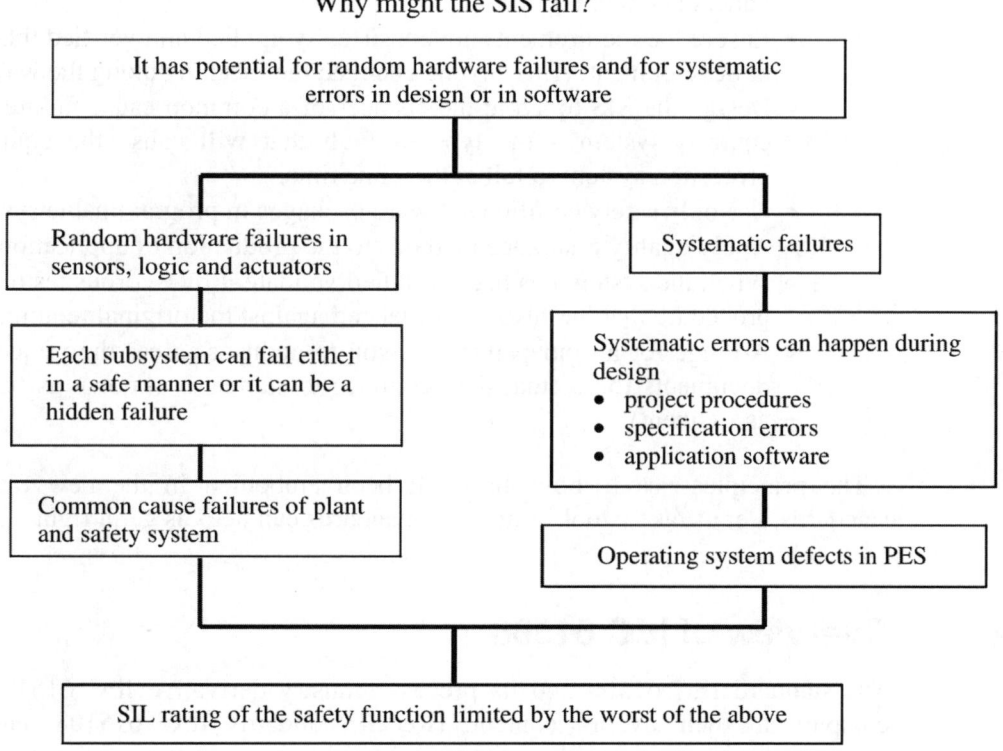

Figure 6.8
Factors affecting the SIL rating

The SIS design project is therefore tasked with minimizing the possibilities of both random hardware failures and systematic design errors. This brings us to the basic techniques used to overcome the failures.

6.3.1   Techniques to reduce failures

For minimizing hardware faults use instruments that:

- Are proven to be reliable
- Are as simple as possible
- 'Fail-safe'; when failure occurs they fail in a defined manner that can be arranged to leave the plant in a safe condition
- May include a high level of internal and external diagnostics (self-checking) to prove at all times that they are working properly
- Arrange for the instrument to shutdown in a safe manner if the diagnostics say it is faulty
- Add redundancy where reliability is suspect to ensure the system can tolerate at least one fault without making the plant unsafe
- Provide for the instruments and the trip function to be functionally tested at prescribed intervals (online or off-line proof testing).

For minimizing systematic errors:

- Ensure the functional requirements are correctly described. (Does everyone understand them?)
- Ensure the requirements are consistently applied and verified though each stage of development. (Has anyone changed the meaning along the way?)
- Design the SIS to avoid the chances of a common cause failure with the basic controls system – the type of fault that will cause the equipment and its protection system to fail at the same time.
- Use only safety certified software packages in programmable systems.
- Apply quality assurance methods to the production of application software.
- When the system has been installed validate it by rigorous testing against a test procedure that has also been checked against the original requirements.
- Arrange for an independent person to audit or assess the project work and its documents to see that they conform to standards (known as functional safety assessment).

The principles listed above have all been embodied in the new functional safety standards. Let us take a look at how the standards can help us get it right.

6.4   Overview of IEC 61508

The standard IEC 61508 and its process industry derivative IEC 61511 are large and complex in their overall content. (Recall standards IEC 615108 and 61511 were introduced in Chapter 3 with an outline of their scope.)

But the essential features can be condensed into a few key items.

- The problems of the systematic errors in design are managed by using a quality assurance system called the safety life cycle.
- The problems of random hardware failure are overcome by a series of design principles that are laid down in detailed hardware design sections of the standards.

- The problems of systematic errors due to software defects are handled by design principles and software quality assurance procedures also laid down in detailed sections of the standards.
- Assurance that whole package will be delivered and maintained properly is provided by mandatory requirements for management of functional safety.

6.4.1   Introducing the safety life cycle

Industry experience has concluded that functional safety can best be achieved by managing all the steps of the design and implementation of a safety system under one managing plan. This plan is known as the safety life cycle.

Figure 6.9 is a simplified version of the safety life cycle. It attempts to show that the stages in the life of a safety system are linked in logical progression and that the working safety system can always be traced back to the original hazard studies and the resulting safety requirements. All steps of the life cycle are underpinned by quality assurance. Any modifications to the safety system involve careful revision of each stage of the life cycle to make sure that no mistakes have crept in due to lack of knowledge of the original design intent. (A good reason in itself for keeping the records secure.)

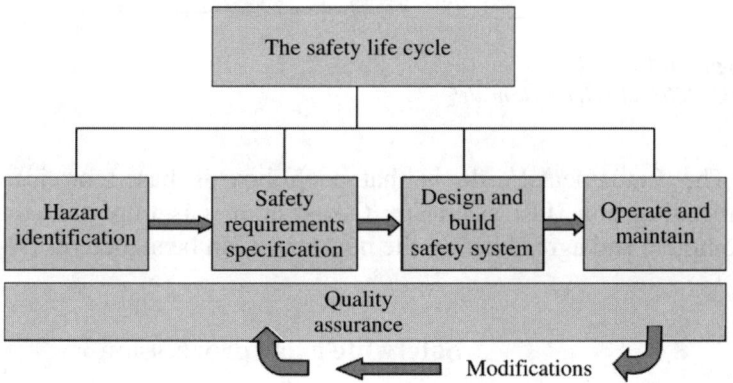

Figure 6.9
The safety life cycle

6.4.2   IEC 61508-safety life cycle

The full safety life cycle model in IEC 61508 is shown in Figure 6.10.

Any project team seeking to comply with the IEC standard is required to plan and document its project activities along the lines set down in this safety life cycle model. The standard provides detailed requirements for the tasks to be performed at the stage of the SLC. These are set down in paragraphs identified against each box on the SLC diagram. Each box is a reference to a detailed set of clauses defining the requirements of the standard for that activity.

The boxes are easy to follow because they are defined in terms of:

- Scope
- Objectives
- Requirements
- Inputs from previous boxes
- Outputs to next boxes.

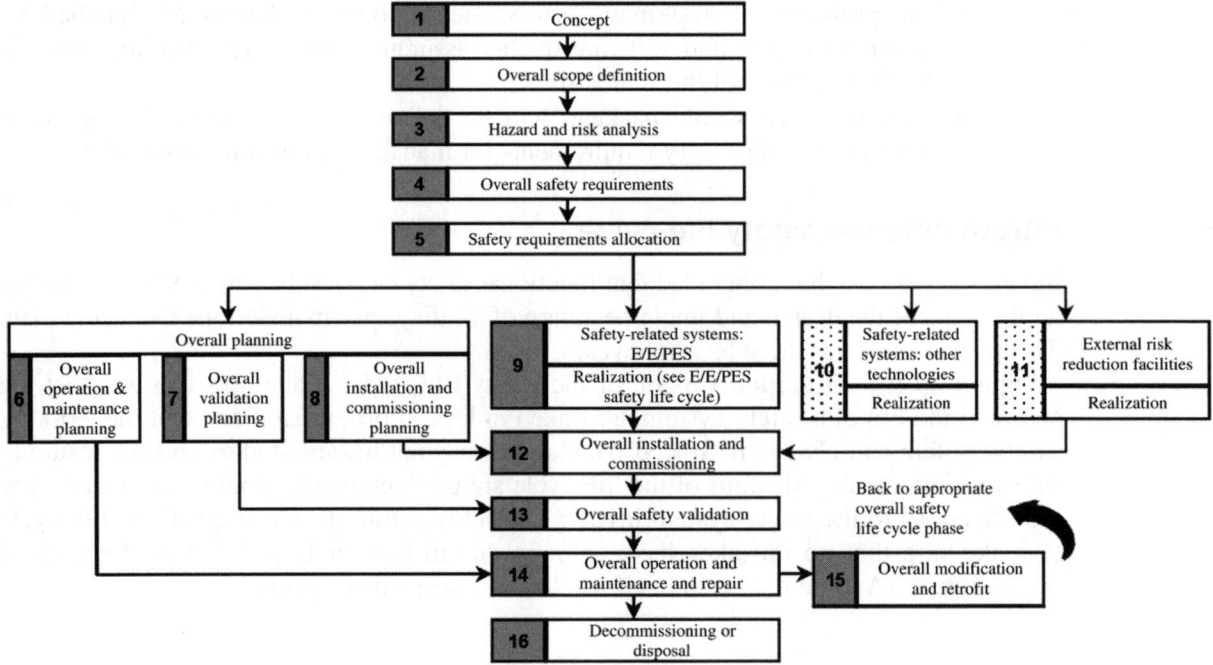

Figure 6.10
IEC 61508 safety life cycle model

The fundamental rule is that each box is based on information contained in the preceding box. Just as in any QA system it is important that each preceding stage is complete and agreed before the next stage can be signed off (see Figure 6.11).

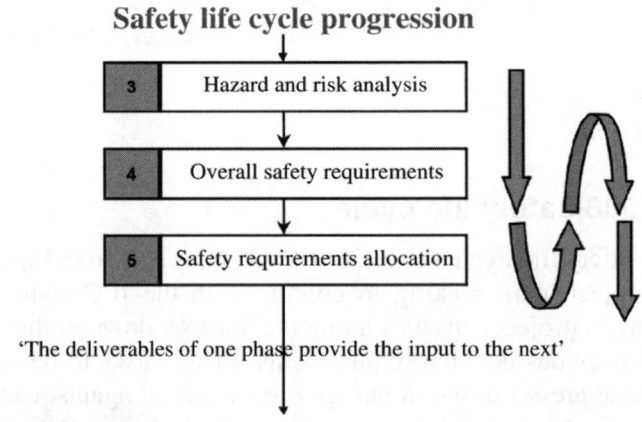

Safety life cycle progression

'The deliverables of one phase provide the input to the next'

It does not require the completion of one activity before starting another: i.e. 'a concurrent design approach can be used'.

Figure 6.11
Safety life cycle progression

Whilst concurrent design is allowed the final result must be that all boxes are aligned and finalized.

The SLC scope includes all stages from initial concepts and hazard studies through to operation, maintenance and modification. The standard covers electrical, electronic and programmable electronic systems and lays down standards of engineering and quality assurance for both hardware and software.

A simplified version of the SLC drawn to show briefly what the project cycle entails is shown in Figure 6.12.

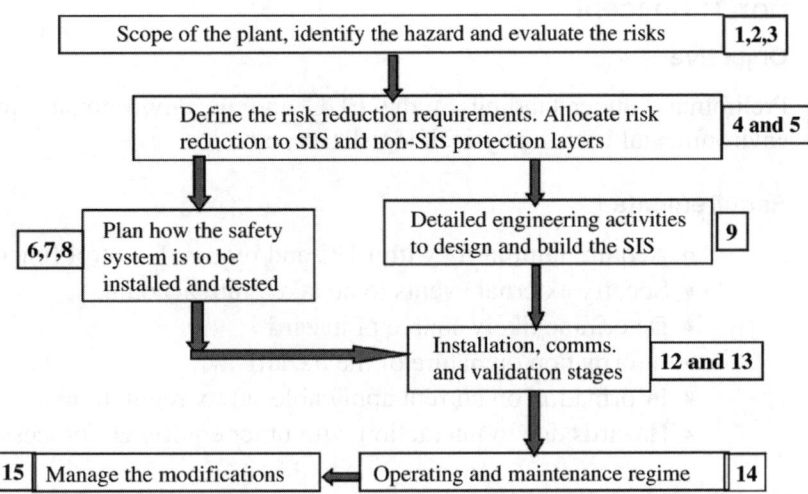

Figure 6.12
Summary of IEC life cycle stages

6.4.3   Key elements of IEC 61508

In addition to its detailed definition of the safety life cycle stages the standard has many other features, only a few key points are noted here:

- *Management of functional safety is just as important as the way we design the SIS:* You cannot claim compliance with the standard unless you can demonstrate that the project has been run under a formalized set of management procedures ensuring everyone knows what they are required to do to support the safety systems at the plant.
- *Technical requirements cover hardware, software, testing:* SIS designs must satisfy constraints on architectures and safe failure fractions. Equipment must have essential proven characteristics. This leads to the certification of devices by third parties, e.g. TUV SIL-3 certification. Self-certification is also permitted.
- *Documentation is mandatory and must be kept current:* Companies must keep a complete set of records of the safety life cycle activities and show that the SIS as installed is valid for the current plant design and its hazard analysis. The testing regimes must match the SIL performance requirements. The testing records must line up with the reliability analysis.
- *Competence of persons:* Individual assigned by companies to work on safety systems must be competent to perform the tasks. The standard does not detail the experience of skills needed but other bodies have developed guidelines.

6.4.4   Getting started with the safety life cycle

The IEC standard requires that we plan to manage the entire SIS project by using safety life cycle steps and by providing documentation for each step. This is very similar in principle to the hazard study series that we saw earlier. The first three levels of process hazard studies align closely with the first three phases of the IEC safety life cycle model.

Let us now look at the IEC Phases 1, 2 and 3 as extracted from the SLC model.

Box 1: Concept

Objective

Preliminary understanding of the EUC and its environment (process concept and its environmental issues, physical, legislative, etc.).

Requirements

- Acquire familiarity with EUC and required control functions
- Specify external events to be taken into account
- Determine likely source of hazard
- Information on nature of the hazard
- Information on current applicable safety regulations
- Hazards due to interaction with other equipment/processes.

Box 2: Scope definition

Objectives

- Determine the boundary of the EUC and the EUC control system
- Specify the scope of the hazard and risk analysis.

Requirements

- Define physical equipment
- Specify external events to be taken into account
- Specify subsystems associated with the hazards
- Define the type of accident initiating events to be considered (e.g. component failures, human error).

Box 3: Hazard and risk analysis

Objectives

- Determine the hazards and hazardous events of the EUC and the EUC control system considering all foreseeable circumstances
- Determine the event sequences leading to the hazardous events
- Determine the EUC risks associated with the hazardous events.

Requirements

- Hazard and risk analysis based on the scope defined in Phase 2
- Try to eliminate the hazards first
- Test all foreseeable circumstances/abnormal modes of operation
- Sequence of events leading to the hazard event + probability
- Consequences to be determined

- Evaluate EUC risk for each hazardous event
- Use either qualitative or quantitative assessment methods.

Choice of technique depends on:

- Nature of the hazard
- Industry sector practices
- Legal requirements
- EUC risk
- Availability of data
- Hazard and risk analysis to consider
- Each hazard and events leading to it
- Consequences and likelihood
- Necessary risk reduction
- Measures taken to reduce the hazard or risk
- Assumptions made to be recorded
- Availability of data
- References to key information
- Information and results to be documented (hazard study report)
- Hazard study report to be maintained throughout the life cycle.

All of which identifies the IEC hazard and risk analysis requirements as being very much the same as the process hazard analysis level 3 or Hazop study.

6.4.5   Alignment of process hazard studies with IEC safety life cycle

Figure 6.13 shows how the new IEC safety life cycle models for SIS correspond to the established process safety life cycle models for hazard studies. The point of departure for the SIS life cycle is ideally at the end of hazard study 3 when the safety requirements specification has been finalized.

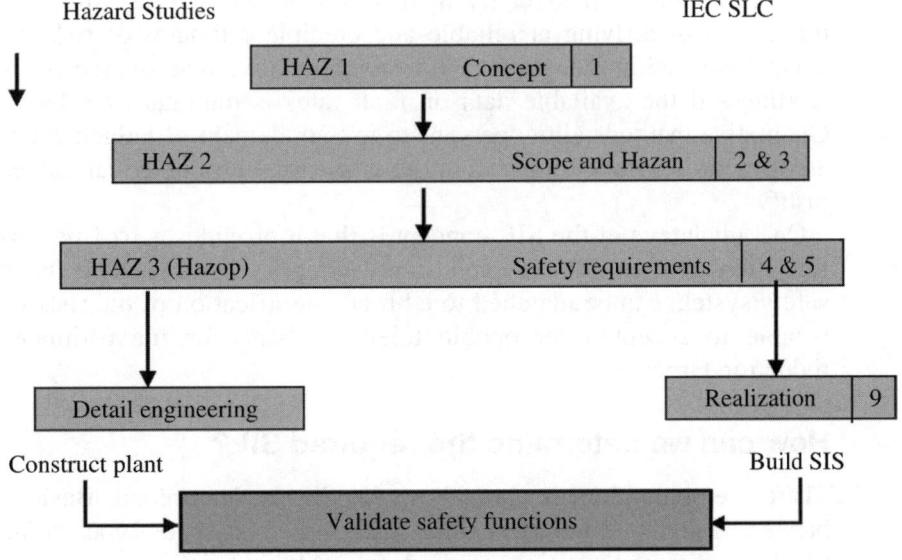

Figure 6.13
Matching hazard studies to the safety life cycle

The plant detail engineering proceeds on the basis of P&IDs that have passed the detailed Hazop study (hazard study 3). The SIS design proceeds on the basis of a safety requirements specification that has also been aligned with the detailed Hazop study.

It seems pretty clear that the first phases of the SLC are similar in principle to the older process safety design life cycle. The main difference is the emphasis in the SLC on risk assessment in the preliminary and detailed hazard studies. This is because risk assessment leads to risk reduction, which in turn defines the SIL requirements of the safety-related system.

So to establish a safety life cycle record for the trip systems all we have to do is refer back to the hazard studied for levels 1 and 2 and perhaps some material from level 3. The handover step is where the hazard study team helps the instrument engineer to generate the safety requirements specification. Let us take a look at what goes into the SRS.

6.5   Determining the safety integrity

As we have seen one of the most important tasks in the SRS development is to specify the safety integrity of each SIS function. This needs to be done fairly early in the development stages to see that our proposed solutions are realistic, achievable and of course affordable. The cost of the SIS will rise steeply with the SIL values. Even if we buy a logic solver that meets SIL-3 the cost of sensors and actuators and engineering work will still be influenced strongly by the SIL rating.

It is important therefore that we have a consistent method of arriving at SIL values within any given organization. To do this we need to consider the most appropriate method of determining the SIL for any particular function. There are at least three recognized methods of doing this and these have been widely documented over recent years and now are built into the ISA and IEC standards.

6.5.1   Diversity in SIL methods

The reason for such diversity in methods of determining SILs is probably due to the difficulties of arriving at reliable and credible estimates of risk in the wide variety of situations faced in industries. Whilst a quantitative risk assessment is desirable it may be worthless if the available data on fault rates is minimal or subject to huge tolerances. Qualitative methods allow persons to use an element of judgement and experience in the assessment of risk without having to come up with numerical values that are difficult to justify.

One advantage of the SIL concept is that it provides a 10:1 performance band for risk reduction and for SIS in each safety integrity level. Hence the classification of the safety system can be matched to a broad classification of the risk and the whole scheme is able to accept a reasonable tolerance band for the estimates of risks and risk reduction targets.

6.5.2   How can we determine the required SIL?

There are some choices about how the SIL is determined. Basically there is a choice between using real numbers (quantitative method) and some variations on fuzzy logic (qualitative methods). One method available is to use the quantitative method with a numerical risk reduction factor we have just been looking at. The other approach is to use qualitative methods, as we shall see here.

6.5.3 Qualitative methods for SIL determination

Many projects find it difficult to agree on hard numbers for the likelihood of a hazardous event and would prefer to work with more approximate descriptors such as 'once in the plant lifetime' or 'possible but unlikely'. IEC 61508 and IEC 61511 provide details of qualitative methods that have been well tried but are not always successful unless handled very carefully. The best known of the qualitative methods is the risk graph or risk parameter chart (see Figure 6.14).

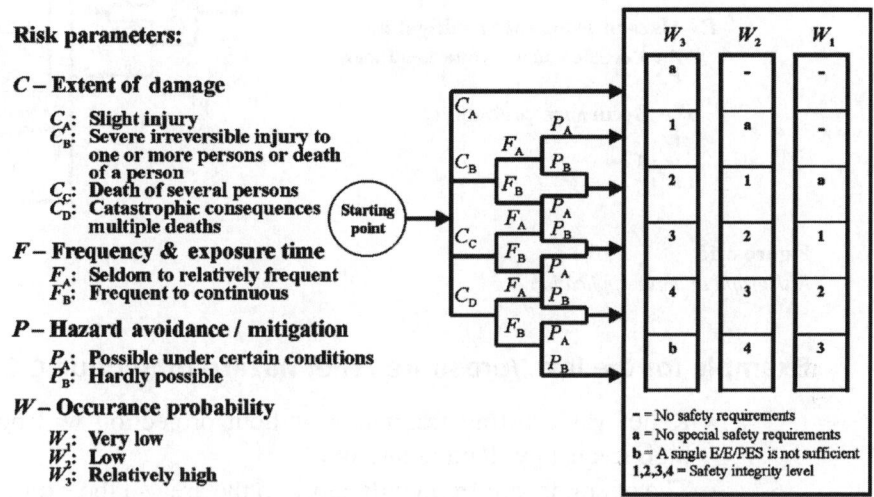

Figure 6.14
Example of a risk graph for estimation of SIL requirements

6.5.4 SIL classification by risk parameters chart

The origin of this method is the German standard DIN V 19250 issued May 1994, which defined a SIL classification code of AK-1 to AK-8. The best descriptions of risk graph methods are to be found in IEC 61511 part 3 Annexes D and E. The descriptions in IEC 61508 part 5 are similar but less detailed.

The difficulty with this method lies in the need to agree a suitable calibration of the risk parameters. The standards provide some guidelines for the parameters but users are free to set up their own interpretations.

In IEC 61511 part 3 Annex D there is a semi-qualitative risk graph where a range of numerical values has been assigned to the parameters. These values are helpful as they represent typical situations found in chemical process plants. This is shown in Appendix H with comments on its usage.

In the example shown marked in on the risk graph shown in Figure 6.15 the hazard analysis team decided:

- *Consequence*:   Cb(Severe injury to one or more persons or possible death of one)
- *Exposure to hazard*:   Fb (Frequent to continuous)
- *Avoidance possibility*:   Pa (Possible to avoid due to early warnings)
- *Frequency of occurrence*:   Wb (Low).

The resulting SIL for the protection system: SIL-1.

Risk Parameters:

C – **Extent of damage**

C_A:
C_B: **Severe irreversible injury to one or more persons or death of a person**

C_C:
C_D:

F – **Frequency & exposure time**

F_A:
F_B: **Frequent to continuous**

P – **Hazard avoidance / mitigation**

P_A: **Possible under certain conditions**
P_B:

W – **Occurance probability**

W_1:
W_2: **Low**
W_3:

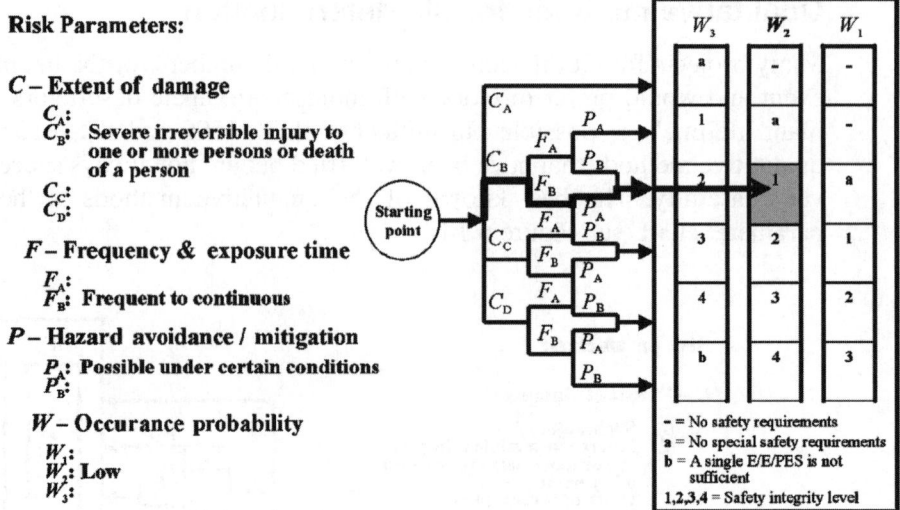

Figure 6.15
Risk graph example application

Example for the level/pressure relief hazard in Section 6.2

- Relief valve lifting frequency without protection will be 1 times per year or high frequency: Parameter: W3.
- The operators are frequently close to the area so the exposure parameter is Fb.
- Being a gas escape we believe the operators will have a good possibility of avoidance so the avoidance parameter is Pa.
- The consequence of exposure is severe injury or death. Consequence parameter is Cb.

The risk graph yields SIL-2 for the overall protection system. If we fence off the area the exposure parameter becomes Fa and the remaining protection (to be done by the SIS) becomes a SIL-1 system. This result aligns with the quantitative method we used in Section 6.1.2.

Whilst this method appears simple it is also very sensitive to the description of the parameters and hence companies are required to test and crosscheck the results carefully. See Appendix H for more notes on this subject.

6.5.5   Summary of methods for determination of SILs

We have already seen that the methods divide into quantitative and qualitative types. IEC 61508 part 5 outlines one qualitative method and two quantitative methods, namely:

- Quantitative method using target risk reduction factors
- Qualitative method using risk graphs
- Qualitative method using hazardous event severity matrix.

For the process industry sector the emerging standard IEC 61511 provides more specific details of the established methods. These are set out in the draft of IEC 61511 part 3 and consist of:

- Quantitative method using target risk reduction factors
- Qualitative method using risk graphs, variations are shown for use where the consequences are environmental damage or asset loss

- Qualitative method using safety layer matrix
- Qualitative/quantitative method using layers of protection analysis (LOPA).

The last two items above are very similar in nature and use formalized definitions of safety layers and protection layers to allow hazard study teams to allocate risk reduction factors to each qualifying layer of protection.

Before attempting to carry out a SIL determination within a project you should carefully read through the material in IEC 61508 part 5 or study IEC 61511 part 3. Both these standards have very helpful information on the basics of SIL requirements determination.

This concludes our brief introduction to methods of determining the SIL targets for a safety system. Now we need to consider some of the design features needed for an SIS to meet the SIL targets that we have set.

6.6   Design essentials to meet SIL targets

IEC 61508 sets down certain essential characteristics that an SIS must have to satisfy a given SIL target. These can be seen in Figure 6.16.

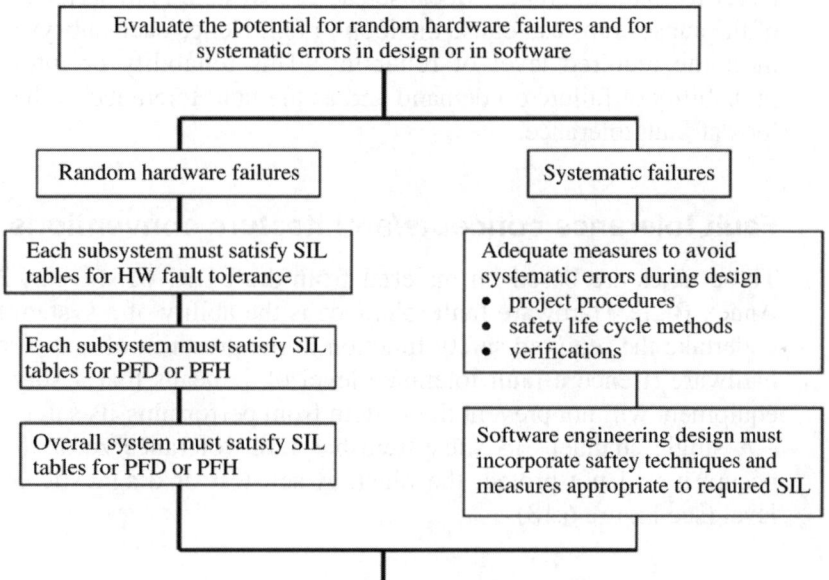

Figure 6.16
Steps in the assessment of achieved SIL

We have discussed the issues of avoiding systematic errors through the use of safety life cycle measures. We have also mentioned the need for software engineering to make special provision for safety in the way the programs are designed and tested. Let us now take a look at the design measures for hardware to meet SIL targets.

6.6.1   Each subsystem must satisfy SIL tables for hardware fault tolerance

We begin by identifying subsystems: Every safety system is characterized by having three subsystems. These are the sensor, logic solver and actuator subsystems (see Figure 6.17).

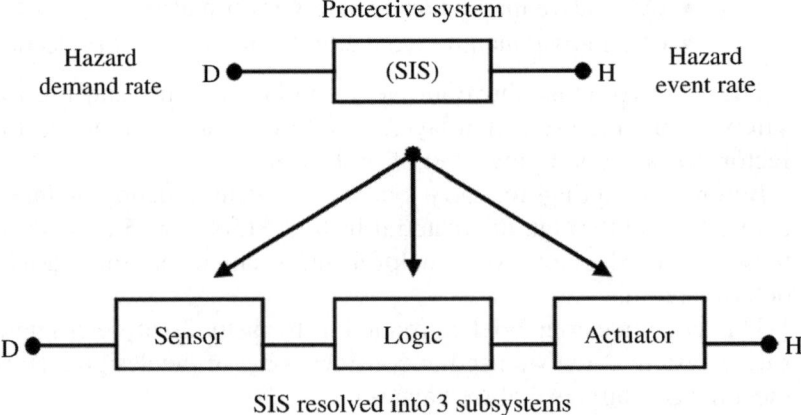

Figure 6.17
SIS reliability block diagram

Each subsystem must now be evaluated individually for its ability to meet the SIL target for fault tolerance. Because each subsystem is in series in the safety loop, any one of the subsystems can cause the loop to fail. Hence each subsystem must be engineered to meet the required level of reliability. This reliability or integrity is expressed as the probability of failure on demand and as the fault tolerance of the subsystem. Let us take a look at fault tolerance.

6.6.2   Fault tolerance concepts/architecture conventions

These notes are based on material from draft IEC 61511 part 2 and IEC 61508 part 6 Annex B.2.2. Hardware fault tolerance is the ability of a system to continue to be able to undertake the required safety function in the presence of one or more dangerous faults in hardware. Hence a fault tolerance level of 1 means that a single dangerous fault in the equipment will not prevent the system from performing its safety function.

A single channel SIS subsystem has fault tolerance zero. If a single dangerous fault occurs it will not protect the plant. A self-test or diagnostic procedure will reduce the level (see Figure 6.18).

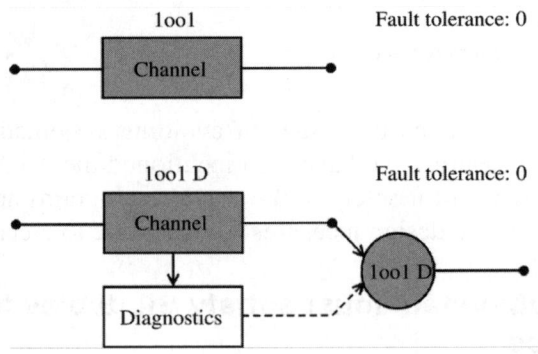

Figure 6.18
Architecture diagram for safety system-1oo1

A self-test or diagnostic procedure will reduce the level of dangerous faults hidden in the system, but the fault tolerance level remains at zero.

A dual redundant channel SIS subsystem will have a fault tolerance level of 1 provided it is arranged to have a 1oo2 voting architecture. A 2oo2 voting architecture will have a fault tolerance level of zero (see Figure 6.19).

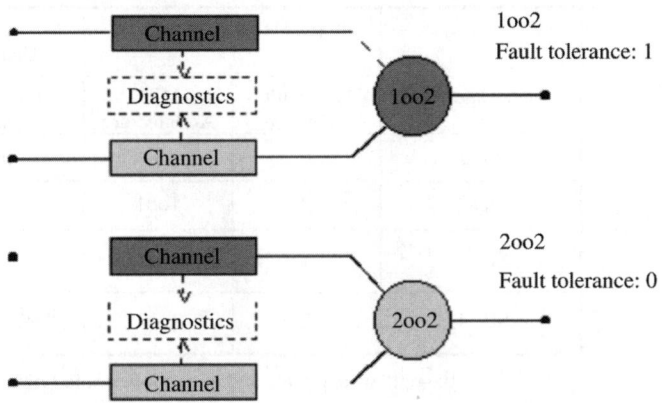

Figure 6.19
Architecture diagram for 1oo2

A 2oo3 architecture has three redundant channels with voting arranged to trip the plant if any two channels command a trip. The fault tolerance level of this system is 1 because anything more than one dangerous fault would prevent the system from tripping (see Figure 6.20).

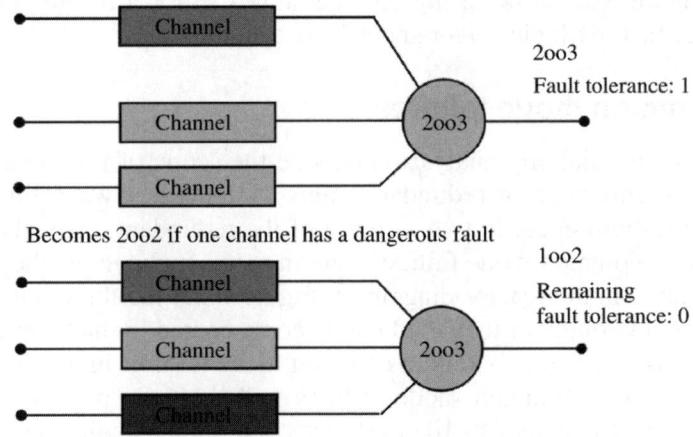

Figure 6.20
Architecture diagram for 2oo3

The conventions shown here are important because IEC 61508 and 61511 have placed constraints on the SIL rating that can be given to a subsystem based on its architecture or fault tolerance level and on the type of device being used. The constraints rules are set out in the following table from IEC 61511 shown in Figure 6.21.

It is important to be aware that using a high SIL value such as SIL-3 requires that sensors, logic and the final actuators should be redundant and in some cases 1oo3 architecture is needed. Note that where devices are considered to be complex, a higher fault tolerance level is required. This typically applies to devices with programmable electronics such as a smart transmitter or a PLC.

IEC requirements: From IEC 61511 – part 1. Clause 11.4.
Applies to sensors and actuators with safe failure fractions 60–90%

	Simple Devices (Non-PES)		Complex Devices (Using PES / Smart etc.)	
Safety integrity	Min. fault Tolerance	Min. Architecture	Min. fault Tolerance	Min. architecture
SIL-1	0	1oo1	0	1oo1
SIL-2	0	1oo1	1	1oo2 or 2oo3
SIL-3	1	1oo2 or 2oo3	2	1oo3
SIL-4	2	1oo3	Special requirements apply, see IEC 61508	

Exceptions apply for higher and lower safe failure fractions

Figure 6.21
IEC fault tolerance rules

Where safe failure fractions can be proven to exceed 90% the fault tolerance level can be reduced by 1. The value of this concession is that redundancy levels can be reduced in such cases leading to substantial cost savings.

If all of this seems a bit too theoretical let us see what a redundant architecture example might look like in practice.

Figure 6.22 shows a high temperature trip for a reactor with 2oo3 voting in the input stage, a 1oo1 logic solver and a 1oo2 actuator stage.

6.6.3   Common mode failures

Detailed reliability analysis is outside the scope of this book but we must take note that the performance of redundant safety systems is always limited by the extent to which failures will occur in two or more of the redundant channels at the same time. These are called 'common mode failures' and in some applications they can be very significant. For example in the reactor diagram in Figure 6.22, the three temperature sensors could all be located wrongly or they could all become coated in thick deposits and all read very low.

Please try listing potential common mode failures in the drench system.

Typically, common mode failures will limit the performance improvements achieved through redundancy to 10 or 20 times better than single channel.

Summarizing the ground rules so far:

- SIL-1 systems provide a basic level of trip protection and can often be engineered from simple single channel designs.
- SIL-2 systems provide a higher level of protection for critical trips but usually require 1oo2 or 2oo3 redundant designs.
- SIL-3 systems employ redundancy and diversity to provide a very high level of protection.
- SIL-4 systems are extremely high performance protection systems that are very difficult to produce. They are reserved for severe and specialized duties.

Redundancy in sensors and acruators

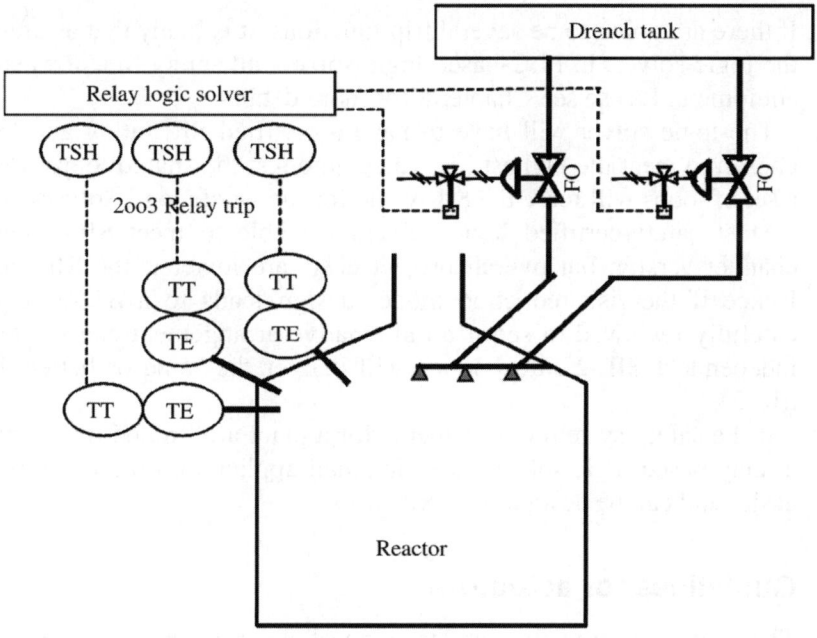

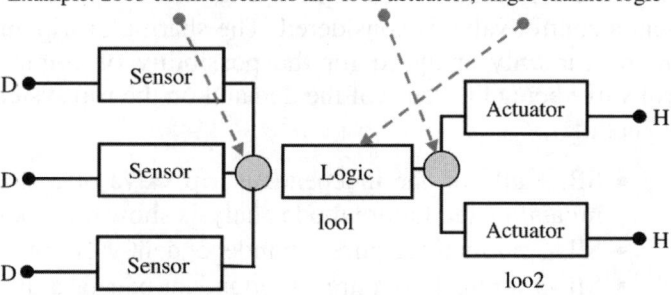

Example: 2oo3 channel sensors and 1oo2 actuators, single channel logic

Figure 6.22
Example of redundancy in a trip system

6.6.4   Selection of sensors

If the process measurement can be done by simple and well-proven devices it is likely that the sensors will be considered as 'type A' or 'simple devices' in the table shown in Figure 6.21 from IEC 61511.

As a general rule for sensors:

- SIL-1 can be achieved with a single type A sensor.
- SIL-2 requires a redundant identical pair of type A instruments.
- SIL-3 requires a 1oo3 voting set or a redundant diverse pair of type A instruments.

If special sensors or unproven sensors are used it is likely that the degree of redundancy may have to be increased.

If programmable sensors (smart transmitters) are used, special arrangements will be required to qualify them as equivalent to type A. These must be used with caution and through consultation with the supplier.

6.6.5   Logic solver

If there are going to be several trip functions, it is likely that a safety PLC will be used for the logic solver. In PLC-based logic solvers all safety functions share the same common equipment. Please see Chapter 8 for more details.

The logic solver will have to have a certified SIL rating and the SIL will have to be chosen to meet the highest SIL value amongst the shared loops. Hence one SIL-3 loop in a set of loops will force the SIL value for the whole logic solvers up to level 3.

Most safety-certified logic solvers are able to meet SIL-3 performance in a single channel version but overall project costs are lower if the SIL rating can be kept to 2. Hence if the risk reduction model design leads to a SIL-3 requirement it should be carefully reviewed to see if an alternative arrangement can be found (for example: two independent SIL-2 safety loops will deliver the same or better risk reduction than one SIL-3).

If the safety system requirements for a plant are limited it may be advisable to consider a relay-based logic solver stage. In small applications these can meet SIL-2 performance easily and can be designed for SIL-3.

6.6.6   Guidelines for actuators

These apply to trip valves and tripping of electrical drives.

In general the comments in Section 6.6.4 for sensors apply equally to actuators. However care is required for SIL-1 and SIL-2 applications where the possibility of tripping a control valve is considered. The sharing of trip and control duties in one valve has to be carefully analyzed for the possibility of common cause failures due to the control valve being the cause of the demand on the trip system.

In general:

- SIL-1 allows one independent trip valve or a shared trip and control valve if reliability and failure mode analysis show the risk reduction can be met.
- SIL-2 normally requires an independent valve or a redundant pair.
- SIL-3 normally requires a redundant pair or a diverse pair of valves or a 1oo3 voting group.

6.6.7   Estimating the spurious trip rate

For the conceptual design stage there is also the question of how to estimate the likely spurious trip rate and see how it compares with the target. Here are some rough guidelines for estimating the spurious trip rate.

Sensors and actuators dominate the fail-to-danger rates of most SIS loops. (Typically 90% of all failures are expected from the field devices.) Field devices have typical safe failure fractions of better than 75%. This means in effect that in a single channel design, the fail-to-safe rate will be four times higher than the fail-to-danger rate.

6.6.8   Effects of proof testing on SIL

We have seen that one of the factors that will determine whether or not an SIS can perform to a risk reduction target (or meet a SIL rating) is the probability of failure on demand, the PFD_{avg}. One of the most important factors in designing the SIS is the effect of proof testing and diagnostics on the PFD for each safety function. Remember, anything that reduces the PFD_{avg} will improve the risk reduction factor of the SIS.

Figure 6.23 shows how proof testing temporarily sets to zero the probability that a trip system has failed. As time passes the probability of failure rises until the next proof test. Hence frequent testing maintains a low average probability of failure.

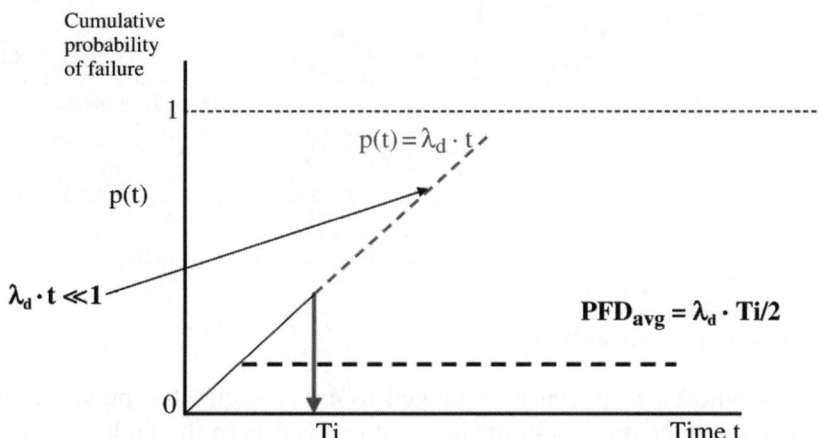

Figure 6.23
How proof testing reduces the probability of failure on demand

6.6.9 Manual testing considerations

The key points for manual testing are:

- The proof test must be good enough to reveal nearly all possible hidden faults.
- Faults not found by the proof test will cause PFD to increase.
- Reducing the proof test interval Ti will improve the PFD_{avg}.
- The test interval should be chosen to ensure that there would be no more than two demands per proof test interval. For example: If the plant has four trips per year the proof test interval should not be greater than 6 months.

Figure 6.24 shows how PFD_{avg} for a typical 1oo1 SIS can be reduced by increasing the frequency of proof testing. The graph is for a single channel SIS subsystem with a fail-to-danger rate of 0.025 per year. If the proof test is carried out every 6 months, the PFD_{avg} value falls into the SIL-2 range. If the system is only required to meet SIL-1, the test interval can be extended to over 3 years.

In this example the online testing work disables the safety function for a period of 4 h. This limits the ultimate improvement in PFD available through frequent testing because the time spent testing contributes to the down time of the SIS. If testing does not disable the SIS, this limitation falls away.

6.6.10 Diagnostic testing considerations

Diagnostic testing has become a significant contributor to improved performance of safety instrumentation since the introduction of programmable systems.

Diagnostic testing involves methods of frequently and automatically testing an SIS subsystem to detect a dangerous fault condition. Provided such tests do not disable the safety function for any significant time (defined by the 'process safety time') they will be invisible to the safety system function.

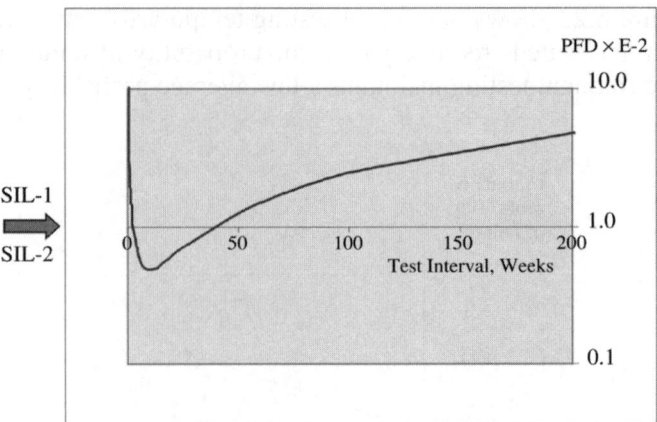

Figure 6.24
Effect of proof testing on PFD

Diagnostic tests can be arranged to either notify the presence of a dangerous fault or shutdown the process immediately on detection of the fault.

Diagnostic coverage

In any subsystem the portion of dangerous faults detectable by diagnostics is called 'Diagnostic coverage'. For example, if 70% of dangerous faults can be found by diagnostic tests, the coverage factor is claimed to be 70%. However the effectiveness of the diagnostic test depends on how often it is performed and how long the repairs take. We saw this in formula set 2 where the term MTTR affects the PFD value derived from diagnostic tests.

If we take our previous example and assume 70% DC and a short test and repair interval (MTTR), the new PFD graph looks like Figure 6.25.

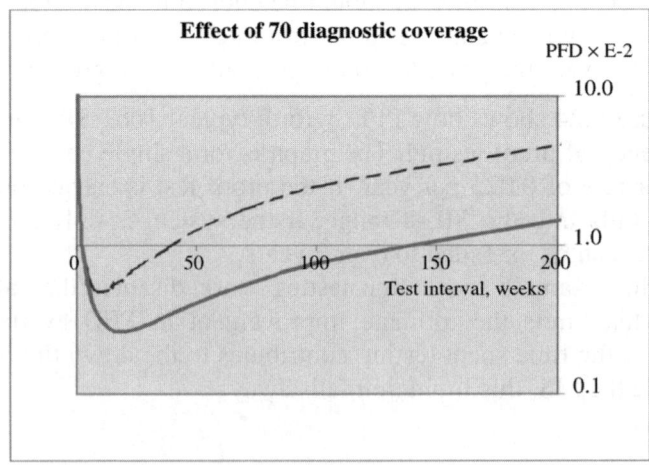

Figure 6.25
Diagnostic tests reduce PFD$_{avg}$

As expected the 70% coverage produces a dramatic improvement in the PFD values and allows the manual proof test interval to be extended without a serious increase in PFD$_{avg}$. Notice how the subsystem can be brought into the SIL-2 range with a test interval of 2 years.

6.6.11 Using redundancy to improve SIL

It is clear from the reliability formulae that identical redundant devices will improve the PFD valus. Even with a 10% common cause allowance the improvements are substantial. In Figure 6.26 the same sensor design has been recalculated for 1oo2 architecture.

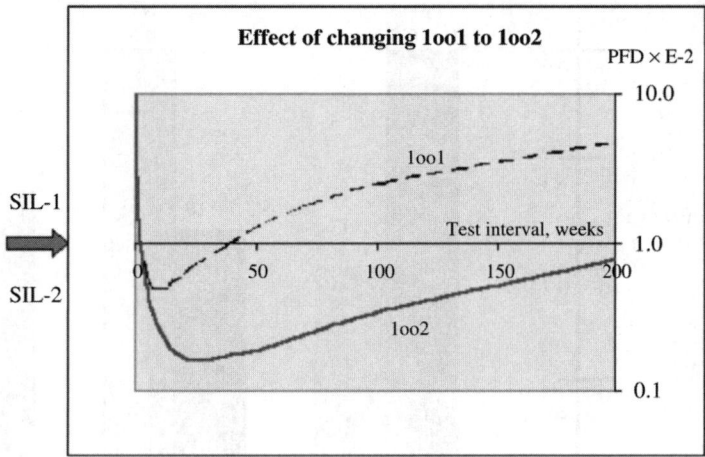

Figure 6.26
Effect of redundancy on SIL rating

6.6.12 Balancing the design between safety and production

In this section we consider the problem of lost production due to nuisance or spurious trips. If the SIS fails to safety every time there is a fault, the safety is not compromised but the plant loses production due to a nuisance trip.

The 1oo2 architectures make the problem worse because there are two devices available to cause trips. Where spurious trips are likely to be a problem the commonly applied solutions are:

- Single channel systems to use high reliability devices
- Proof testing must be designed to avoid causing trips
- Use 2oo3 architecture so that one channel can fail-safe without tripping
- Use 1oo2D architecture supported by logical control as follows.

The 1oo2D architectures are designed to keep working on one channel whilst one faulty channel is being repaired. This is commonly used for logic solvers and requires intelligent diagnostics to switch out the faulty channel and reconfigure the voting logic without tripping. It is also possible to apply this method to sensors and actuators through the design of testing methods and with the aid of diagnostics.

The 2oo3 architectures allow one channel to fail-safe whilst leaving two channels still protecting the plant. This method is simple to apply for sensors but is not attractive for trip valves due to cost and complexity. Figure 6.27 shows the benefits of redundancy for avoidance of nuisance trips.

6.6.13 Conclusions on SIS design

It is clear that the SIL target should be chosen very carefully because this is where the project team has to achieve a balance between safety and cost. Setting an SIL target that is higher than necessary invites severe cost penalties and operating difficulties. Several methods exist to help with the decision process and using more than one method is a good way of verifying the result.

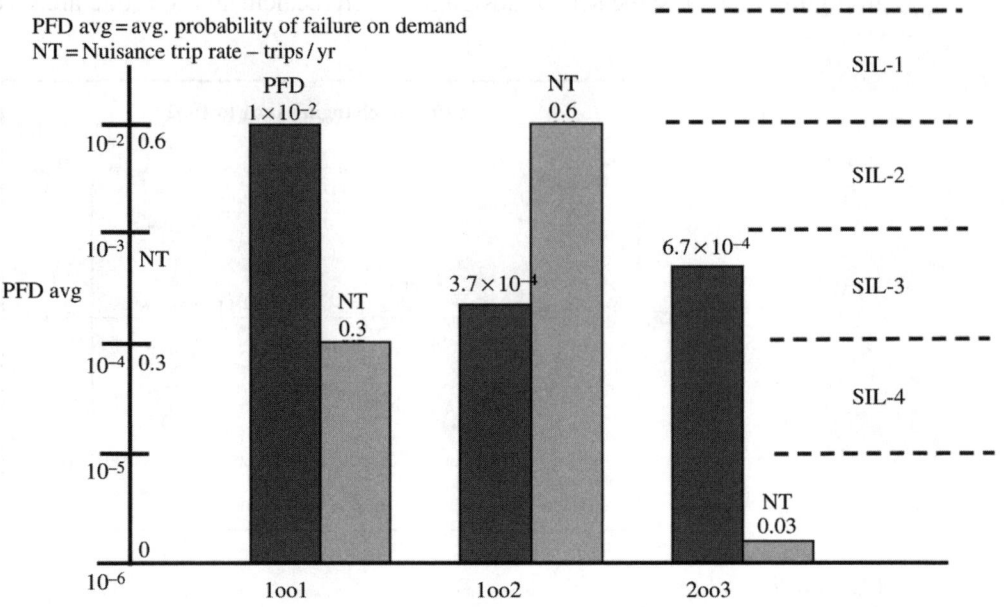

Figure 6.27
Comparison of redundancy options

IEC 61508 and 61511 have set clear requirements for meeting SIL targets. These involve systematic analysis of the devices and the architectures. It is necessary to have some knowledge of equipment failure rates to carry out the calculations but generic failure rate date is often sufficient due to the wide margins allowed in the SIL bands.

The PFD calculation methods allow designers to adjust testing rates, diagnostic coverage and redundancy levels to meet the performance targets.

6.7   Specifying the SIS requirements

Finally in this section we need to show how the SIS should be formally specified by the design team so that the engineering of trip systems can proceed accurately from the basis of the hazard studies. This is done by developing the safety requirements specification.

6.7.1   Components of the SRS

It is generally convenient to structure the safety requirements specification (SRS) into three main components as shown in Figure 6.28.

6.7.2   SRS input section

The input information section:

- Provides information on the process and the process conditions
- Defines any regulatory requirements
- Notifies any common cause failure considerations
- Lists all safety functions covered in the SRS.

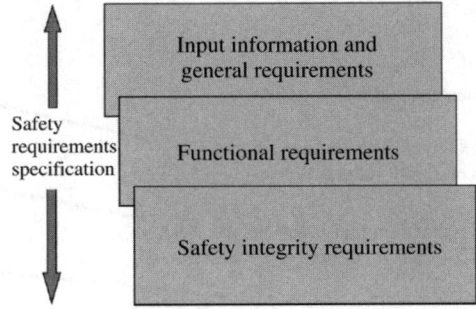

Figure 6.28
Structure of a safety requirements specification

6.7.3   SRS functional section

The functional requirements specification defines for each function:

- The safe state of the process
- The input conditions leading to response
- The logical responses
- The output actions to bring the process to a safe state
- The process safety time or speed of response.

6.7.4   SRS integrity requirements section

The integrity requirements specification defines for each function:

- The required risk reduction and the SIL
- Expected demand rate and the low or high demand mode
- Energize or de-energize to trip
- Spurious trip constraints
- Proof testing strategy
- Environmental conditions.

6.7.5   Development of the SRS

The objective of an SRS is to capture all the information necessary for the future design and continuing support of a safety function. Two versions of SRS are created during the development stage.
   These are:

- Overall safety requirements
- SIS safety requirements.

In IEC 61508, Phase 4 defines an overall safety requirements specification. This describes the risk reduction requirements for overall safety including the non-SIS and external protection functions. The safety allocations phase (5) then allocates risk reduction duties to both SIS and non-SIS protection layers for each safety function.

The SIS safety requirements are then defined at the start of Phase 9 on the basis of the known requirements for all risk reduction contributors. The structure of the two levels of SRS is the same and we can start by looking at the overall SRS and then adding in more specific details for the SIS (see Figure 6.29).

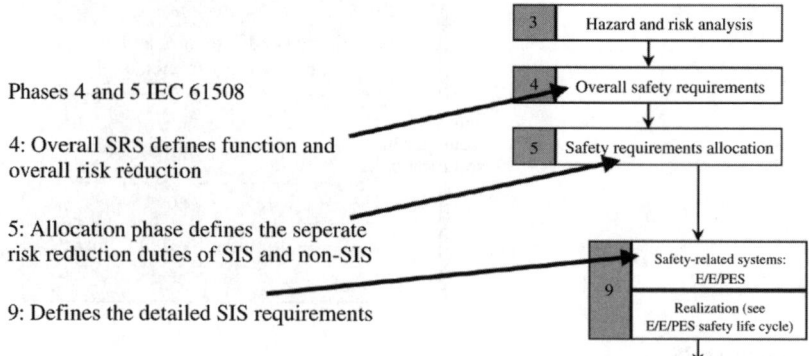

Phases 4 and 5 IEC 61508

4: Overall SRS defines function and overall risk rèduction

5: Allocation phase defines the seperate risk reduction duties of SIS and non-SIS

9: Defines the detailed SIS requirements

Figure 6.29
Development steps to define the SRS

The SIS safety requirements specification should be developed in parallel with the non-SIS protection layers. This is the reason why the IEC 61508 standard calls for overall safety requirement to be defined in Phase 4 followed by an allocation Phase 5 which defines sharing of protection duties across the layers of protection. The final SRS for the SIS is then part of the 'realization' Phase 9.

Phases 4 and 5 often involve looking ahead to the conceptual design stage since we need to have a good idea of what is achievable when the SRS is being firmed up. The development of the SIS and non-SIS requirements therefore has a relationship to the hazard studies that can be seen in the design sequence diagram shown in Figure 6.30.

6.7.6   Design sequence

Details of what should be done to specify risk reductions and safety allocations are set out in the IEC standard. Here is a summary of a typical sequence of events as shown in Figure 6.12.

The diagram indicates the sequence of activities that are expected to follow from the initial identification of a hazard during either level 2 hazard study or level 3 (Hazop) study. These activities will lead to an outline or 'conceptual' design of the trip system that has to be tested for acceptability in the hazard study team.

If the design is unacceptable for any reason, typically cost or complexity, the allocation of risk reduction duties would have to be adjusted until a satisfactory balance between SIS and non-SIS measures has been found. Often the sequence of activities will be performed quickly to create draft designs for consideration by the hazard study team. Revisions to the design will loop around the sequence until an approved final version of the design is accepted.

The activity steps are summarized below:

- The hazard study team identifies a hazard by the normal analysis process used in hazard studies.
- The team estimates the consequences of the hazardous event if there are no protection devices installed and if there are no measures such as evacuation alarms to mitigate the consequences.
- The team estimates the likelihood of the hazardous event either as a frequency (e.g. 0.1 per year or once in 10 years) or as a qualitative term such as 'probable' (see Note 1). This frequency is known in IEC 61508 as the unprotected risk frequency F_{NP}.

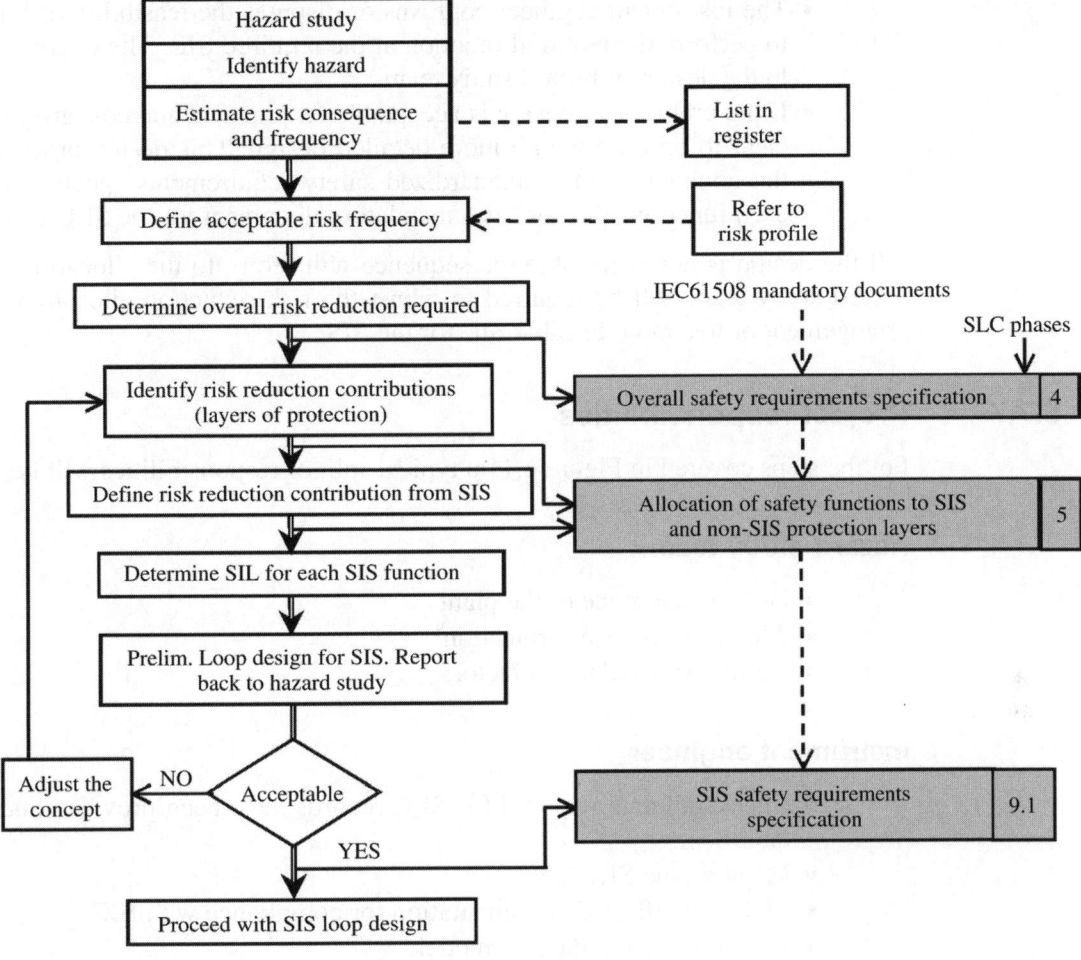

Figure 6.30
How the SRS is developed from the hazard study information

- Using guidelines such as a risk profile chart the hazard study team defines the 'acceptable or tolerable risk frequency'. IEC 61508 calls this F_T.
- The overall risk reduction requirement for the hazardous event is then defined as F_{NP}/F_T. This value and the justification for it must then be documented into the safety life cycle record for Phase 4 (see Guideline No. 2).
- The process designers or safety officers have the responsibility to define the expected protection measures that will contribute to the required risk reduction. These are known as 'layers of protection' (see Note 2).
- Each layer of protection contributes to reducing the risk. If the hazard study team or designers call for an instrumented trip system (SIS), it will be necessary for its risk reduction contribution to be nominated (e.g. a risk reduction factor of 100).
- All risk reduction factors for each safety function will then be listed and described into the safety life cycle record for Phase 5 (see Guideline No. 2).
- The instrument engineer uses the risk reduction target allocated to the SIS to determine its safety integrity level (SIL) and to begin the conceptual design of the trip system (see Note 3).

- The instrument engineer confirms or disputes the feasibility of building an SIS to perform the required function at the required SIL. The engineer reports back to the design or hazard study team.
- If the preliminary design is acceptable on practical and cost grounds, the way is clear to proceed with a more detailed design. This design process begins with the completion of a standardized safety requirements specification (SRS) for each function. The SRS is a mandatory document for the SLC record.

If the design is not acceptable the sequence will return to the allocation phase and the hazard study team will be required to adjust the risk reduction allocations to a feasible arrangement or to a more feasible SIL for the SIS.

6.7.7   Project responsibilities

For the steps covered in Figure 6.11 a typical split of responsibilities will be:

Hazard study team

- Defines the scope of the plant
- Allocates layers of protection
- Defines risk reduction factors.

Instrument engineer

- Checks all data required for SLC records have been provided and records it in table 1
- Manages the SLC steps
- Maintains all SLC documentation for compliance with IEC
- Prepares a risk reduction model
- Calculates risk reduction factors and SIL for each loop
- Does conceptual design
- Reports back to hazard study
- Describes the process and its intended basic control systems (EUC and EUC control system)
- Identifies possible hazards due to EUC and EUC control system
- Estimates risks in terms of frequency and consequences
- Defines target frequencies for acceptable risk
- Proposes layers of protection which may include an instrumented trip (SIS).

The key point to note here is that an individual within the hazard study team must be appointed to manage the safety life cycle activities required for an SIS to achieve compliance with IEC 61508. That person will also have to ensure that the documented records are completed and validated for each step of the safety life cycle.

6.7.8   Input requirements

Basic data from the preceding stages of the SLC. (This could be the hazard study report.)

- SRS subject, title, ref. number, etc.
- Process data sheets, chemical hazard data sheets

- References to process descriptions, flow sheets or P&IDs correctly defining the scope of the plant
- Regulatory requirements applicable to the plant or the process
- List of safety functions to be included in the SRS
- Common cause failure possibilities.

6.7.9   Functional requirements

The first thing we need to be clear about is that each individual safety function has to be specified. This means that there has to be a separate specification sheet for each function. Possibly the best way to deal with this is to have a safety functions list in the common input section of the SRS, with a data sheet for each function.

6.7.10   Safety integrity requirements

In the overall SRS the safety integrity is simply the required overall risk reduction for each listed function.

In the allocation phase the safety integrity is the risk reduction allocated to each of the SIS and non-SIS contributors. There are some noteworthy points in the IEC standard at this phase:

- The IEC standard requires that we take into account the skills and resources available at a particular location when working out a suitable safety integrity value.
- In some applications a simpler technology solution may be preferable to avoid complexity.
- Iterative design may be needed to modify the proposed architecture of the SIS to ensure it can meet the required SIL.
- For each function, state if the target safety integrity parameter is to be based on high or low demand mode of operation.
- Allocation to use techniques for combination of probabilities. This includes quantitative and qualitative methods for determination of SILs.
- Allocation to take into account the possibility of common cause failures between the protection functions.
- As the allocation proceeds the SIL values of the SIS for each function are to be listed. Basis of high or low demand to be stated.
- Shared equipment to be specified to the highest SIL function.
- No single SIS to have a SIL of 4 except by special demonstration/experience.
- No SIS to have a SIL of less than 1.
- All information to be documented.

Using the above requirements the completed allocation phase will provide a well-documented baseline for the completion of the detailed safety requirements specification for the SIS. The safety integrity requirements for the SIS are then developed to conform with the following requirements summarized from Phase 9.

For each SIS safety function:

- The required SIL and also the target risk reduction factor where quantitative methods have been used
- High or low demand mode

- Requirements and constraints to enable proof testing of the hardware to be done
- Environmental extremes
- Electromagnetic immunity limits needed to achieve EMC.

6.7.11 Conclusions on the SRS development

We have seen how the safety requirement specification has been expanded from overall safety requirements through allocation into a detailed SRS for the SIS. The result of these activities will be a fully documented record of the design requirements and the history of where the requirements have come from. We are now in a position to consider some documentation aids for the SRS.

6.8 Documenting the SRS

What tools can be used to capture the requirements efficiently?
The main needs are:

- A specification sheet blank form listing all the technical data items required to complete an SRS
- A means of representing the functional requirements that is accurate but flexible enough to accommodate changes
- A systematic means of determining the target SILs.

Tools for the first two items are suggested here. Methods for SILs are considered in the next section.

6.8.1 Contents checklists for the SRS

On the basis of the preceding notes we can now set out a checklist plan as shown below. The normal practice in an operating or engineering company will be to develop a company-specific SRS template for use in all projects. The following lists can be used to assist with the task of collecting the information needed to complete the SRS. All items in the lists would also go into a company specification form.

Contents checklist for the general section of SRS

General Section Items	Notes
Title and subject of the SRS	SRS normally covers all safety function on the plant. Can be split into areas
Spec. no./date/rev no./author	Document management details
P&IDs	Ref. nos of applicable P&IDs
A list of the safety functions required	Each function separately identified for plant area, unit and function number
Cause and effect matrix and/or logic diagrams	Ref. numbers of drawings describing the functional logic of the SRS
Process data sheets	Reference to existing data sheets
Hazard study report references	Links to Phase 3 SLC records

General Section Items	Notes
Process information (incident cause, dynamics, final elements, etc.) of each potential hazardous event that requires a SIS	Should be in Phase 3 SLC records
Process common cause failure considerations such as corrosion, plugging, coating, etc.	Needed for this spec.
Regulatory requirements impacting the SIS	Reference to Phase 1 SLC
Spec. to be derived from the allocation of safety requirements (IEC Phase 5) and from safety planning	Doc. refs to be given to overall SRS (Phase 4) and allocation phase (Phase 5) records
General requirements for the design of the SIS	For example, a centralized programmable logic solver with communications interface to existing DCS. Standard process sensors and valves to be used except where found to be unsuitable
Power supply requirements and action on loss of power to the SIS	Power failure response and restart rules for the SIS
Definition of any individual process states which when occurring concurrently create a separate hazard	For example, overload of a vent scrubber system
Application software safety requirements: general features common to the SIS logic solver subsystem	General requirements for the application software are applicable here. These would typically be specified in a separate document that will be linked to the SRS. A Sasol standard defining essential features of the application software would be appropriate here

Contents checklist for the functional requirements section of SRS

The following list will apply to each individual safety function listed in the general section. Hence, an individual checklist should be used for each function.

Functional Section Items	Notes
Details sufficient for the design and development of the SIS	This is general requirement from IEC standards. All requirements should be clear and simple for third parties to use
Is control to be continuous or not?	For example, a 24 h continuous process or a batch process
Is the demand mode low or high? What are the assumed sources of demand?	Refers to the basis of demands being less than twice per proof test interval (low demand) or more than twice per proof test interval (effectively a high demand mode)

Functional Section Items	Notes
The definition of the safe state of the process, for each of the identified events. See also concurrent hazard item in general section	Required to ensure the SIS design knows what safe condition to go to
Method of achieving safe states	Detail of the process plant actions
Interfaces with other systems	Upstream/downstream or service unit interactions need to be highlighted
All relevant modes of operation of the EUC	Safety function needs to be working for which states? e.g.: plant start up, normal production, storage mode, etc. Also reasonably foreseeable abnormal conditions
Any specific requirements for SIS start up/restart?	
The process inputs to the SIS and their trip points	Process and operator input variables, tag numbers if available
The normal operating range of the process variables and their operating limits	
The process outputs from the SIS and their actions	Description of valve and trip actions
The functional relationship between process inputs and outputs, including logic, math functions and any required permissive logic	The functional logic. Supplied at least as narrative description, preferably as a functional logic diagram or cause and effect matrix
Requirements for overrides/inhibits/bypasses	May be included in functional logic
Selection of de-energized to trip or energized to trip	Some applications may be better suited to energize to trip designs. SIS design would have to cater for this
Consideration for manual shutdown	Define what if any manual shutdown facilities are needed
Application software requirements (for details on this item refer to clause 12 of IEC 61511 dealing with software SRS). See also IEC 61508 part 3 clause 7.2	Any features specific to the function would be defined here and would form part of the application software SRS
Action(s) to be taken on loss of energy source(s) to the SIS	Power failure response and restart rules for the SIS
Response time requirements for the SIS to bring the process to a safe state	This derives from the hazard study and process info. Overall response time of SIS including valve stroke times to be matched against the response time available
Response action to any fault detected in the SIS	Declare if plant is to trip on detection of SIS fault. Alternatives to be declared where applicable. Human factors to be considered, e.g. can operator be expected to cover the situation?
Any possible combination of output states of the SIS that could be dangerous	A check on possible combinations to be done (originates from IEC 61511)

Functional Section Items	Notes
Human–machine interfaces requirements	Alarms, indicators, message and logs needed per trip function. Specific operator interface items per function
BPCS (i.e. or Control PLC) interface requirements	e.g.: Control loop resets or sequence aborts needed to assist control system stability during a trip
Reset functions	Actions allowed or forbidden, etc.

Contents checklist for the safety integrity requirements section of SRS

The following list will apply to each individual safety function listed in the general section. Hence an individual checklist should be used for each function.

Each function will be evaluated in Phase 5 for its individual SIL requirement (see Guide No. 1). The common logic solver subsystem (i.e. relay panel or PLC) for any set of functions will have a SIL value at least as high as the highest SIL of the functions it serves.

Safety Integrity Requirements Items	Notes
The required SIL for each safety function	As determined in Phase 5
Low or high demand mode for each function	See general section. Check this for each function
Estimated demand rate	Basis of calculations for hazard rate done in Phase 5
Maximum tolerable spurious trip rate	Usually based on production loss rates. Basis for redundancy design options. See also next item
Reliability requirements if spurious trips may be hazardous	Spurious trips may contribute to hazardous events at the plant. Hence from a safety integrity viewpoint they may need to be avoided. This is independent of the cost of lost production
Minimum and worst case repair times feasible for the SIS	Impacts redundancy design
Environmental extremes	Environmental conditions impact the ability of instruments to meet SIL targets. These must be declared at the start of design
Requirements for diagnostics to achieve the required SIL	Difficult sensor applications or valves that are prone to problems may require a high level of diagnostics to be built into the design. These provide frequent confirmation that sensors and valves are fully functional

Safety Integrity Requirements Items	Notes
Requirements for maintenance and testing to achieve the required SIL	Defines the intended proof test intervals. Defines the amount of testing that can be done off-line during shutdowns vs online. Describes facilities required for proof testing

6.8.2   Defining the functions

We need a reliable method of defining the functional relationships between inputs and outputs in the SIS. Usually we have to circulate the information among many parties and it has to be available at the plant for use in operations, training and in future studies. There are various ways of graphically presenting the information.

We illustrate examples of three widely used methods comprising:

- *Matrix table for cause and effect*: Use a matrix table of inputs and outputs making up a simple trip system. This method of defining functions is good for placing on P&I drawings and for explicitly defining interlocks and basic trip logic. It is not so good for defining shutdown sequences.
- *Trip Logic Diagrams*: Trip logic diagrams are widely used to define safety functions. They are useful because a complete scheme can be followed on one drawing and also because many graphical programming systems use the same format (see Figure 6.31).

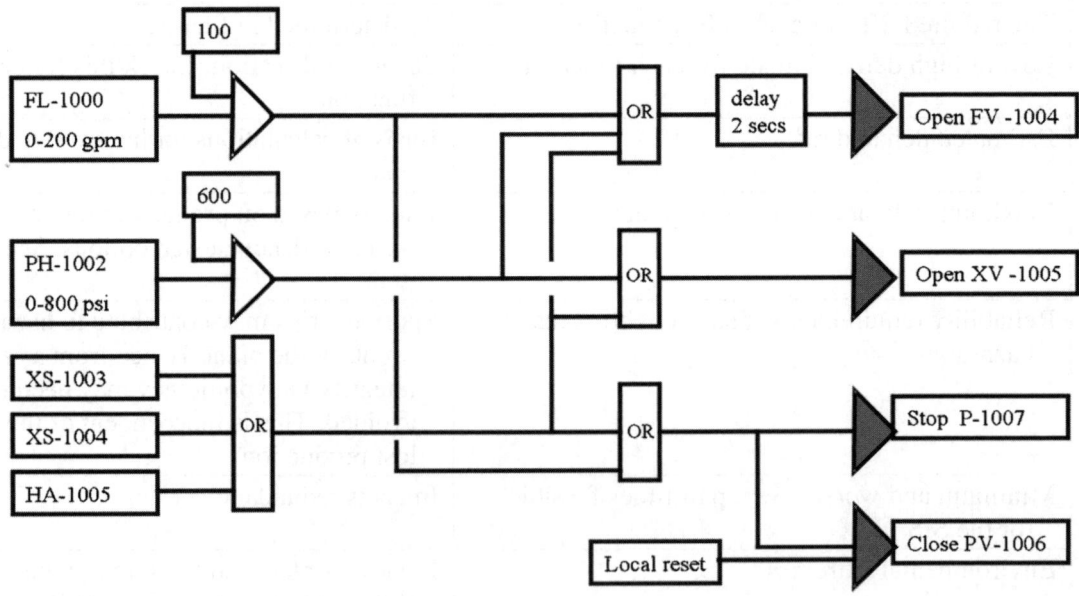

Figure 6.31
A trip logic diagram example

- *Trip link-up diagram*: This method allows a complex trip function to be represented on a multiple set of engineering drawings. This type of diagram is very useful when used as a means of defining the overall scheme to a design contractor because the task can be broken down into easily identified small sections (see Figure 6.32).

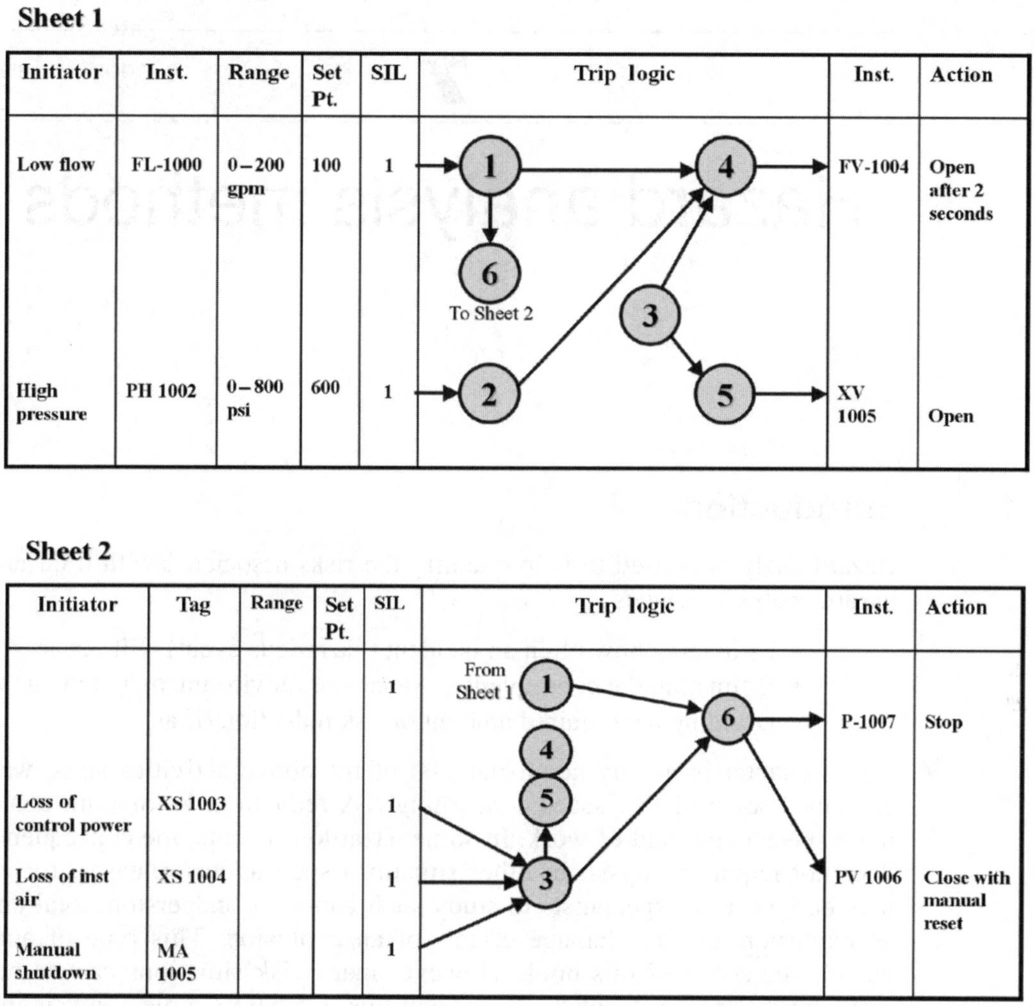

Sheet 1

Initiator	Inst.	Range	Set Pt.	SIL	Trip logic		Inst.	Action
Low flow	FL-1000	0–200 gpm	100	1			FV-1004	Open after 2 seconds
High pressure	PH 1002	0–800 psi	600	1			XV 1005	Open

(Trip logic: nodes 1 → 4 → FV-1004; 1 → 6 (To Sheet 2); 2 → 3 → 4; 3 → 5 → XV 1005)

Sheet 2

Initiator	Tag	Range	Set Pt.	SIL	Trip logic		Inst.	Action
				1	From Sheet 1		P-1007	Stop
Loss of control power	XS 1003			1				
Loss of inst air	XS 1004			1			PV 1006	Close with manual reset
Manual shutdown	MA 1005			1				

(Trip logic: 1, 4, 5, 3 → 6 → P-1007; → PV 1006)

Figure 6.32
Example of trip link-up diagram

6.9   Conclusions

The ground rules for design of safety instrumentation have become well established over many years of experience in major operating companies. We have seen an outline of these rules and have highlighted some of the factors to be considered when calling up a trip function.

It is particularly important to be aware of the relationship between cost and safety integrity levels in the SIS. The hazard study provides a good opportunity to get the SIL ratings right at the early stages of the project.

International standards have captured the working procedures needed to reduce the chances of systematic errors in design by setting down the stages of the safety life cycle. We have seen how the early stages of the life cycle match up to hazard study procedures. The methodology for transferring the trip requirements from the hazard study stages into the design stages has been thoroughly defined in the new international standards. The procedures follow basic quality assurance methods and can be set down for use by companies in a systematic manner. It is reasonable to expect that these procedures will improve the linkages between hazard studies and the management of safety systems.

7

Hazard analysis methods

7.1   Introduction

Hazard analysis is used to help quantify the risks associated with a hazard. The task of hazard analysis includes:

- Estimating how often an incident (hazardous event) will occur
- Estimating the consequences to persons, environment and plant
- Deciding the required amount of risk reduction (if any).

This chapter is mainly about the first of the above activities since we have already examined some of the issues concerning risk reduction. Estimating consequences is a more specialized field of work. In some hazardous events, the consequence is clear and does not require analysis. In other situations such as gas releases, for example, there may be a need for specialists to study such factors as dispersion, total energy levels in an explosion and the damage effects of an explosion. This type of analysis work is outside the scope of this book. However many risk situations can be handled just by reducing the frequency of the event and this is where simple analysis methods can be valuable.

7.1.1   What is quantitative risk assessment used for?

In summary, quantitative assessment of risk is valuable for:

- Knowing how often we can expect an event to occur ... this assists basic communication with other parties
- Ranking of risks for priority or severity on a risk matrix chart
- Comparing risks with tolerable risk targets
- Deciding on quantitative risk reduction requirements (risk reduction factors)
- Deciding on SIL requirements for protection systems.

There is another side to the quantification of failures. We also have a need to evaluate how the risk reduction measures are going to perform. Two areas in particular are relevant here.

Predicting or justifying the failure rates of safety system components. For example: field instruments, valves and logic solver systems. Many of the devices used in SIS applications are subject to rigorous failure mode and effect analysis (FMEA) as part of the qualification programs used to certify the instruments as suitable for safety system duties (see Safety certified instruments in Chapter 8). Sometimes FMEA methods are used in evaluating the likely failure rates of control systems as part of the EUC risk analysis.

Showing the effects of risk reduction measures on a hazard situation. We shall see in a moment how risk reduction can be modeled very effectively into a fault free analysis.

7.2   Outline of methods

Two methods of hazard analysis are widely used. Failure mode and effects analysis looks at possible component faults and tabulates their impact on risks. Fault tree analysis (FTA) looks at a hazard event and resolves the causes into basic events. If the event rate can be stated the hazard rate can be predicted. Choosing the right method for the job depends on a basic knowledge of the methods and seeing how they fit the application.

7.2.1   Failure mode and effect analysis

What is FMEA?

It is a failure mode and effect analysis method that is used to:

- Identify equipment or system failures arising from component faults
- Evaluate the effect of failures
- Prioritize the effects of failures according to severity of defects.

Compared with Hazop, the FMEA techniques do not begin with suggested deviations from design intent; they work by listing all possible failure modes of devices or components in the plant followed by an evaluation of the effect of each failure mode.

Some of the failure effects (consequences) will be harmless and some may be dangerous. Hence any method of ranking the failure effects will allow the user to find the most severe problems and give them first consideration.

The prioritization or risk ranking is done mainly by using:

- Risk matrix with risk priority numbers being generated as we saw in Chapter 1
- Criticality analysis, where the method is known as FMECA, abbreviation for Failure Mode Effects and Criticality Analysis.

7.2.2   Reasons for using FMEA

It is used in many industries and different versions exist for machinery, design, systems and products. Our interest here is mainly in process FMEA where it can be used to:

- Identify specific accident situations
- Evaluate safety measures
- Quantify risks
- Evaluate hazards at preliminary design or due to operating procedures
- Improve reliability of a process
- Meet regulatory requirements (evidence of suitable hazard study and risk assessments)
- Perform a systematic hazard evaluation.

7.2.3   When to use FMEA?

The hazard study 2 and 3 phases are good times as soon as sufficient design detail is available to identify the key components of the process. For machinery studies a greater level of design detail is probably needed.

Failure mode and effect analysis studies can be kept current for the life of the process and are fairly easy to revisit at periodic intervals due to the simple listing nature of the records. They are therefore suited to safety management requirements and can be part of re-validation studies.

Failure mode and effect analysis is often recommended by regulatory bodies for dealing with complaints and calls for corrective actions.

7.2.4   What to use it on?

Mechanical equipment such as pumps, compressors, etc. where component failures are considered to be a problem.

Systems for which there are few drawings but where individual components are readily identifiable.

Safety instrumented system components such as transmitters, valve controls and safety PLC hardware systems. In these devices an assessment of the fail to danger rate is critical and needs to be evaluated from the basic failure rate data of the components.

7.2.5   What is FMEDA?

Safety critical instruments achieve their high safe failure fractions partly through the use of internally driven diagnostic routines designed to detect dangerous failure modes before the instrument is asked to perform a trip. The detection capabilities of the device are factored into the risk ratings, and the FMEA then becomes FMEDA 'Failure Mode Effect and Diagnostic Analysis'.

7.2.6   Standards for FMEA

The reference and original standard for FMEA is the US MIL STD 1629. Recently IEC have published IEC 60812 'Analysis techniques for system reliability – Procedure for failure mode and effects analysis. FMEA'.

7.2.7   Outline of methodology

For process FMEA the following steps apply:

- Collect the information on the process including P&IDs and all available operating information
- Define the scope and purpose of the study to set the boundaries
- Structure the plant into items of plant that have a defined function or can be managed reasonably (examples: cooling system, heat exchanger, part of a distillation column).

Set up a worksheet for each item of plant that has the following typical columns.

Column 1: name

This column describes the name of the component in consideration (definition of chapter/component).

Column 2: code

This column describes the code number or reference of the component.

Column 3: function

This column describes the function of the component. This is important for anybody who writes or reads the document to understand the way in which the system works. It is an easy way to get a good understanding of how the different components work and how eventually the system works.

Column 4: failure mode

This column describes the failure modes of the component. For example, leakage, fail or close for valves, fracture, wear out, etc.

Column 5: cause

This column describes the cause of the failure mode of column 4. For example, aging, overload, misuse, etc.

Column 6: failure effect on function level

This column describes the effect of the failure on function level of the component.

Column 7: failure effect on next sub level; criticality

This column describes the effect of the failure one level higher than in column 6. Depending on the complexity of the system, it is possible to consider more levels. This gives a good insight in how the system works.

Column 8: safeguards

Identify the current safeguards in the plant item that may exist to prevent or mitigate the consequences of this failure effect.

Column 9: failure rate

In this column the failure rate of the current failure mode is given, usually in failures per unit time. If no failure rates are available, it may be possible to use the information of standards or databases. Estimates should be noted as such. A range of values may have to be tested. Alternatively, use the occurrence ranking scales.

Column 10: remarks / action

This column is reserved for any comment that is of importance. For example, the action to be taken to prevent a failure mode like this, changing the design, condition monitoring, better education of employees, etc.

Now additional columns are needed for the risk ranking of the failure modes:

- *Column 11:* Severity ranking. Use the scale of consequences
- *Column 12:* Occurrence or likelihood ranking
- *Column 13:* Detection ranking. This is the scale that modifies the likelihood by a factor based on the likelihood of detection before the fault becomes critical.

These three columns allow a ranking to be stated for each event by multiplying them together. This is called the 'Risk Priority Number'.

Individual users of FMEA will need to construct their ranking tables using descriptors appropriate to the application. One approach is to use ascending scales of 1 to 10 for severity (S) and for occurrence (O) and then a descending scale for the detection(D).

Severity scale

Effect	Rank	Criteria
None	1	Might be noticeable by operator
Very slight	2	No effect on process, negligible effect on product, no hazard
Very severe	9	Hazards capable of injury to one or more persons
Catastrophic	10	Severe hazard capable of causing death of more than one person

Occurrence scale

Occurrence	Rank	Criteria
Improbable	1	Not expected in 1000 years
Remote	2	Not expected in plant lifetime
Very likely	9	More than 1 per year
Frequent	10	More than 3 times per year

Detection scale

Detection	Rank	Criteria
Extremely likely	1	Controls will almost certainly detect this fault as soon as it occurs
Very likely	2	Controls expected to normally detect this fault
Very unlikely	9	Detection not expected
Never	10	Cannot be detected

Notice that this last scale does not take into account the speed of the hazardous event; hence detection may or may not help to avoid the event. Additional influence factors can be added as needed.

The above example creates a scale from 1 to 1000. Risk priority numbers above a defined threshold will then qualify for risk reduction treatment or at least reliability improvements with improved detection. Again this illustrates the benefits of diagnostics.

7.2.8   Conclusion

Failure mode and effect analysis studies and the results tables are likely to be very effective in component-centered problems but may be difficult to apply in some process situations. Hence, selection of this method needs careful evaluation before assuming it will be suitable for a hazard analysis task.

7.3 Fault tree analysis

Fault trees are a valuable aid to risk assessment and the development of protection schemes. They are normally used once the Hazop study has identified a potentially hazardous event and the team has requested some analysis of the likelihood and consequences.

According to an Instrument Society of America (ISA) technical report (ISA Tr84.03) FTA was originally developed in the 1960s to evaluate the safety of the Polaris missile project and was used to determine the probability of an inadvertent launching of a Minuteman missile. The methodology was extended to the nuclear industry in the 1970s for evaluating the potential for runaway nuclear reactors. Since the early 1980s, FTA has been used to evaluate the potential for incidents in the process industry, including the potential for failure of the Safety Instrumented System (SIS).

Failure tree analysis begins with a 'top event' that is usually the hazardous event we are concerned with, for example, 'explosion'. The 'tree' is then constructed by developing branches from top-down using two basic operators: 'AND gate' and 'OR gate'. The logic gate symbols are shown in Figure 7.1.

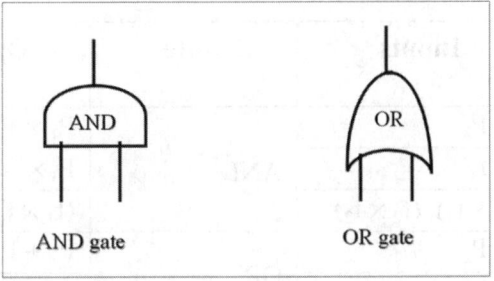

Figure 7.1
Logic gate symbols for FTA

The logic gates allow the contributing causes of the top event to be set out and combined according to the simple rules of AND or OR, for example, see Figure 7.2.

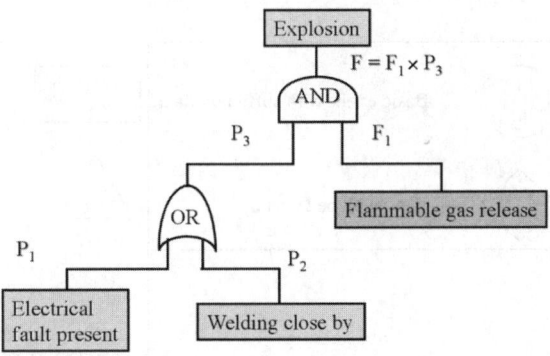

Figure 7.2
Example of basic fault tree

Functions of the gates

- 'AND' gates are used to define a set of conditions or causes in which all the events in the set must be present for the gate event to occur. The set of events under an 'AND' gate must meet the test of 'necessary' and 'sufficient'.
- 'OR' gates define a set of events in which any one of the events in the set, by itself, can cause the gate event. The set of events under an 'OR' gate must meet the test of 'sufficient'.
- – 'Necessary' means each cause listed in a set is required for the event above it to occur; if a 'necessary' cause is omitted from a set, the event above will not occur.
- 'Sufficient' means the event above will occur if the set of causes is present; no other causes or conditions are needed.

The information about each event is described as either:

- P = Probability of the event occurring, or
- f = Frequency of the event, or
- f × t = Duration of the event.

From which the following combinational rules are obtained:

Inputs	Gate	Operation	Output of the Gate
P_1, P_2		$P_1 \times P_2$	P
P_1, f_1	AND	$P_1 \times f_1$	f
$(f_1 \times t_1), (f_2 \times t_2)$		$(f_1 \times f_2)(t_1 + t_2)$	f
P_1, P_2	OR	$P_1 + P_2$	P
f_1, f_2		$f_1 + f_2$	f

The combinational rules allow the information known about each individual event to be combined to predict the frequency of the top event and the intermediate events.

Event symbols

Event symbols used in FTA are shown in Figure 7.3.

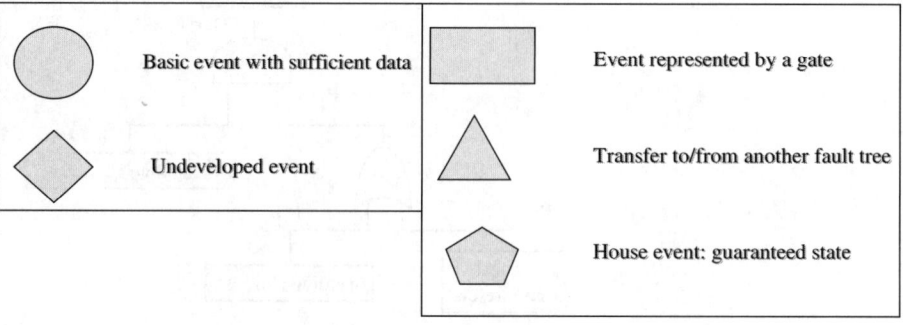

Figure 7.3
FTA: event symbols

These provide a means of classifying events:

- Basic event is the limit to which the failure logic can be resolved. A basic event must have sufficient definition for determination of an appropriate failure rate.
- Undeveloped events are events that could be broken down into sub-components, but, for the purposes of the model under development, is not broken down further. An example of an undeveloped event may be the failure of the instrument air supply. An undeveloped event symbol and a single failure rate can be used to model the instrument air supply rather than model all of the components. Fault tree analysis treats undeveloped events in the same way as basic events.
- Rectangles are used above gates to declare the event represented by the gate.
- Transfer gates are used to relate multiple fault trees. The right or left transfer gates associate the results of the fault tree with a 'transfer in' gate on another fault tree.
- House events are events that are guaranteed to occur or guaranteed not to occur. House events are typically used when modeling sequential events or when operator action or inaction results in a failed state.

The fault tree construction proceeds by determining the failures that lead to the primary event failures. The construction of the fault tree continues until all the basic events that influence the top event are evaluated. Ideally, all logic branches in the fault tree are developed to the point that they terminate in basic events.

FTA example

The example shown in Figure 7.4 is based on a batch processing plant where a PLC is used to sequence the operations of a large pressure vessel. During the batch cycle, a stage is reached where hydrogen must be supplied into the vessel and its pressure must be gradually raised to 6 bar.

The problem is that initially the vessel has a large vapor space that is filled with air. To avoid a dangerous gas mixture the vessel must first be evacuated to remove the air. There is also the possibility of leaks into the vessel whilst it is under vacuum. The PLC therefore carries out a leak test before starting the hydrogen feed. This test also reduces the risk of hydrogen leaks from the vessel once the pressure is positive.

The EUC risk (IEC terminology) is then due to a range of mechanical defects and PLC control/instrument defects. The fault tree shows how these defects are grouped under certain types of hazards all leading up to the potential for an explosion. The failure rates are estimated values but they do comply with the IEC requirement that normal control systems should not be credited with a failure rate lower than 10^{-5}/h.

Summary of rules for constructing fault trees

From an internal ICI Publication *Guide to Hazard Analysis* by J. L. Hawksley.

Aims: To do as simple a study as possible i.e. get the maximum benefit from the minimum of work.

This requires: the correct logic, the most significant causes to be identified, a 'broad brush' fault tree to be drawn initially and only the areas of significant concern developed in greater detail.

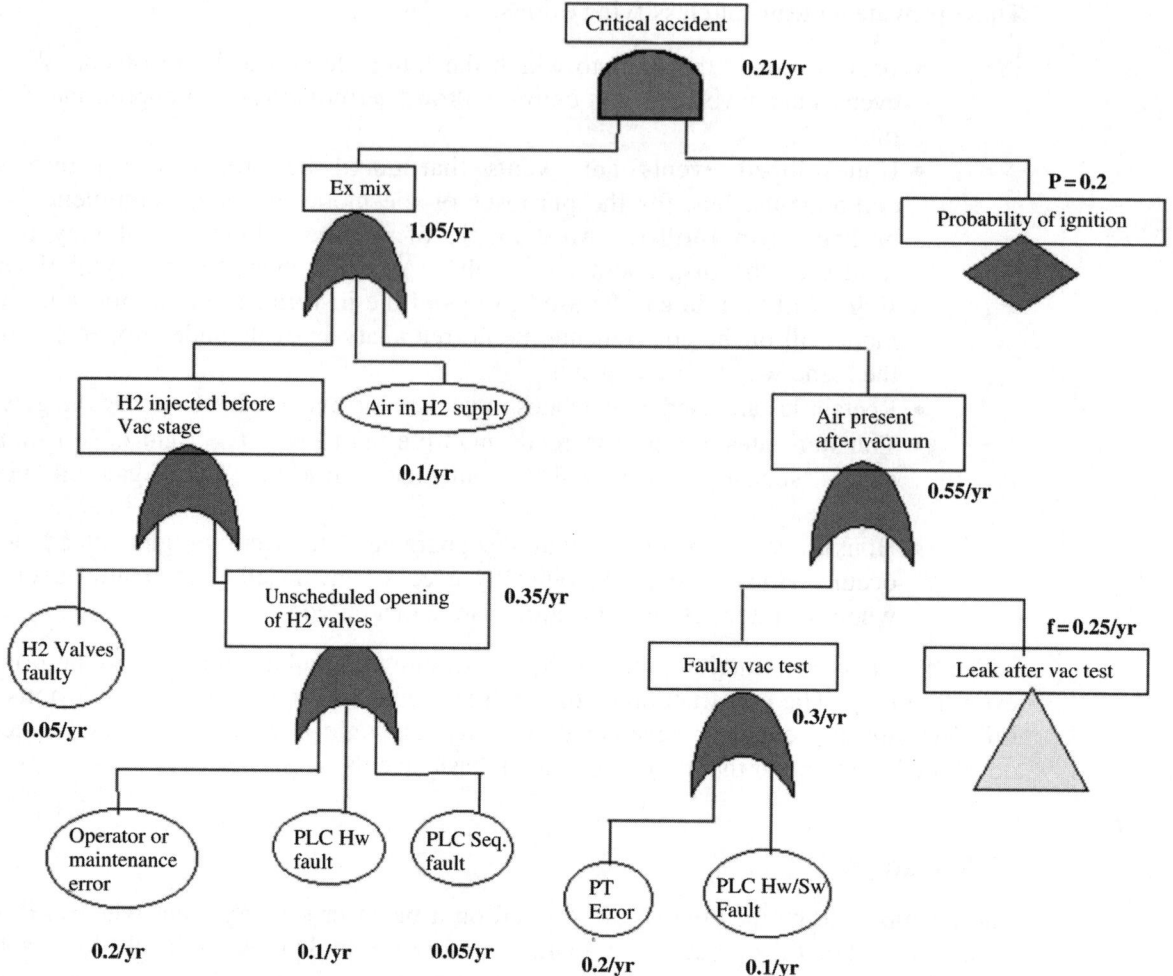

Figure 7.4
Fault tree for batch explosion diagram shown without SIS protection

Requirements:

- A physical description of the system
- A logical description of the system
- A clear definition of the hazard of concern
- A plant visit (if the system exists).

Key points for drawing up the fault tree:

- The fault tree is composed of events joined by 'AND' and 'OR' gates
- Start with the hazard or 'Top event'
- Draw a demand tree first, and then add the protective systems
- Think in small steps
- Think in terms of physical properties and relative physical positions of equipment
- Check very carefully for common mode effects
- There is not necessarily a single 'correct' fault tree
- Have a box containing a reasonable and adequate description at every gate
- Check the logic of a completed tree by going along each branch to the top event.

Practice with fault trees

The best way to become familiar with the basics of fault trees is to work through a simple example. The material shown in Figures 7.5–7.9 is based on examples given in pages 49–53 of the Hazan Manual issued by I Chem E and written by Trevor Kletz. This manual is a very helpful guide to basic practices in the field of hazard studies.

In the examples below the fault trees begin on the left with the hazardous event (Trevor Kletz regards a free meal as 'a common industrial hazard'). Note how frequency and probability are multiplied at the AND gate.

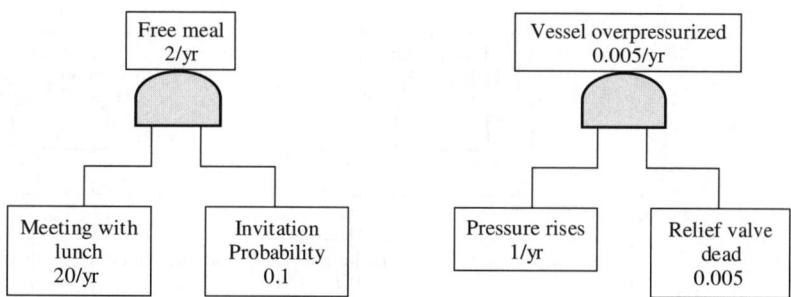

Figure 7.5
Introductory practice with fault trees
Acknowledgements: Trevor Kletz

Extending the fault tree

Fault trees lend themselves to breaking out the problem into specific detail. In Figue 7.6 the meetings with lunch are not all with salesmen. Hence an OR gate is used to split the events.

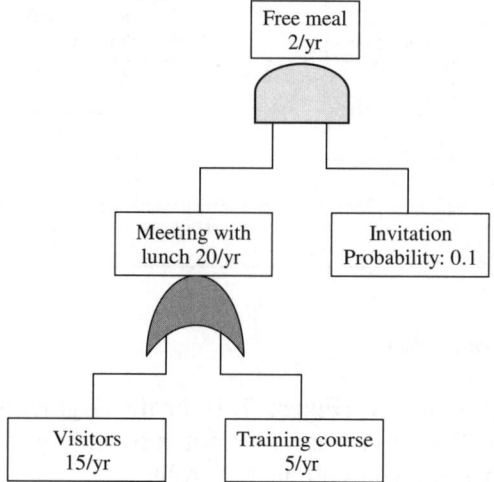

Figure 7.6
Illustration of fault tree analysis

Fault trees are extended until we have the data for the *frequency* or *probability* of each event on the left (or bottom) of the diagram. The improved resolution of the data leads to a modified frequency for the top event.

The fault tree then provides a model for evaluating the effect of other factors (Figures 7.7 and 7.8). In this case the lunch frequency increases due to a new training course.

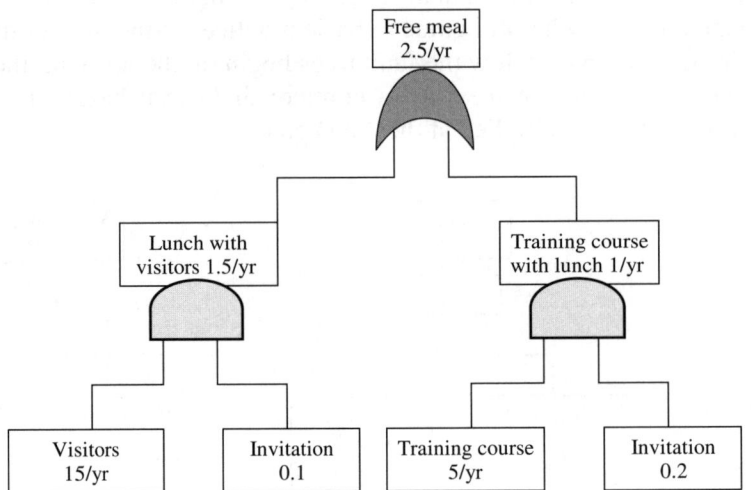

Figure 7.7
Extension of fault tree analysis

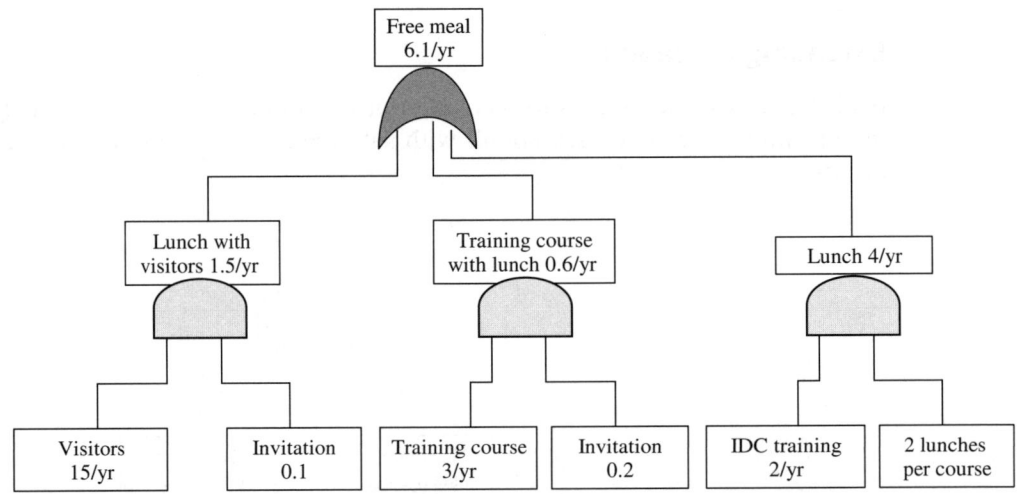

Figure 7.8
Adding more events to the FTA

In the next example (Figure 7.9), again originating from the Hazap manual, the fault tree for a car failing to start is set out. Try adding in some estimated frequencies to see it predicts how often you will have a problem.

7.3.1   Adding risk reduction measures in FTA

So far we have seen how useful fault tree analysis can be for anglicizing the risk of a known top event. The next step is to build in the possible risk reduction measures and predict the new risk frequency for the top event. It is easy to do this using the general approach shown in Figure 7.10.

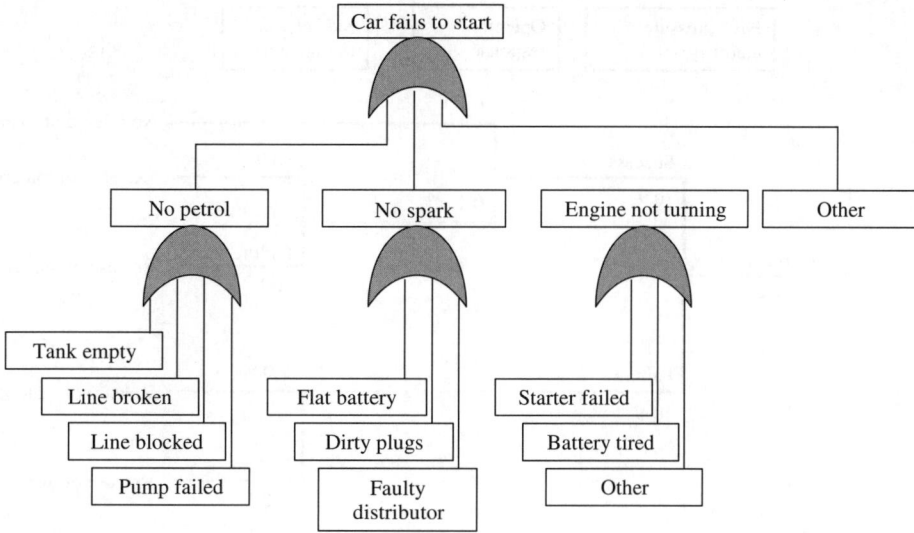

Figure 7.9
FTA for fault finding

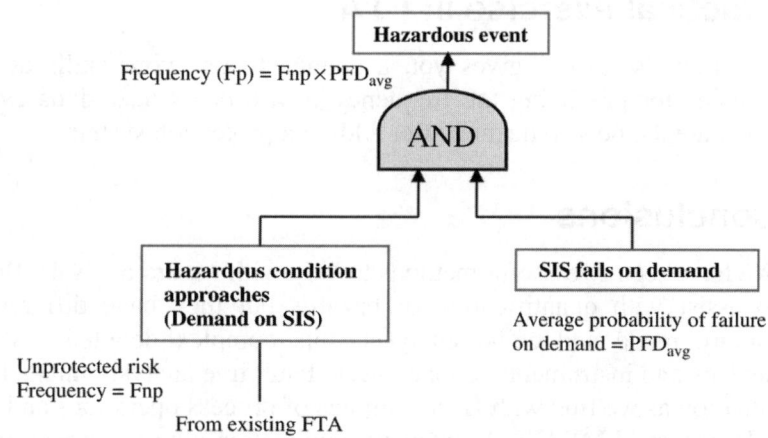

Figure 7.10
Generic method for adding SIS protection into a fault tree model

In the diagram the protection in the form of a SIS has a proposed probability of failure on demand (PFD_{avg}). This probability feeds into an AND gate with the predicted frequency of the hazardous event (demand rate F_{np} per year) assuming no protection. The additional AND gate reduces the frequency of the hazard tp $F_{np} \times PFD_{avg}$.

7.3.2   Event tree analysis

Event tree analysis is a widely used modeling method that is closely related to FTA. In effect this method takes the top event and divides the possible consequences into as many options as may exist and defines the probable splits.

In the event tree diagram (Figure 7.11) the success or failure of the protection responses defines the choice of route to the end event. This clearly provides a consequence tree and uses the PFD of each protection measure to calculate the frequencies of each outcome.

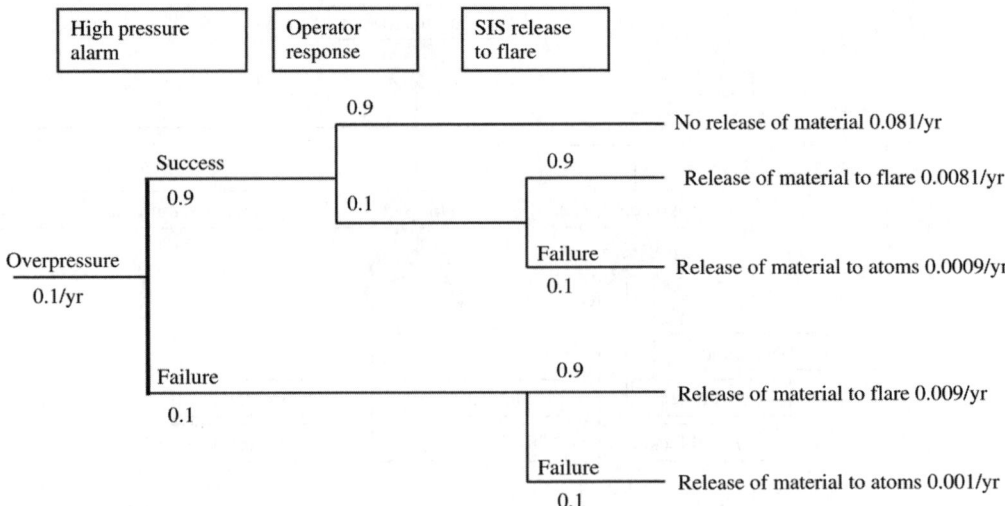

Figure 7.11
Event tree analysis example

7.4   Practical exercise in FTA

Our Exercise No. 6 gives you a chance to test your skills at developing a fault tree analysis for predicting the frequency of a process hazardous event. It also asks you to calculate the new frequency after adding a protection system.

7.5   Conclusions

We have seen two useful methods to help with hazard analysis. Both methods can be used to assist with quantification of hazards but they have different fields of application. Failure mode and effect analysis suits complex detailed systems such as electronic devices and instruments or machinery. Fault tree analysis suits a broader and more simple situation as we find with Hazop studies of process operations and the protection systems.

In general, FMEA finds extensive application with equipment manufacturers wishing to predict the reliability of their products. While FTA is best suited to end-users evaluating the hazard problems or the performance of protection systems. As a graphical method, it is ideally suited to reporting and presenting hazard analysis results and protection system proposals to the wider audience.

8

Factors in the choice of protection system

8.1 Introduction and objectives

This chapter considers some of the factors affecting the choice of designs for the alarm and trip protection systems. It does not attempt to deal directly with the issue of choosing non-SIS modes of protection such as relief valves or pressure relieving system based on process equipment. However, a project engineer or hazard study team may well be faced with striking a balance between investing in say, for example, mechanical pressure relieving systems with associated absorber systems and providing a trip system to reduce the frequency and severity of the demands on pressure relief. You may recall from Chapter 6 that balancing the allocation of risk reduction measures between SIS and non-SIS is the 'allocation' activity described by step 5 of the safety life cycle. This leads to the typical questions:

- Do we have a practical choice of protection methods available?
- Is a trip system going to be cheaper than a design change?
- Is a better trip system going to be cheaper than a better mechanical protection system?
- How does the SIL affect the cost?

The answers to these questions will involve someone doing a feasibility study on the technical choices (if any real choice exists) and they will be asked to estimate the costs. The key questions are likely to be:

- What will the trip system cost?
- What type of equipment is suitable?
- What will affect the choice of SIS equipment?

The objective here is to inform hazard study participants on the key factors affecting the SIS design and its costs and to show how the setting of SIL targets in a systematic and efficient manner will benefit the project.

8.1.1 Equipment cost factors

What are the 'big ticket items' in SIS? It may help to first take the basic model of an SIS and look at the components (see Figure 8.1).

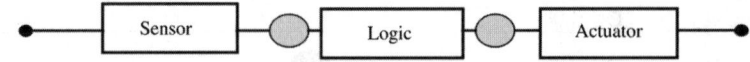

Figure 8.1
Basic SIS configuration in block form

The rules for meeting a target SIL rating for any particular function require that each subsystem (sensors, logic, actuators) shall be capable of meeting the SIL performance target (refer Chapter 6 for clarification here if needed).

So it is convenient to look at each subsystem from a performance and cost viewpoint and then add in the associated costs of engineering design and maintenance including proof testing:

- *Sensing instruments* are expensive and the total cost must include the installed cost. They may need to be redundant and diverse to meet the SIL requirements.
- *Logic solver systems* can be built very simply for simple applications but the real cost of performance must include the engineering upkeep and the costs of lost production through spurious trips. Programmable systems offer great performance advantages as SIS functions and complexity of SIS functions increases but their cost can be high and the costs must include for software licenses and programming tools.
- *Actuators* can range from simple tripping of electrical drives to very expensive high performance valves. The temptation is sometimes to use the existing control valve as a trip device to save cost and complexity but in many cases this will not satisfy the SIL requirements.
- *The engineering manpower costs* of a trip system can be one of the biggest budget items in a large installation. Cost items include specification time, documentation and validation work, programming and testing of logic, reliability calculations, training and the cost of periodic proof testing.
- *Maintenance costs* occur because trip systems require periodic proof testing and accurate calibration. The logic solver programs require skilled supervision and diligence in the control of any detailed changes. Therefore the trip system has a substantial call on the instrument maintenance costs for manpower and training. On larger SIS installations, software licenses and hardware service contracts are also large budget items.
- *Production losses* can be the largest cost item if the trip system affects a large continuous production process or one that has a risk of plant damage if an emergency shutdown has to take place. The cost of one spurious or avoidable trip may exceed the value of the entire SIS.

Underlying all of the main cost items is the major influence of the target SIL require from the protection system. We have seen that SIL-1 performance delivers risk reduction in the range of 10–100. We have also seen that SIL-3 at the top of the scale delivers RRFs in the range of 1000–10 000. Because of the very demanding nature of an SIL-3 safety system it is likely to cost 5 or 10 times the amount of a SIL-1 system performing the same function. All of the cost factors identified above will increase sharply as the safety integrity needs rise.

Here is a short guide to the architecture requirements of SILs 1 to 3. The arrangements shown are a starting point for the actual designs because variations can be obtained according to each application and within the guidelines of IEC 61508/61511.

8.1.2   SIL architecture guide

In all examples the approach is to select a suitable arrangement for each of the three typical subsystems for an SIS, i.e. sensor, logic solver and actuator.

SIL-1 safety system

Single channel architectures are acceptable for typical process instrument applications as shown in Figure 8.2 provided the instruments are 'proven in use' or certified suitable for safety applications.

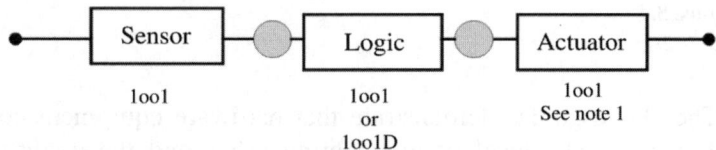

Note 1: It may be acceptable to trip the control valve but if the control valve contributes to the hazard condition a separate trip valuve must be used.

Note 2: If sensor reliability is poor a 2oo3 voting group may be justified.

Figure 8.2
SIL-1 trip

SIL-2 safety system

Dual channel architectures are needed for most sensors; logic solvers can be single channel if the diagnostic coverage is high, but see Note 2. Actuator must be independent, may have to be 1oo2 if reliability is poor (see Figure 8.3).

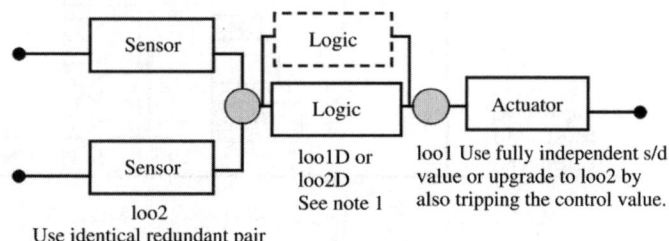

Note 1: If a PES logic solver used a high diagnostic coverage version is acceptable for SIL-2. However spurious trip requirements normally dictate the use of a 1oo2D redundant pair.

Note 2: If sensor reliability is poor a 2oo3 voting group may be justified.

Figure 8.3
SIL-2 trip

SIL-3 safety system

Dual channel architectures are needed for sensors, logic solvers and actuators. Diverse measurement principles should be used. In some cases 1oo3 voting may be justified. Do not use 2oo3 without analysis as this will not improve safety performance (see Figure 8.4).

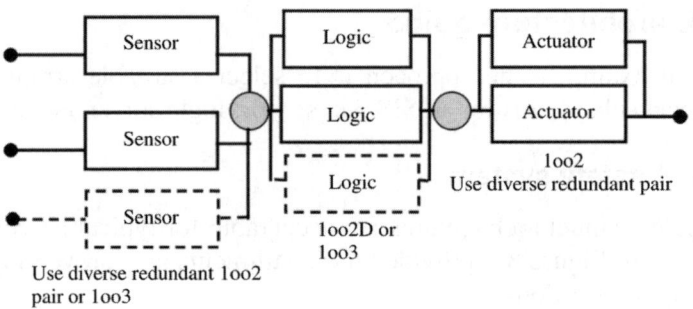

Figure 8.4
SIL-3 trip

The above guides demonstrate that hardware equipment cost will obviously rise with SIL values. The level of engineering effort and the quality of all documentation and validation work will also rise steeply with SIL ratings. This is because of the contribution of systematic error possibilities to the failure rates of the safety function. Hence we shall see a cost rise in all areas leading to the situation indicated in Figure 8.5 which is a very rough indicator of the cost vs SIL position based on the sharing of logic solver stage across at least ten safety functions.

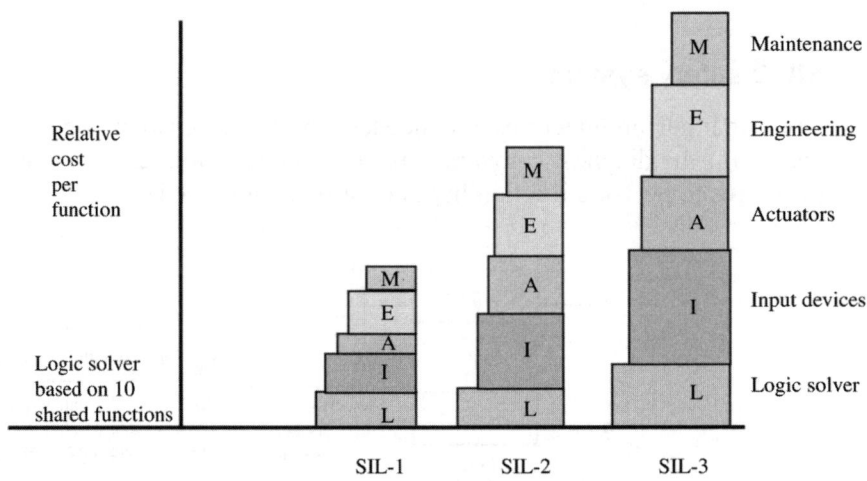

Figure 8.5
Relative costs of a typical safety function vs SIL

8.1.3   It pays to get the SIL right!

Indications from SIL determinations and reviews conducted by Assett Integrity Management Ltd, an experienced UK consultant, are that of the entire trip and interlock functions found on several large oil and gas installations, some 40–50% do not require any SIL rating at all. This is because they have a low impact on safety or losses. Of the classified trips, over 95% are found to be SIL-1 with just a few SIL-2 and SIL-3 rated safety functions. This spread will reflect the fact that many SIL-1 functions serve to protect against minor hazards or equipment damage items such as lube oil failures or overheated bearings. We should also note that most safety-related alarms would be engineered as SIL-1 systems.

The tendency is for users to overspecify the SIL requirements perhaps because of the need to work with a high safety margin where doubt exists on the true scale of risk. The cost penalties for doing this become very large when applied systematically to a large facility as the SIL targets are raised.

So it is important that right from the start the project should try to get the SIL requirements defined for each safety function. This is another reason why it helps a lot to get the formal safety requirements specification drafted as soon as the need for a trip function has been identified.

8.2 Equipment selection

Let us now spend a few minutes to become familiar with some characteristics of the types of SIS equipment available and see what factors may affect the choice of equipment.

Deciding on a particular type of SIS technology is likely to be an iterative process at the beginning and it usually starts with the decision on the logic solver technology. The tendency is to move from relay-based, hardwired panels to PLC-based programmable systems. The step up to PLCs is a big step, but once made, the presence of a safety PLC in the design tends to influence the details of how the rest of the safety functions are engineered. Let us first look at the logic solver choices.

As stated in the previous section, the SIS can be considered as three subsystems: sensor, logic solver and actuator. The general-purpose model for the SIS looks as shown in Figure 8.6.

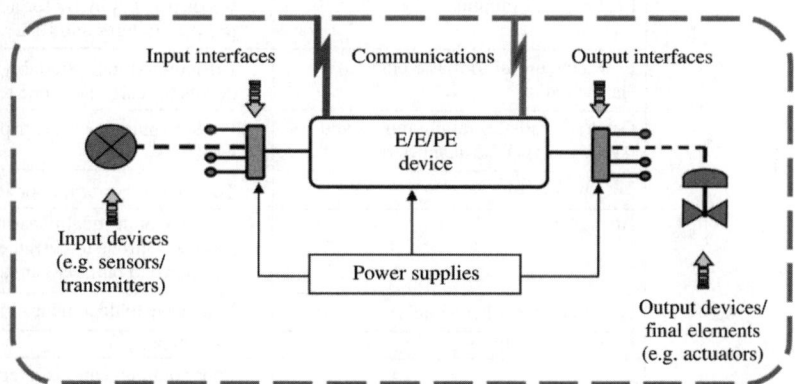

Figure 8.6
Safety-related control system

The procedure for selecting the equipment usually involves deciding each of the subsystems separately. Usually the first decision is for the logic solver stage because the capabilities of this stage will to some extent dictate the details of the sensor and actuator designs. Most PES logic solvers will be purchased for at least SIL-2 performance and are often certified for SIL-3. Hence SIL rating has only a moderate effect on the cost. The entry-level costs of some systems are high but are not sensitive to SIL ratings. The payback will lie in greatly improved management of the overall system.

8.2.1   Deciding the logic solver

The key issues in logic solver selection include the following:

- Choice of technology: relay, solid state or programmable
- SIL ratings
- Total package cost and the number of functions that can be served
- Available skills and resources
- Communication requirements with process controls and control room.

Here we briefly consider some of the technical factors influencing the choice between the use of relay-based, solid state or programmable logic solvers. This section describes some basic features of each type of logic solver.

8.2.2   Relays

Relay-based shutdown systems were the mainstay of the process industry for all the decades up to the 1980s. There are many good relay-based logic solver systems in service around the world and there are still many applications where they are the most appropriate choice for a shutdown system. There are however several disadvantages to using relays including their dependency on trip amplifiers and signal processing devices for any computing functions. The following chart lists the merits and demerits of relay-based systems.

Good Points	Bad Points
Simple and inexpensive for small systems	Complexity grows quickly
EMI and RFI immune	Needs trip amplifiers for analog signals: prone to failures and RFL
Low tech, useful for basic plants, Hardwired	Difficult to modify reliably: wiring details obscure and prone to errors
Generally fail-sfe: failure modes and failure rates can be quantified	Prone to nuisance trips, trip amps are a weakness
Easy to service	Too easy to corrupt, poor security
Adaptable	Custome builr: quality assurance problems, prone to design errors, expensive to build larger systems
Easy to design when simple	Expensive to document
Easy to manually test	No diagnostics
	Poor communications/interfacing

8.2.3   Solid-state systems

At the same time as the need for more complex and reliable shutdown systems became pressing, the availability of smaller and smarter electronics increased. The era of solid-state systems as an alternative to relay-based systems probably began in the late 1970s and ran until the establishment of really attractive PES solutions in the mid-1990s. From the evidence of the applications still being installed it looks as if the best of the solid-state systems is going to continue to be used and improved for many years to come.

Figure 8.7 shows the configuration of a typical solid-state SIS with its input signal processing stage, the logic solver function performed by standardized electronic function blocks mainly AND gates, OR gates, logic inverters and timers.

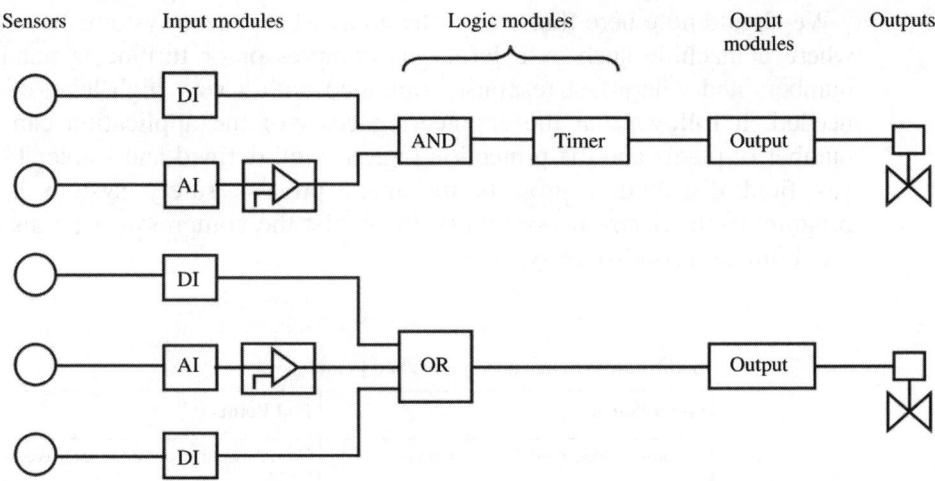

Figure 8.7
Solid-state system

Considering the merits and demerits of solid-state systems they have essentially the same characteristics as relay-based systems with the advantage of using purpose-built components such as multi-channel input signal processing boards and logic solver blocks. Early versions suffered from a substantial disadvantage over relays because they lacked fail-safe capabilities. It is possible for a static logic element to fail high or low or just stop switching. The failure may remain undetected and hence presents a high fail to danger risk.

The answer to the failure mode question was to use dynamic logic. The modules of the logic solver are operated in a continuous switching mode transmitting a square wave signal through each gate or circuit. Diagnostic circuits on board each modules then immediately detect if the unit stops passing the pulses. The detectors in turn link to a common diagnostic communication module that reports the defect to the maintenance interface (see Figure 8.8). Normally the detection of a failed unit will lead to an alarm and sometimes a trip of the plant.

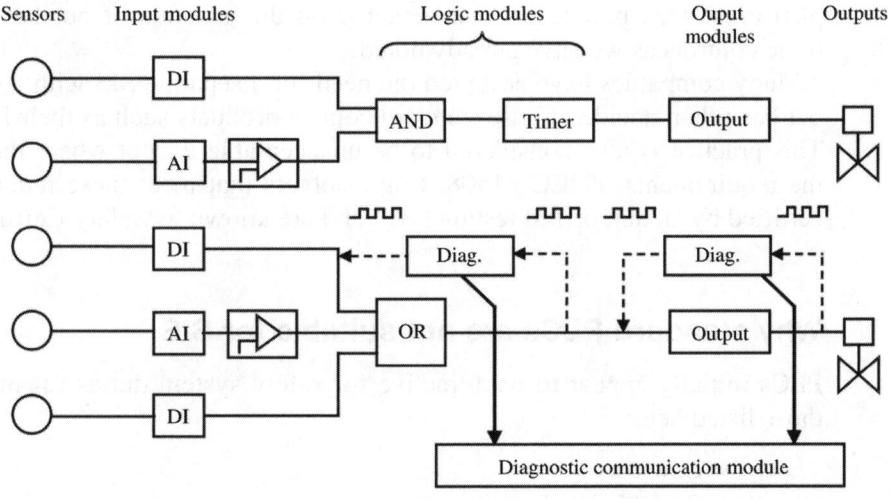

Figure 8.8
Solid-state system with diagnostics

We should note here that one of the areas where these systems have been successful is where a machine such as a large gas compressor or turbine is manufactured in large numbers and where fast response combined with a very high level of safety integrity is needed. It follows that the engineering costs of the application can be spread over a number of plants and the protection logic is well defined and stable. Thus for example in gas field distribution projects the main process safety system is likely to use a programmable electronic system (PES) whilst the compressor stations may be built with fixed function solid-state systems.

Solid-state systems: good points/bad points

Good Points	Bad Points
Purposes-made modules and panels	Complex construction, expensive to build
EMI and RFI immune through design	
Hardwired logic, good security against corruption, avoids software woories	Complex wiring, difficult to modify, long lead times
Good for machinery protection and high integrity repeat versions, E.g. Burner management, compressor	Inconvenient for plants with changing requirements
Generally fail safe: failure modes and failure rates can be quantified through testing	Redundancy and diagnostic features not so good as PES
Good testing facilities	
Fact acting, parallel logic	

8.2.4   Programmable systems

For logic solvers the use of PES presents obvious attractions. Most basic control systems utilize PLC- or DCS-based equipment and even single-loop control modules are PES devices. Many plants have placed their SIS logic functions into their basic control system platforms. This practice is unacceptable on the grounds of not having separation from basic controls as we have already noted.

Many companies have accepted the need for a separate SIS with a separate logic solver but have then decided to use standard control products such as their favored PLC models. This practice is also considered to be unacceptable except where the logic solver meets the requirements of IEC 61508. Logic solvers that meet these requirements are usually certified by an authorized testing body and are known as Safety Certified PLCs.

8.2.5   Why standard PLCs are not suitable for SIS

PLCs initially appear to be attractive for safety system duties for many reasons such as those listed here:

- Low cost
- Scalable product ranges
- Familiarity with products
- Ease of use

- Flexibility through programmable logic
- Availability of good programming tools
- Good communications.

The PLC fits in easily to the SIS model as shown in Figure 8.9.

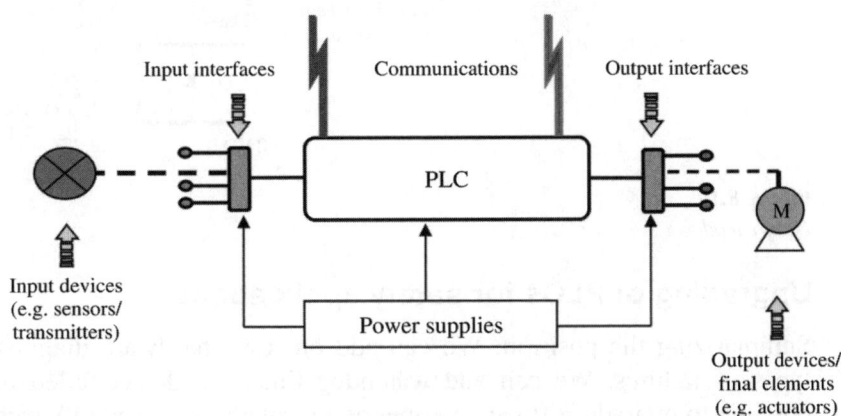

Figure 8.9
PLC as safety controller

But there are significant problems:

- Standard PLCs were not designed for safety applications
- Limited fail-safe characteristics
- Unpredictable response in the presence of hardware faults
- High risk of covert failures (undetected dangerous failure modes) through lack of diagnostics
- Reliability of software (also stability of versions). Absence of safeguards against program malfunctions in the operating systems
- Flexibility without security
- Unprotected communications
- Limited redundancy structures.

Here is a simple example of the covert failure problem (Figure 8.10). The output stage of the PLC operates a fail-safe solenoid or motor trip relay. It has to stay energized for weeks but we will not know if it is shorted until it has to trip the function. This is an unrevealed fail to danger condition or 'covert fault'. The broken wire fault is an 'overt fault' or revealed fault, which will fail-safe but creates a 'nuisance trip'.

The problem appears to be that there is no way of ensuring a consistently high standard of engineering both hardware and software across the wide product range of PLCs. It is too easy for end-users to make mistakes in the software applications and this, combined with a relatively low level of hardware diagnostics, means a low level of confidence that the PLC will always be on duty and will always respond when finally asked to do so.

PLCs require active diagnostic testing to overcome the risk of undetected failures in a dangerous state and leaving the plant undefended. If all failures were certain to be failsafe this would not be necessary, as the plant would always be protected in the event of PLC failure.

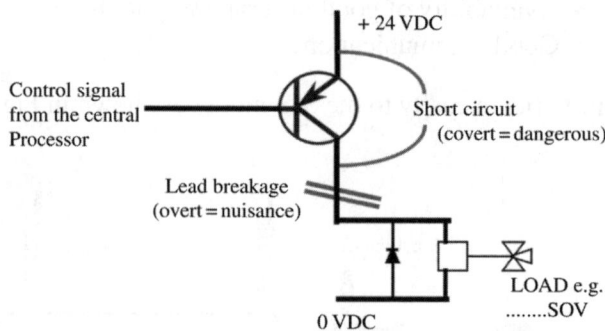

OVERT FAILURE AND COVERT FAILURE

Figure 8.10
Failure modes

8.2.6   Upgrading of PLCs for safety applications

Summarizing the position: We can add our own hardware diagnostic devices for some types of failures. We can add watchdog timers to detect failed operating cycles. It is possible to upgrade software engineering through improved QA techniques. It is possible to consider dual redundant standard PLCs in hot standby mode. However, the standard PLC does not lend itself to covering all the possible failure modes through the normal fault detection systems. Even if our special measures are effective, at the end of all these extra efforts we have the problem that we have built a special application that needs to be carefully documented and maintained. And then we have the problem of proving it to others or certifying it for safety duties.

That is basically the case for vendors to produce a special purpose PLC built specifically for critical safety applications. Let us look at what it takes.

8.2.7   Characteristics of safety PLCs

The answer to the problem of undetected faults in PLCs lies in the concept of fault coverage and fault tolerant systems. The answer to the problem of hidden defects in software is high quality embedded software combined with strictly limited user programming facilities.

- Automatic diagnostics continuously check the PLC system functions at short intervals within the fault tolerant time of the process.
- A very high level of diagnostic coverage is achieved through internal self-checking hardware and software.
- When a dangerous fault is detected the PLC is automatically shutdown.
- Redundant hardware options are available to provide uninterrupted operation even if one channel has failed.
- All faults are notified for attention and repair.
- Online hot replacement of failed modules can be provided.
- Application software is written within a protected framework of proven instructions or approved function blocks.
- All application software are updated transparently to redundant channels.

8.2.8   1oo1D safety PLC

Figure 8.11 shows the basic architecture of the PLC upgraded to include for diagnostic devices embedded in the construction of the PLC. This unit is able to overcome the objections listed for standard PLCs and is now the basic module concept for several safety control system manufacturers. Essentially this unit in single channel configuration will trip the plant if any of its diagnostic functions finds a fault in any of the stages: input, CPU, power supply or output.

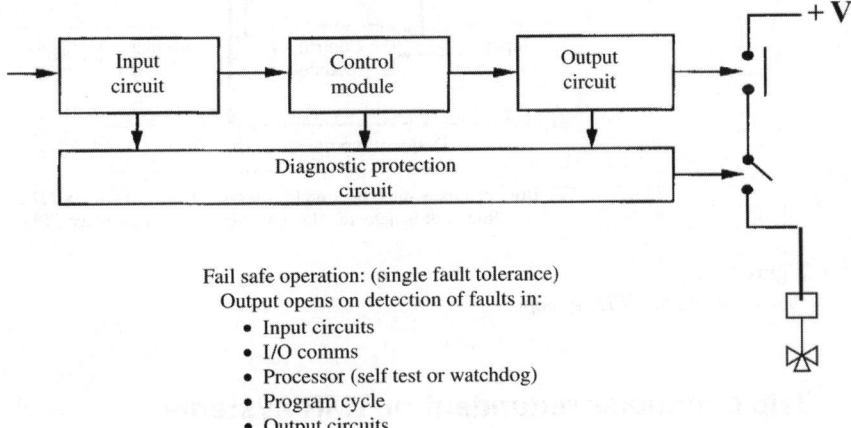

Fail safe operation: (single fault tolerance)
Output opens on detection of faults in:

- Input circuits
- I/O comms
- Processor (self test or watchdog)
- Program cycle
- Output circuits

Figure 8.11
Single channel safety PLC

The term 1oo1 comes from the notation that 1 fault in the 1 channel system will not cause a failure of the safety function. The D denotes diagnostics protection. The high diagnostic coverage ensures that this PES device is acceptable for SIL-1 and SIL-2 applications in single channel format.

Even though many process applications operate in low demand mode the designers of safety PLCs will want to ensure that their products are qualified for the continuous mode. All self-diagnostics are executed typically within a 1-s time frame.

TUV class 3 specifications allow a 1-s interval to execute the following:

- Diagnostics to ensure very high coverage of possible faults
- Application software executed at least twice
- Operate a secondary means of de-energization (SMOD) of all the affected outputs in the event of diagnostics finding a fault
- Diagnostic coverage must cover a high percentage of possible faults – typically 95–99.5%

8.2.9   1oo2D safety PLC

In this version the entire logic solver stage from input to output is duplicated and if one unit fails its diagnostic contact will open the output channel and remove that unit from service. The SIS function then continues to be performed by the remaining channel (see Figure 8.12).

The notation 1oo2D applies because the system will still perform in the presence of 1 fault amongst 2 units. The parallel connection of the two units substantially improves the availability. Note that diagnostic performance is further improved by cross-linking between the CPU of one channel and the diagnostics of the second channel.

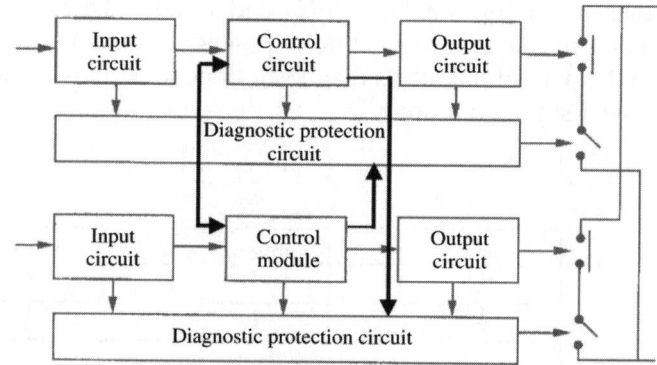

Both channels must operate to trip output. Reverts to 1oo1D if
module fault is detected. Diagnostics must check other CPU

Figure 8.12
Dual channel safety PLC example

8.2.10   Triple modular redundant or TMR systems

The safety systems built on the 1oo2D modules have found a strong market in the general
area of process plant applications. In some of the most demanding safety areas including
offshore oil and gas and in the nuclear field these systems have to compete with an
alternative architecture based on the principle of 2 out of 3 voting. These are known as
triple modular redundant (TMR) systems. Figure 8.13 illustrates the principle.

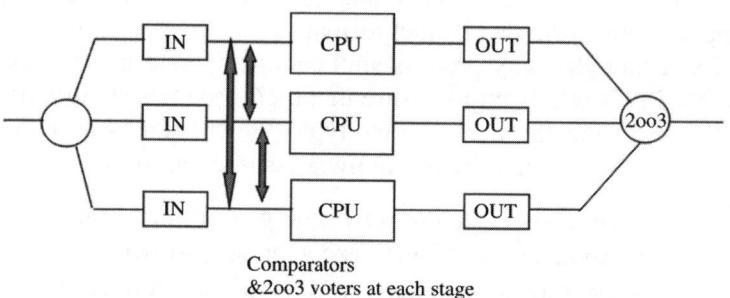

Comparators
&2oo3 voters at each stage

No single point of failure. High safety integrity. High availability

Figure 8.13
Basic arrangement of TMR safety PLC

These units have the advantage of not having to use such complicated diagnostics as the
1oo2D systems because every stage of the PES can be built up with 2 out of 3 majority
voting for each function.

For example, the three input stages operate in parallel to decide which is the correct
value to pass to the processor. The three processors compare data and decide which is the
correct action to pass on to the output stages, etc. All internal communications can use a
triplicate bus. A good example of this type is the Triconex range of safety controllers. The
advantage of 2oo3 systems is that if one channel switches off the remaining two channels
still qualify for SIL-3 performance. This allows plants to keep running safely whilst the
one channel is under repair.

8.2.11   Quad 2oo4D safety PLCs

Some safety PLC designs have moved from 1oo2D architectures to so-called Quad processor designs where two CPUs are mounted on one assembly. This makes the single channel PLC system into a 1oo2 operating pair that can satisfy SIL-3 requirements in one module. High availability safety PLCs are then built with two redundant modules. As with the TMR designs, if one module switches off due to an internal fault the redundant channel (1oo2) remains in service and can satisfy SIL-3 performance indefinitely.

8.2.12   Communication features of safety controllers

Before we move away from the high-tech end of the safety systems we need to touch on the issue of communications. The points are summarized:

- Strong need for communications between the SIS and the plant control systems
- For operator information and co-ordination with control
- For tidy-up of DCS or PLC controller states or sequences arising from action of the SIS
- For event recording
- For I/O status and status of the SIS itself
- Security is required in communications to prevent incorrect writing of data into the SIS
- Communications and data formats need to be compatible with DCS/PLC vendor standards or open standards
- Growth of certified interfaces.

8.2.13   Classification and certification

Having looked at some safety PLCs it should be clear that the technology is well established to meet the objectives of achieving high safety integrity using PES. The remaining obstacle is the task of proving the performance of the product. This is where the earlier DIN 19250 standard and the new IEC 61508 standard make a major contribution by setting out engineering procedures and technical performance standards that must be met in order to achieve designated SIL performance levels.

We have noted that certification of a device for SIS duties must cover the hardware and diagnostic performance capabilities as specified by IEC 61508. Certification must also cover the software engineering life cycle activities leading to the embedded software as well as the programming tools supplied for the end-user.

The combination of hardware and software certification means that the end-user can obtain a fully certified product for the logic solver section of the SIS. This does not of course relieve the end-user of any obligations as far as the quality of his software application/configuration is concerned.

Certification is a comprehensive and specialized task undertaken only by well-established authorities such as TUV. Details of testing guidelines and practices are outside the scope of this book, but it is helpful to note that TUV publish details of their testing programs on their website. In particular their website publishes a list of type approved systems so that the progress of certification for each manufacturer's product can easily be checked.

Because IEC 61508 requires overall management of functional safety to be treated as an essential factor for the development, supply and operation of safety systems, much attention is now being paid to Functional Safety Capability Assessment (FSCA). This relates to the overall capability of a company, whether it is a supplier or end-user to properly perform all the tasks involved in supporting functional safety systems. In the future we should look for FSCA approvals for suppliers and end-users as part of the quality assurance package associated with SIS products.

8.2.14   Cost considerations for safety certified PLCs

Safety PLCs can bring many advantages to an SIS installation. These include:

- Ease of configuration of the signal processing and logic functions
- Very high performance for safety through diagnostics
- Secure management of the safety function through software
- Automatic documentation of the safety logic
- Ability to support diagnostic methods for sensors and actuators
- Cost saving through sharing of many safety functions in one unit
- Very low risk of spurious trips through redundant architectures.

The entry-level costs of most safety PLC systems are high and usually the PLCs will be purchased for at least SIL-2 performance and are often certified or upgradeable to SIL-3. This is done to ensure that they can service the highest SIL rating of the loops they serve. Provided that plant has a fairly large number of safety functions the contribution to cost per function is not such a large item. The payback will lie in greatly improved management of the overall system.

This completes our brief review of logic solvers. The next paragraphs consider the choices for sensors.

8.3   Key points about sensors and actuators

Some important points to keep in mind when dealing with sensors and actuators are summarized here:

- Sensors and actuators are the most critical reliability items in an SIS.
- Separation, diversity and redundancy are key features for success.
- Safety-related instruments must have a proven record of performance. IEC 61508/61511 have specific requirements.
- Logic solver intelligence and communication power will help to provide diagnostic capabilities to assist field device reliability.
- Failure modes and programming errors are potential problems for intelligent instruments.
- Sensors and actuators dominate reliability issues.

Consider the first point in the above list: A typical reliability table will illustrate why the field devices are the major contributors to the possible failures of most safety functions.

Item	Fail to Danger (Rate per year)	PFD$_{avg}$ (3 Month Proof Test)	PFD$_{avg}$ (% of Total)
Input sensor loop	0.05	0.006	32
SIL-2 Logic Solver PLC		0.0005	3
Output Actuator loop (solenoid + valve)	0.1	0.0125	65
Total		0.019 (SIL-1)	100

The field devices taken together contribute 97% of the PFD for this example. Note also that the PFD figures for the field devices are affected by environmental conditions and maintenance factors such as test procedures, calibration methods and the effects of mechanical plant maintenance activities.

All of this means that once an adequate logic solver has been chosen for the SIS the bulk of the engineering and maintenance effort should concentrate on the field devices. This fact is sometimes masked by all the effort that has gone into improving the performance of logic solvers. Possibly this is because the field systems are application specific rather than product specific as is the case with PLCs, i.e. vendors can improve the PLCs as products but end-users have to look after the field instruments as applications.

Note that the PES logic solvers include the benefits of auto-diagnostics to improve the PFD figures. In the past it has been difficult to do the same for sensors and actuators but in cases where this can be done there will be major improvements to the PFDs.

8.3.1   Sensor types

There are only two basic categories of sensors: switches and transmitters (see Figure 8.14).

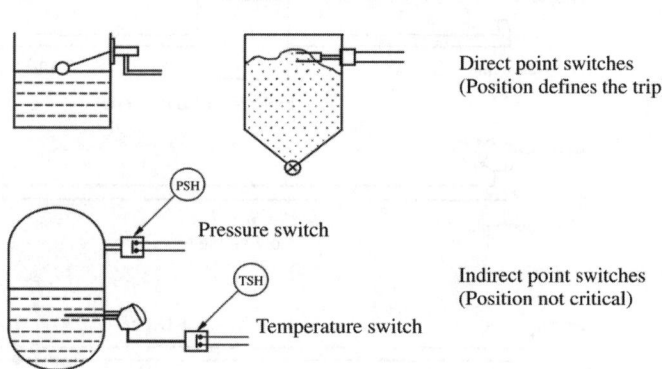

Figure 8.14
Types of sensors

Switches

Simple electrical or pneumatic switches that produce a state change when the limiting condition is exceeded.

These can be divided into sub-classes:

- Direct point switches where the sensor's physical location defines the limit value, e.g. limit switch or proximity switch, a fixed level switch such as the float switch commonly used on small boiler drums.
- Indirect point switches such as pressure switch or temperature switch where the basic measurement is converted mechanically or electrically to a signal, which has an adjustable limit value.

Transmitters

These employ sensing elements where a signal representing a process parameter is continuously sent to the logic solver. The logic solver rather than the sensor then determines the limiting condition.

These can be subdivided into some typical variants as shown in Figure 8.15 and described here.

- 4–20 mA current transmission to an input A/D converter stage in the logic solver. The field instrument is powered by a portion of the loop current supply
- 4–20 mA current transmission or a voltage signal typically 1–5 V transmitted from a separately powered converter stage either control room- or field-mounted
- Low-level signal (from thermocouple or RTD element) directly to an input converter stage in the logic solver
- Digital signal transmission using a bus protocol such as Fieldbus or Profibus from a field-mounted transmitter to a bus converter stage at the logic solver.

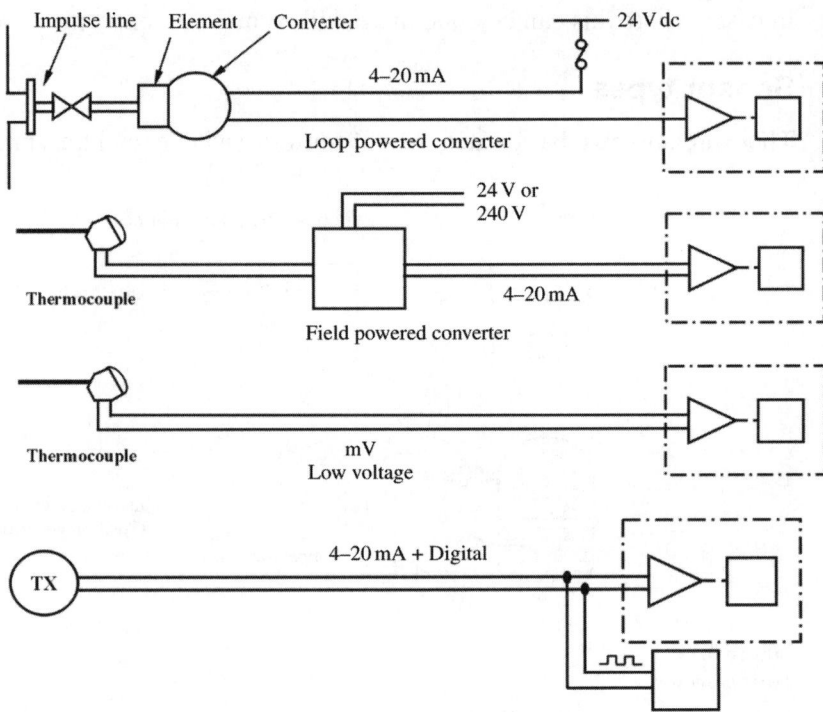

Figure 8.15
Transmitters

8.3.2   Advantages of analog transmitters over switches

All the standards for safety systems advocate the use of analog transmitters in preference to switches or sensors for safety applications.

The advantages are listed here and should be noted carefully when planning a safety system:

- Good reliability and accuracy
- Signal present at all times
- Potential for diagnostics, easier to detect faults
- Possible to compare signal with other parameters
- Trending and alarming available
- Multiple set points
- Competitive pricing
- Rationalized spares.

8.4 Guidelines for the application of field devices in the SIS

The application of any measurement and control device for duties in an SIS must take into account two primary considerations:

1. The device should be applied using the best design techniques to minimize failures.
2. The selected device should meet the qualifying requirements of IEC 61508 or IEC 61511.

8.4.1 Design techniques to minimize failures

The ground rules for design to minimize dangerous failure rates include the following techniques.

Fail-safe design

Design sensors and actuators to result in fail-safe responses to their most likely failure modes. Then review spurious trip rates to see if they are acceptable. For example, a temperature transmitter serving a high temperature trip function will have the element burnout option set to 'upscale'.

Separation

Ensure physical and functional separation between BPCS and SIS sensor/actuator systems as far as practicable.

Diagnostics

Search for ways of introducing diagnostics to frequently confirm the healthy operation of the device.

Redundancy

Use redundancy where a reduction in fail to danger rate is needed or where a low spurious trip rate is essential.

Apply redundancy to meet fault tolerant architecture constraints of IEC 61508 for SIL-2 and SIL-3.

Diversity

Search for diversity of sensors where the risk of common cause failures is significant.

8.4.2 Design for fail-safe operation

- Sensor contacts closed during normal operation
- Tx signals go to trip state upon failure (normally < 4 mA)
- Broken wire = trip

- Output contacts closed and energized for normal operation
- Final trip valves go to trip (safe) position on air failure
- Drives go to stop on trip or SIS signal failure.

The points summarized here are fundamental to good SIS design practices. For a comprehensive guide to the design practices for SIS sensors and final elements we would suggest:

- ISA S84: Appendix B provides guidance on many topics in SIS design
- IEC 61511 part 2 (when released).

There are, however, no set rules as each installation must be evaluated to see if the majority of possible failure modes will lead to a safe condition for the process.

8.4.3   Separation of sensors from BPCS

One of the ground rules of SIS is 'Avoid sharing sensors with the BPCS (Basic Process Control System) except under specially reviewed conditions'.

The reasons for this should be fairly obvious but it is surprising how often the rule is broken due to financial or physical constraints.

Do not share sensors because it:

- Violates the principles of independence for the SIS from the BPCS
- Creates potential for common cause failure
- Does not create a separate layer of protection
- Procedures for maintenance and testing and device protection may not be adequate.

The separation rules for sensors are summarized here based on guidance contained in IEC 61511 part 2 par.: 11.2.4:

- Sharing of sensor between SIS and BPCS only allowed if safety integrity targets can be met. This would require sensor diagnostics and is only likely to be possible for SIL-1.
- Separate sensor is allowed to be copied to BPCS via isolator.
- SILs of the levels 2, 3 and 4 normally require separate sensors with redundancy.
- SIL-3 and SIL-4 normally require separation and diverse redundancy.

Figures 8.16 and 8.17 illustrate the sharing and separation of sensors in a typical boiler drum level application.

In Figure 8.16, 'fail high' state of the level transmitter will cause the boiler drum level to fall to a dangerous empty condition and the safety function will fail for the same reason.

Note how separation includes separate connections into the process, a contribution to reducing common cause failure potential (Figure 8.17). Figure 8.18 presents a fault tree analysis comparing the effects of shared and separate sensors on the damage rate.

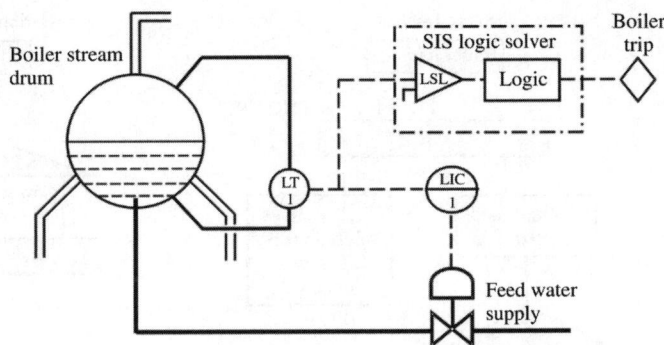

Figure 8.16
Shared sensors

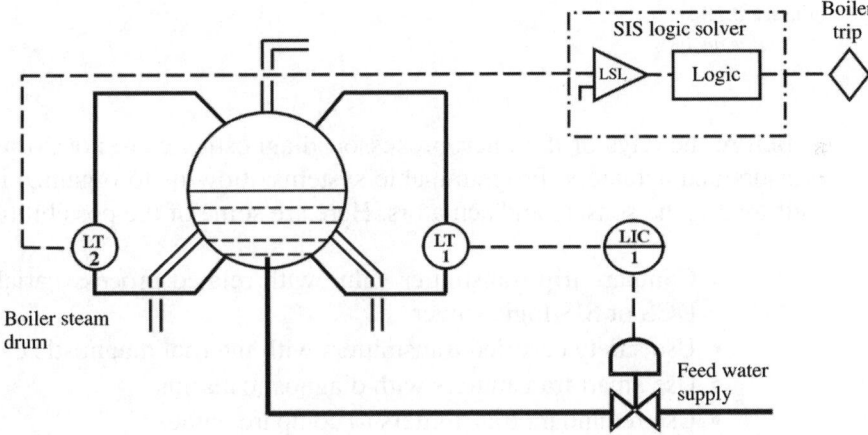

Figure 8.17
Separate sensors

8.4.4   Fault tree analysis for boiler low-level trip

Note that the fault tree on the left side of figure 8.18 separates the faults due to feed water failures from those due to the transmitter failure. No protection is provided against transmitter failure, which will cause a low-level condition through its action on the level control loop LIC-1. Hence the boiler damage rate will be high compared with the correctly designed version on the right-hand side.

8.4.5   Sensor diagnostics

Diagnostics, as we have seen before, involve the monitoring at frequent intervals of the operating condition of a device such as a PES or a sensor. Thus a high level of diagnostic coverage will convert many potentially dangerous failure conditions into a safe condition with an alarm or safe shutdown. Potentially large savings can be found by reducing the frequency of manual proof testing and by reducing the levels of redundancy needed to achieve the target SIL.

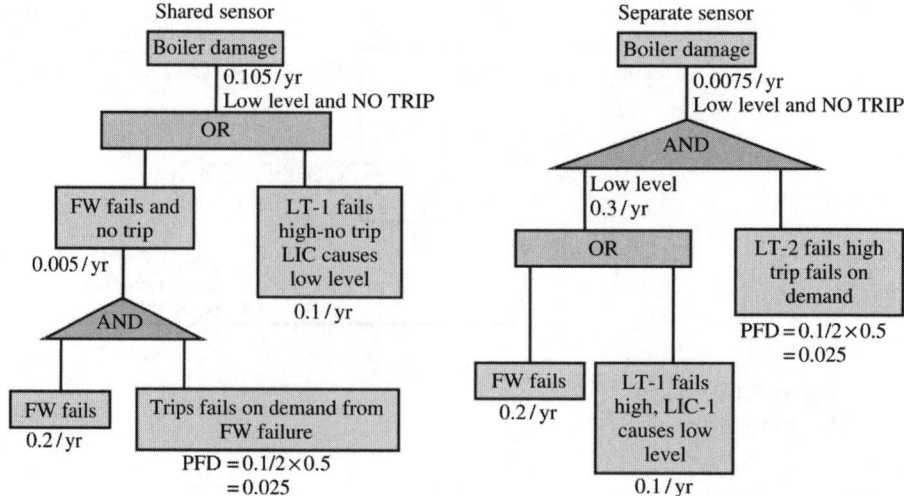

Figure 8.18
Fault tree analysis

Before the days of the microprocessor, diagnostics were not normally applied at all to sensors and actuators. Programmable systems allow us to organize improved methods of self-testing the sensors and actuators. Here are some of the possibilities:

- Compare trip transmitter value with related process variables. Can be done in DCS or SIS logic solver
- Use safety certified transmitters with internal diagnostic capabilities
- Use smart transmitters with diagnostic alarms
- Use redundant transmitters to compare values
- Smart and safety certified transmitters provide diagnostic monitoring of their own electronics and these can be arranged to switch the transmitter output to a safe mode value.

The question is what level of diagnostic coverage is achieved? There is still scope for measurement error due to impulse line problems or changed process conditions. It is the end-users' responsibility to evaluate what portion of potential failures are detectable by the diagnostics and this portion is then factored into the reliability model for the SIS loop.

8.4.6   Redundancy and separation issues in sensors and final elements

It is often quite difficult to arrive at a decision on redundancy in field device applications. There are conflicting demands between safety reliability, cost and regulatory requirements. Let us look (Figure 8.19) at the constraints and find some guidelines. Here again is the constraints table we saw in Section 6.6.2.

Constraints applied by IEC standards

IEC 61511 sets down minimum hardware fault tolerances where the safe failure fraction (SFF) of an instrument is between 60% and 90%.

IEC requirements: From IEC 61511 – part 1. Clause 11.4.
Applies to sensors and actuators with safe failure fractions 60–90%.

Safety integrity	Simple Devices (Non-PES)		Complex Devices (Using PES/Smart etc.)	
	Min. fault tolerance	Min. architecture	Min. fault tolerance	Min. architecture
SIL-1	0	1oo1	0	1oo1
SIL-2	0	1oo1	1	1oo2 or 2oo3
SIL-3	1	1oo2 or 2oo3	2	1oo3
SIL-4	2	1oo3	Special requirements apply, see IEC 61508	

Exceptions apply for higher and lower safe failure fractions – see text.

Figure 8.19
Fault tolerance of subsystems

Exceptions

- Reduce FT by 1 if SFF > 90% (e.g. use diagnostics to achieve this). OR
- Reduce FT by 1 if the instrument is 'Proven in Use' AND has a limited configuration facility AND is password protected AND SIL is below 4
- Increase FT by 1 if SFF < 60% (normally applies to a non-fail-safe device, to be avoided anyway).

We should note the distinction between simple (non-PES) and complex (PES) devices. This places a penalty on instruments where there is less confidence about the failure modes due to the use of PES.

The following table indicates a greater requirement for redundancy where PES devices such as smart transmitters are used but exceptions apply that effectively to allow us to use most types of smart transmitters just as if they were conventional analog transmitters.

For the meaning of 'Proven in Use' please refer to the next subsection on selection of instruments.

Hence when considering the choice of sensor, bear in mind that if it does not meet the best credentials it may be necessary to install redundant units to meet the SIL targets.

Constraints applied by SIL target failure rates

If the device on its own has a poor PFD figure (i.e. relatively high compared to the SIL target value) it may be necessary to use a redundant 1oo2 or 1oo3 configuration to meet the target.

Constraints applied by spurious trip target rates

The spurious trip rate of a 1oo2 pair can often be ten times the dangerous failure rate. If this is too costly for production losses the logical thing to do is install a 2oo3 configuration. These configurations are very commonly seen in process plant applications, but we must always keep in mind the extent of common cause failure potential or systematic design errors such as wrong calibration of all three transmitters. Reduction of common cause requires more diversity.

Redundancy summary

Architect	Features
1oo1	Basic single channel for SIL- and SIL-2 with simple non-PES devices. SIL-2 is difficult
1oo1D	Diagnostics applied to improve safe failure fraction. Allows PES devices to be used in single channel if SFF > 90%
1oo2	Basic dual channel for SIL-1, SIL-2 and SIL-3 with simple non-PES devices
1oo2D	Able to tolerate 1 fault and revert to 1oo1D during repair (SIL-2). Allows PES devices to meet SIL-3 if safe failure fraction exceeds 90%. Does not satisfy diversity for SIL-3 if sensors are identical. Reduces spurious trip rate, good alternative to 2oo3
2oo3	Alternative to 1oo2 and provides much improved protection against spurious trips. PES devices do not satisfy SIL-3 unless SFF > 90%

The advantages of diagnostics become clear in this table. If the SFF can be increased above 90% by using internal and external (i.e. in the logic solver) diagnostics, the redundancy rules allow one less level of redundancy. In particular the 1oo2D configuration used with sensors means that spurious trips can be avoided by allowing the system to degrade to 1oo1D when a defect is reported in one of the sensors. The defective sensor is then repaired whilst the plant continues to operate under increased surveillance.

8.4.7   Common cause failures

In many cases the gains in reliability achieved through redundancy will be limited by common cause failures unless we take special measures to reduce them. There is always the danger that all our sensors or all our valves will suffer the same defect at the same time.

ISA S84 and its technical report Tr 84 part 2 have valuable guidance on common cause failures.

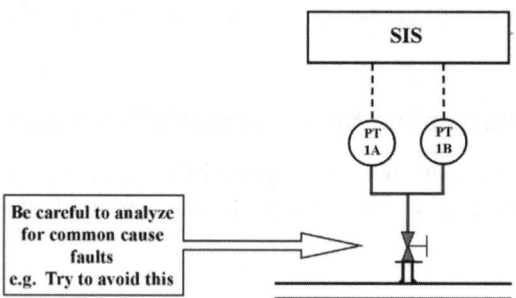

Figure 8.20
Example of potential common cause failure

Some specific examples of common cause failures and systematic failures are as follows:

- Shared process connection (example shown in Figure 8.20)
- Process chemistry disables safety function of final element (valve plugs or valve corroded)

- Valve leaks due to corrosion AND this leak is not detected by mechanical integrity inspection
- Solenoid valve fails due to incorrect installation AND this is not detected by testing
- Solenoid vent port is plugged by dirt daubers or plugged by insulation AND is not detected by testing
- User application logic errors
- Poor communication of the SRS functional specification to SIS designer and installer
- Transmitter calibrated incorrectly (wrong specification, bad calibration standard, technician makes a mistake)
- Wrong specification device (transmitter, solenoid valve, shutdown valve, etc.) installed.

8.4.8   Diversity

Diversity means using different types of sensors and actuators or different measurement and operating parameters to achieve the same result in a safety system. The objective is to eliminate or minimize the possibilities of common cause failures or systematic errors. Figure 8.21 shows a typical example where the same limiting condition, in this case pressure in an ammonia storage vessel, can be detected by pressure and by temperature (which determines the saturation vapor pressure of the ammonia).

Example of diversity

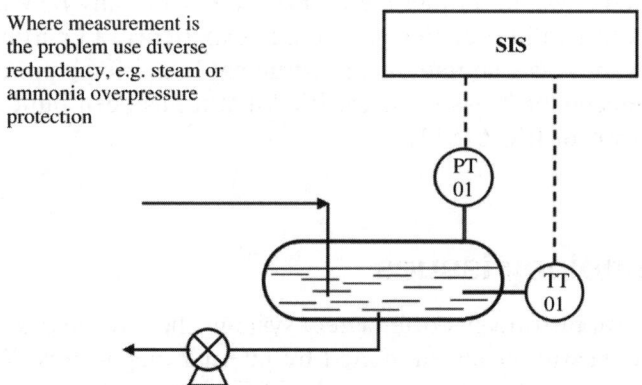

Figure 8.21
Diverse redundancy in sensors

As SIL values rise so does the need for diversity in the protection systems. At SIL-3 the IEC standards expect us to provide redundant and diverse safety instrumented systems. At SIL-4 design teams go to great lengths to avoid the risk of common cause failures by having redundant and diverse sensing systems connected to redundant and diverse logic solvers.

8.5 IEC 61508 requirements for field devices

Firstly we must remember that whatever instrument we use it must be qualified for use in our SIS.

When introducing the paragraphs on selection of components IEC 61511 has this clause:

'11.5.2.1 All components and subsystems necessary to achieve a safety instrumented function shall either be designed in accordance with IEC61508-2 and IEC 61508-3, as appropriate, or the requirements for a component to be considered as proven-in-use shall be satisfied (see 11.5.3).'

This basically offers us the following choices:

- Justifying that the instrument meets IEC 61508 essential characteristics (not too difficult if it is a simple device). Or
- Having a certificate of compliance from a testing authority (e.g. TUV: SIL-3). Or
- Justifying the instrument as 'proven in use'.

8.5.1 Proven in use

What does 'proven in use' mean? Here is the definition found in IEC 61511:

A component may be considered as proven-in-use when a documented assessment has shown that there is appropriate evidence, based on the previous use of the component, that the component is suitable for use in a safety instrumented system

IEC 61511-1 describes the requirements in some detail in 10 sub-clauses. The assessment can be achieved provided the user company can find the evidence of successful application and can show that the supplier's QA system will deliver repeat versions of the originals.

All of the above is intended to make sure that users do not use 'just any old instrument' in a safety system. In fact the problem lies with 'any new instrument' since it is difficult to establish valid previous operational experience. In particular we need to be aware that if we want to use an intelligent instrument in a safety application it has to meet the above requirements or it has to be certified that it has been built and tested in accordance with IEC 61508 or IEC 61511.

8.6 Technology issues

Intelligent instruments offer safety systems the advantages of being able to perform better quality measurements supported by internal diagnostics. We have seen how self-testing and reporting will help increase the SFF of a field device.

There are disadvantages for safety systems. Firstly there are the general reservations about the risks of programmable systems in safety applications. These relate to:

- Potential for systematic errors in the software
- User configurations may create new untested versions of the instrument
- Unauthorized in-service changes to settings, zero, range, mode, etc.

One of the main purposes of IEC 61508 was to address these types of issues and find ways of dealing with them. Hence with the aid of safeguards based on IEC 61508 it becomes possible to use intelligent instruments in a safety system provided we stick to the rules.

In brief the answers to the above possible problems are:

- Instruments using PES should be manufactured using hardware and software engineering procedures in accordance with IEC 61508.
- Limited software instructions are made available to the end-user to program the instrument within a tested range of configurations.
- The program of the instruments is password protected.

We have seen that unless the above requirements are met, the fault tolerant rating of the instrument is loaded down by comparison with a non-PES version.

8.6.1 Example of safety critical transmitter

Figure 8.22 shows a schematic from the Moore Safety Critical Transmitter. Note how a redundant and diverse measuring system is used to compare the internal values of the pressure sensor. Diagnostics driven by the PES section will shutdown this transmitter into a fail-safe state if a defect is found.

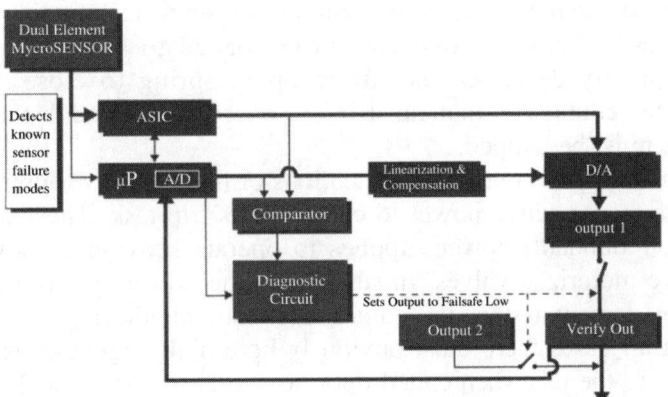

Figure 8.22
Safety critical transmitter

The hardware and software engineering of the transmitter system has been certified by TUV to be in compliance with IEC 61508.

8.6.2 Benefits of a certified safety critical transmitter

The benefits given below are those that we identified earlier in this Chapter based on the high SFF and limited access by users to the software:

- Internal diagnostics with high coverage factor
- Very low PFD_{avg} values. Saves on proof testing, etc.
- Certified for single use in SIL-2 (instead of dual channel)
- Certified for dual redundant use in SIL-3 (instead of 1oo3).

8.7   Guidelines for final elements

The devices used by the SIS to actuate the protective function are called final elements and are often called actuators (see Figure 8.23).

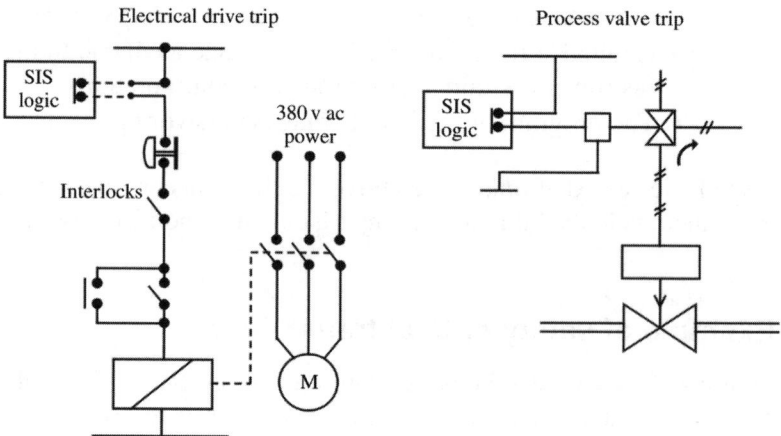

Figure 8.23
Final control elements

Final elements are most commonly seen as process valves with air or hydraulic power to actuators that are spring-loaded to close on release of the actuator fluid. (Typically described as 'air to open, spring to close'.) Or they are simply motor starter contactors that must be de-energized to break the power supply to the drives that must be tripped.

In large power control applications or in large valve applications the final elements may have to use active power to carry out its trip task. These applications require backed-up (i.e. redundant) power supplies to operate heavy-duty power isolating contactors or to drive motorized valves. In such cases the power system becomes an integral part of the final element and would require diagnostic monitoring.

In all cases there must obviously be a high degree of assurance that the final element will do the job when called upon to act. Hence there will be an emphasis on diagnostics and regular proof testing.

8.7.1   Separation rules: final elements

For shutdown valves, it is acceptable for SIL-1 systems to use an existing process control valve to trip the process but only if the hazard analysis shows that faults in this valve are unlikely to place a demand on the SIS function.

IEC 61511 specifically calls for the trip solenoid to be arranged as shown in Figure 8.24 where the SIS trips the basic control valve. The arrangement ensures that positioner defects do not prevent the trip action from being executed by the solenoid.

For SIL-2 applications the preferred method to achieve separation is to install a separate shutdown valve specifically for shut-off or venting duties. A ball or gate valve often meets this duty.

The redundancy requirements of SIL-2 and even SIL-3 can often be met by also tripping the control valve, leading to the arrangement shown in Figure 8.25. This arrangement is also potentially flawed due to the possibility that hazardous eventscaused by faults in the control valve.

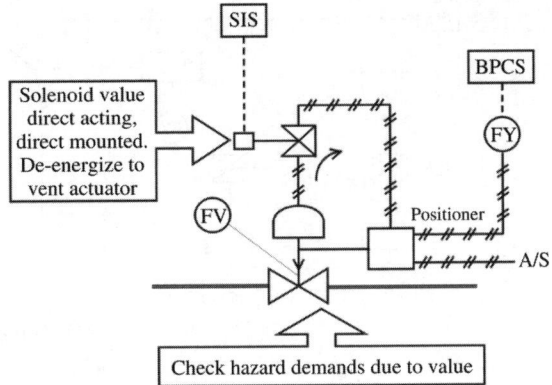

Figure 8.24
Tripping arrangement for share control valve

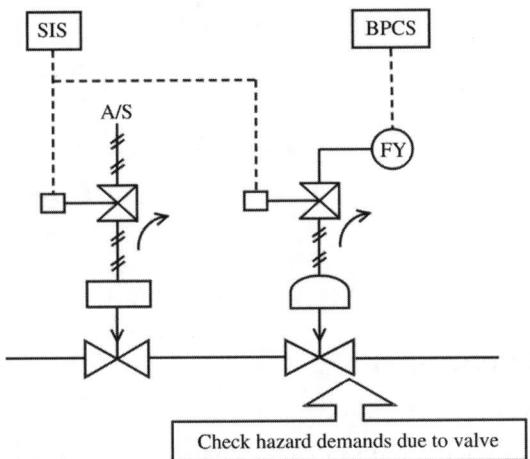

Figure 8.25
Diverse separation of control and shutdown valves

When this situation occurs the choice lies between fitting another redundant valve and demonstrating that the design is still adequate to meet the SIL targets. We shall look at how self-testing devices may help with this problem later. For the moment we can show how the performance of the above valve arrangements can be evaluated. In most cases the decision on how to arrange the final elements will have a major cost and performance implication. This is an area where careful application of fault tree analysis can be very helpful. Here is an example.

8.7.2   Evaluation of an SIS output stage

This is an exercise in evaluating the safety integrity of a proposed design for a trip system output stage where options have to be considered for a control valve to be used as a trip valve or as a redundant back-up to a trip valve.

Option 1   Find the trip failure rate for a SIS designed as per Figure 8.26 where V1 is trip valve and V2 is a control valve with the trip command forcing it to close. Do this by constructing a fault tree analysis diagram. See guidelines below in the starting information.

Option 2   Deduce the failure rate if V1 is removed from the design.

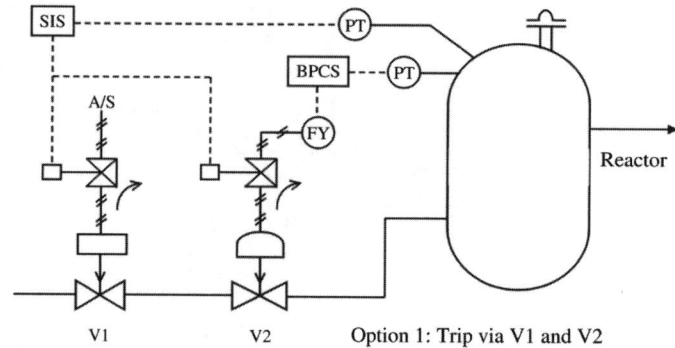

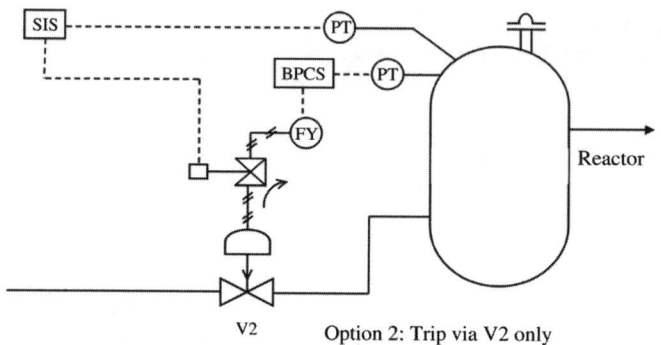

Figure 8.26
Options 1 and 2

Starting information

The data given is as follows:

- λdd is the fail to danger rate of the combined solenoid valve, actuator and valve for V1 or V2. There are no undetectable failures
- The proof test interval is Ti = 0.5 year
- λdd for V1 = 0.2 per year, PFD_{avg} for V1 operating alone = 0.05
- λdd for V2 = 0.2 per year, PFD_{avg} for V2 operating alone = 0.05
- PFD for V1 and V2 in 1oo2 pair. PFD_{avg} = 0.008 (based on common cause failure factor, β = 10% and manual proof test only, Ti = 0.5 year)
- Trip demand from process and control excluding failures of V2 is estimated to be 1 per year
- The combined PFD_{avg} for the SIS sensors and logic solver is PFD_{avg} = 0.1.

Method

- Draw a fault tree for option 1 starting with top event = 'hazardous event'
- Split the causes of failure into two branches
- V1 and V2 both fail on demand from process
- V1 fails on demand from trip demand caused by failure of V2
- Develop both branches downwards to show the fault logic
- Insert the faults per year and PFD data on the fault tree and derive the trip failure rate for the top event

- Revise the diagram and figures to show option 2 and option 3
- Compare the results and comment on the value of installing extra valves.

Information only reliability formulae are not covered in the scope of this book.
Equations used for PFDs
for single valve:

$$PFD_{avg} = \lambda dd \cdot Ti/2$$

for V1 + V2 in 1oo2 operation:

$$PFD_{avg} = \{1 - \beta\} \{ (\lambda dd)^2 \cdot (Ti)^2 \}/3 + \beta \cdot \lambda dd \cdot Ti/2$$

The solution for Option 1 looks as shown in Figure 8.27.

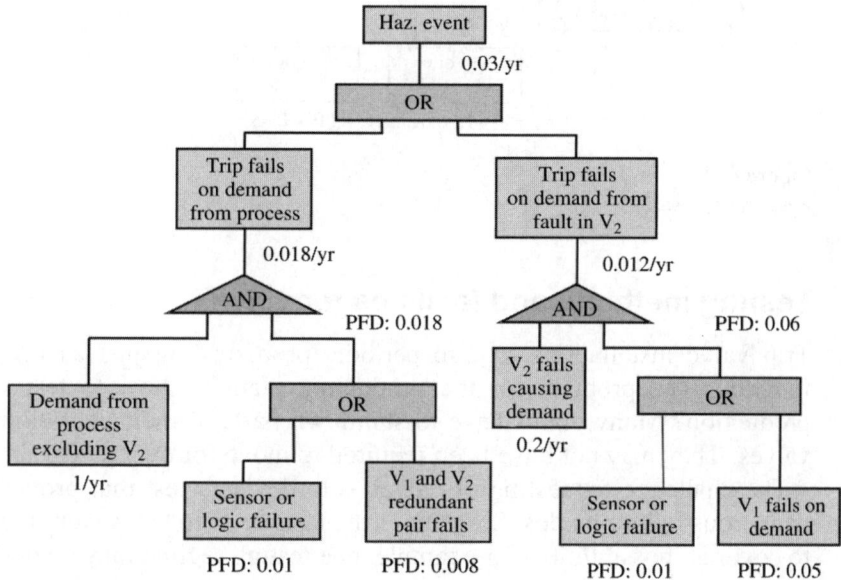

Figure 8.27
Fault tree Option 1

In this fault tree model there are two failure scenarios:

1. The pair of valves has failed when the process runaway occurs.
2. Valve V2 fails open causing a pressure runaway and the SIS fails to respond due to a sensor, logic or V1 fault.

The results added together predict an accident rate of 0.03 per year.
The solution for Option 2 looks as shown in Figure 8.28.
In Option 2 there is no independent valve to protect the process if V2 is the cause of overpressure. The result is a very high-predicted accident rate.

8.7.3   Diversity in ESDVs

In shut-off valves diversity can involve installing two different types on the same line. For example a gate valve and a ball valve will both provide shut-off but it is unlikely that both valves would stick open during the same period. Similarly it is unlikely that both have sealing or seating failures at the same time. However, IEC 61511 says you should not use diversity if it means installing an unfamiliar or untried valve for the shut-off duty.

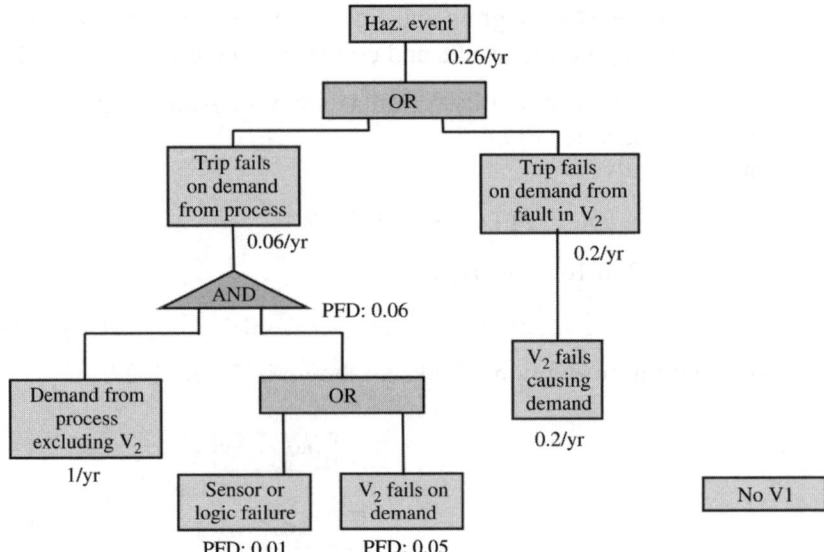

Figure 8.28
Fault tree Option 2

8.7.4   Testing methods and facilities for valves

Trip valve installations require periodic proof testing just as for the rest of each SIS function. The problem for the production plant is: how to test valves without losing production? Many plants have to shutdown parts of their operation to test the shutdown valves. They may not have been required to move for over a year or more.

The challenge for testing of valves is to devise a test that properly covers all possible dangerous failure modes. Sometimes the tests have to be broken down into different types to cover all possibilities, for example, one testing regime may be defined as follows:

- Partial closure tests at short intervals to prove that the valve is not stuck and that the solenoid will work
- Full closure test online with bypass valve open to prove full movement
- Full shut-off test against line pressure at the time of plant shutdown to prove the valve has the ability to close against line pressure
- Maintenance check for leakage to be done during plant shutdown.

Unless a plant has short production runs the cost of testing may have to include the cost of plant downtime whilst the full closure tests are being done. This can be a major cost factor unless the test period can legitimately be extended to, say, once every 2 years.

8.7.5   Bypass arrangements

Figure 8.29 illustrates a permanent bypass arrangement for testing a shutdown valve online. Note the bypass warning 'ZA' alarm. This arrangement is used as a standard feature in some companies; others will avoid this as being risky, inadequate for proving or adding too much cost. It offers the advantage of avoiding downtime costs for scheduled proof testing.

Always provide alarms or alerting procedures for the operator to know there is a bypass in force.

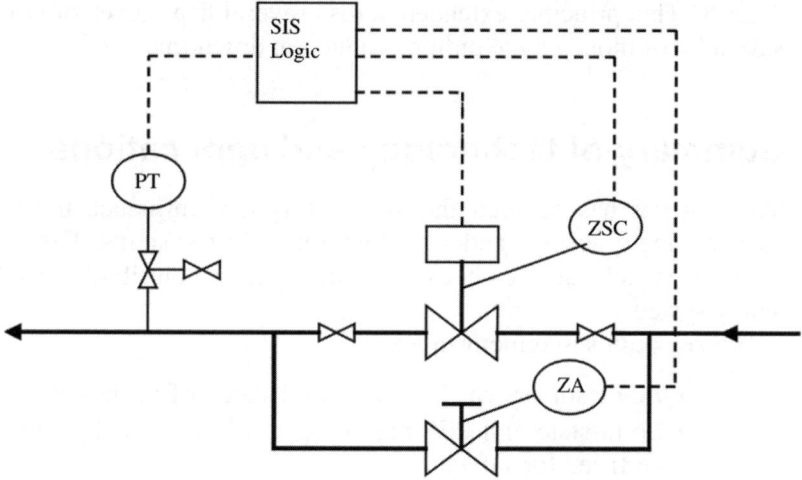

Figure 8.29
Bypass arrangements

8.7.6 Using smart valve controllers and diagnostics to save cost of testing

There is a strong incentive to find diagnostic methods for ESDVs that can reduce the frequency of full closure tests. Recently manufacturers have been able to offer the safety certified smart positioner or digital valve controller that is capable of performing automatic partial closure testing on demand. Figure 8.30 illustrates the method.

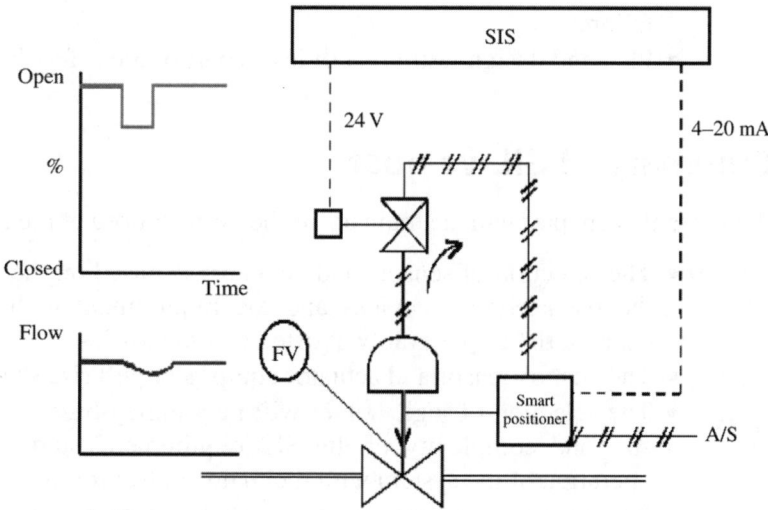

Figure 8.30
Partial closure testing using 'smart' valve controllers

Usually a partial closure test can be done safely whilst the valve is in service without disturbing the process. If this is done say once per week by simply pushing a button, the need for a full closure test can be reduced from typically once per year to once every

3 years. This principle extended across several trip valves on a plant has the potential to save a lot of money by avoiding production downtime.

8.8   Summary of technology and applications

This chapter has covered the key factors that influence the cost and arrangement of sensors, logic solvers and actuators for safety systems. Great care is required in the application field devices because of their major contribution to the potential failure of a safety system.

The key points to remember are:

- Logic solvers can be relay, solid state or PES based
- Solid-state and PES-based logic solvers (usually called safety PLCs) must be certified for safety
- Certification of PES logic solvers includes hardware, software and communication interfaces
- PES logic solvers bring the benefits of diagnostic abilities for themselves and the sensors
- Instruments must be well proven and have well-defined failure modes
- Instruments should be properly qualified for safety applications through certification or proven-in-use records
- Intelligent instruments are treated as PES systems under IEC 61508
- Follow well-established rules of Separation, Redundancy and Diversity for sensors and valves
- Strive to obtain good diagnostic coverage through logic solver power or self-testing schemes
- Sensor and actuator subsystems require careful analysis of common cause failures
- Plan and design testing facilities as part of the safety life cycle activities.

8.9   Summary of SIL vs cost

We close this chapter with a summary of the points raised at the start:

- The selection of sensors and actuators is based on the need for separation from the basic control devices and the requirement to have dependable fail-safe characteristics preferably assisted by diagnostics.
- The cost of sensor and actuator equipment will rise sharply with SIL ratings.
- The selection of logic solver will be a policy-based decision depending on the size and complexity of the SIS requirements and the need to integrate the operation of the system with the main control room facilities.
- Engineering cost rises steeply with SIL rating due to the increased levels of verification and validation required by IEC 61508.
- Maintenance costs will be higher with high SIL ratings because of the greater attention to detail required to ensure no possible errors have crept into testing and record-keeping. Also because high SIL requires redundant and diverse sensors and actuators, all of which have to be proof tested.
- Production costs through spurious trips need not be affected by high SIL values. Whilst the fail-safe rate of dual redundant subsystem doubles over the

single system these problems can be overcome by 2oo3 voting systems or by 1oo2D voting in the logic solver. For a modest investment in sensors the spurious trip rate can be reduced to a very low value.

- Major cost benefits can be achieved by making sure that SIL targets are not set unnecessarily high.
- If the SIL requirements are systematically defined at the time of the hazard studies there is a better chance of getting the SIL targets agreed at realistic levels.

9

Exercise in specifying an SIS from the Hazop

Objective

To consolidate the methods described earlier in this book for deriving the safety systems from the Hazop.

9.1 Introduction

This chapter is an exercise in trying to convert Hazop study recommendations into a safety system requirements specification. The idea is to test out some of the methods described in the book. This is not a comprehensive exercise but it is intended to highlight some of the thinking that has to go into the development of a safety system from the time that the Hazop identifies a problem.

This exercise is directed at the key points we have been following in the book:

- Hazop reports the identification of a hazard and calls for additional safeguards.
- Hazard analysis determines the target SIL of the safety function.
- The design team defines safety function in alignment with the Hazop.

The product of the exercise is a safety requirements specification. In our exercise we will simplify the specification a bit for brevity.

9.2 Process description

We use a hypothetical case study that is similar to a typical problem seen in polymer production by a batch process.

Figure 9.1 shows an autoclave used to carry out polymerization batch operations very approximately as used in the manufacture of PVC. In this example we are going to suppose there are four large batch autoclaves in the whole plant. The batches run in staggered phases.

The autoclave is filled with monomer in a water mix and a catalyst is added. Heating is applied by the TIC control loop until reaction starts to occur and the process becomes exothermic. The TIC loop automatically applies cooling as the temperature approaches the set point and stabilizes the temperature of the contents over a period of several hours as the contents convert to polymer.

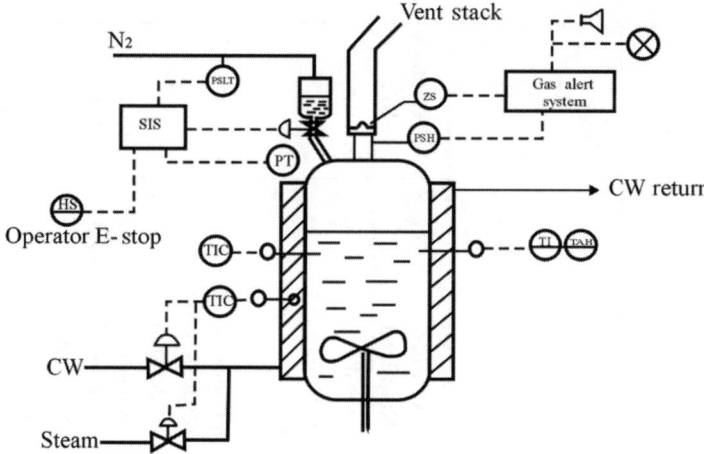

Figure 9.1
Autoclave arrangement for batch polymer production

The risk of a runaway reaction occurs due to failures in the cooling system, a stoppage of the stirrer and faults in the control system or in chemical composition. When this occurs a high temperature alarm is used to warn the operator that the reaction is going too fast. Operator is expected to respond with manual override actions on the cooling system to try to recover control. Figure 9.2 illustrates the stages of reaction runaway that may occur if the alarm and operator action fail to solve the problem.

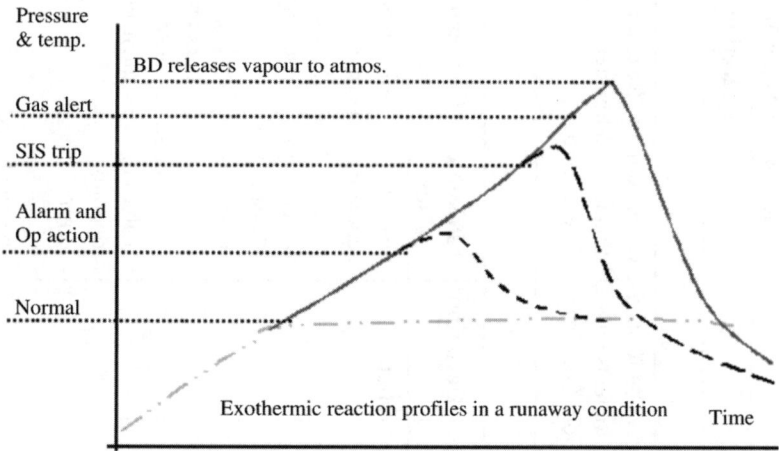

Figure 9.2
Temperature and pressure profiles as reaction runaway develops

If the reaction still gets away the pressure will rise rapidly in the autoclave and would lead to a vessel rupture with the risk of injuries and fatalities as well as the release of a large toxic vapor cloud and possibly a flammable vapor. The primary safeguard is a large aperture bursting disk that will be sized to ensure rapid discharge of the vapors to atmosphere via a vent pipe.

The Hazop study has recognized two additional layers of protection proposed by the design team and so has recorded the position in the Hazop worksheets as follows:

Hazop study title: Autoclave batch operations, example only		Sheet: 1
Drawing no.: IDC diagram 1	**Rev no.:**	**Date: 18 Nov 2002**
Team composition		**Meeting date:**
Part considered: Autoclave unit		
Design intent	Material: VCM /Catalyst/Water Activity: Reaction stage after addition of catalyst Source: Contents only Destination: No transfer Description: See Chapter 9 description	

No.	Guideword	Element	Deviation	Possible Causes	Consequences	Safeguards	Comments	Actions Required	No.	Action by:
A 1	Temp.	Autoclave	High	Loss of coolant	High reaction rate, pressure rises rapidly	Low coolant pressure alarm	Limited options to increase cooling	Request stopped alarm	A1.1	Inst design
				Stirrer stops causing loss of heat tfr	See high pressure conse-quences	High temp. alarm	Rapidly leads to runaway and overpressure event	Operator responses to be defined. Consider situation due to common failure of CW	A1.2	Process eng.
				Control fault, valves or operator error		Emergency power for stirrer			A1.3	Hazan study team
				Chemical mix incorrect		Dual flow meters	NIS is the primary safeguard and should be used in response to alarms	Finalize requirements for automatic trip of NIS	A1.4	Hazan study team
						Neutralizer injection system (NIS)				

Hazop study title: Autoclave batch operations, example only		Sheet: 2
Drawing no.: IDC diagram 1	Rev no.:	Date: 18 Nov 2002
Team composition		Meeting Date: 01 Nov 2002
Part considered: Autoclave unit		
Design intent	Material: VCM /Catalyst/Water Activity: Reaction stage after addition of catalyst Source: Contents only destination: No transfer Description: See Chapter 9 description	

No.	Guideword	Element	Deviation	Possible Causes	Consequences	Safeguards	Comments	Actions Requiered	No.	Action by:
A 2	Pressure	Autoclave	High	Runaway reaction caused by high tempera-ture For causes see 'High Temp.'	1. Vessel overpressure, Rupture of vessel with injuries and possible fatalities LOC with toxic vapor cloud 2. With BD safeguard installed the consequence of BD relief is a hazard of toxic exposure for persons in the area	1. Bursting disk relief to vent stack 2. NIS trip on high pressure 3. Gas alert with evcuation to safe room	Worst case is 3 autoclaves in reaction with CW failure Safeguards are acceptable subject to hazard analysis confirming risk targets can be met	Hazan to report on event frequencies for vessel rupture SIL rating for NIS	A2-1	Hazard study team

The Hazop worksheets identify the major hazard of vessel rupture with resulting loss of containment and recognize the safeguard of the bursting disk and vent stack. The second level of hazard is toxic exposure if the bursting disk is forced to release the autoclave contents to atmosphere.

This analysis requires that the safeguards meet two risk targets. The overall safety requirements for the vessel rupture event are likely to be defined as a very low frequency for the event. A higher frequency can be tolerated for the toxic exposure even though this may cause some injuries to persons.

Figure 9.3 is a simplified risk reduction diagram for the situation where all the safeguards are in place.

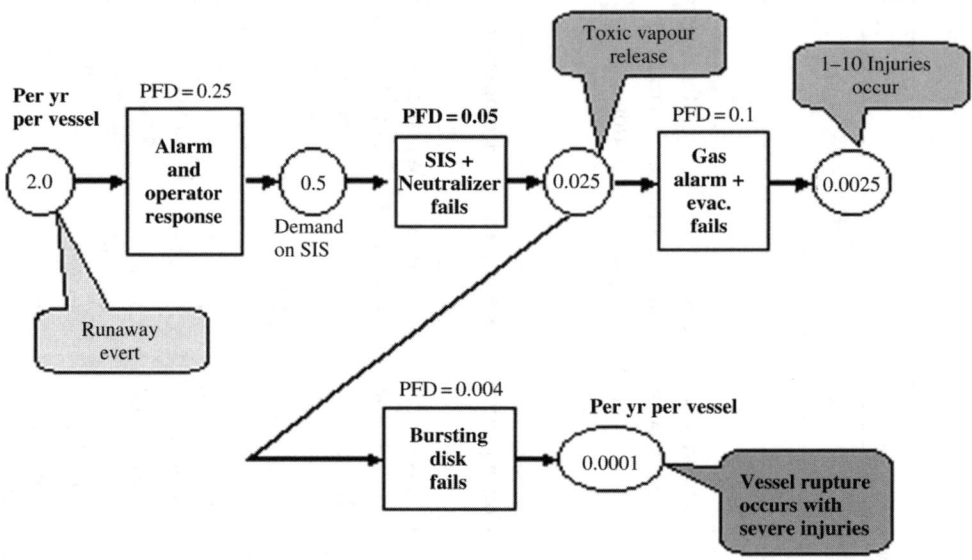

Figure 9.3
Risk reduction diagram for a single autoclave runaway

This diagram has captured the information produced by the hazard analysis team. It indicates that a runaway event is expected on the average twice per year for each autoclave. The first level of response is credited with a low success rate (PFD = 0.25) leading to a demand of 0.5 times per year on the safety system operating the neutralizer injection system. If this fails to stop the runaway reaction the bursting disk will operate as pressure rises and there will be vapor release via the vent, but the vessel will be safe. Hence, the exposure risk to persons can be seen on the diagram.

If the bursting disk should fail to relieve the pressure, the vessel rupture event will occur as shown. The failure probability of the disk and vent has been shown as $PFD_{avg} = 0.004$ (1 in 250). The diagram predicts the vessel rupture rate at 0.00005 or 1 per 10 000 years.

Finally we have to allow for the fact that there are four autoclaves on this plant. This means that the event frequencies for the plant will have to be multiplied by 4. The predicted rate for the vessel rupture event then becomes 1 per 2500 years and if we assume 1 potential fatality for this event this will determine the risk to the operators.

If the bursting disk works and if the gas alarm system works with persons retreating to a refuge we would have to estimate the number of persons likely to be injured due to still being in the way of vapors. If it is 1 person we see that the injury rate becomes 0.025 per year or one person injured per 40 years. Failure of the gas alarm would increase this figure to possibly 10 persons for the site or average 1 person per 4 years.

9.2.1   Determination of the SIL for the neutralizer trip

Risk ranking tables or profiles can be used to decide if these figures will be tolerable if all safety systems are installed. We are going to assume they are acceptable in this exercise. In which case our diagram defines the risk reduction allocations and sets the PFD for the neutralizer functional safety system at 0.02. This figure suggests that neutralizer injection system can be an SIL-1 system. The problem for the designers is that there are two failure contributions: The SIS portion and the chemistry portion.

$$SIS + Chem = 0.02$$

If the neutralizer chemistry has a $PFD_{avg} = 0.01$, the PFD for the SIS will be: $0.05 - 0.01 = 0.04$. This allows to SIS portion to still qualify as SIL-1.

Using the alternative approach to SIL determination we could try the risk graph. The overall safety system when run on a risk graph indicates SIL-3 for the total protection requirement. With the bursting disk in place the risk graph for the neutralizer function also indicates SIL-1.

9.3   Safety requirements specifications

9.3.1   Overall safety requirements

We will not detail the overall safety requirements specification here but it will comprise a statement saying that the risk of serious injuries or a fatality has been predicted to be as high as 2 events per year for the whole plant if no safety measures are in place. The target of 1 vessel rupture event per 2500 years is considered to be tolerable and as low as reasonably practicable. Therefore an overall risk reduction of 5000:1 is required.

This is to be achieved by a combination of two layers of protection:

- Neutralizer injection to reduce the frequency of runaways
- Bursting disk to relieve the pressure if runaway occurs.

In addition, the risk of gas exposure by persons on the plant is to be reduced by providing an alarm system with an evacuation procedure leading to a toxic gas refuge.

9.3.2   EUC risk

In the risk reduction diagram we see that there is an alarm for high temperature in the autoclave and also alarms warning of loss of cooling water pressure. The Hazop has detected that if the stirrer stops, there is also a risk of runaway. It has called for an alarm if the stirrer stops. The alarms and the possible reactions of the operator have been credited with a risk reduction factor of 4 (PFD = 0.25). These alarm functions make a contribution to reducing the demand on the protection systems, but because the RRF is below 10 it is likely that these will not be treated as safety-related and they can be handled in the main plant control system.

9.3.3   Safety allocations

Risk reduction allocations will be at least 250:1 for the bursting disk and 20:1 for the neutralizer system incorporating a SIS. The risk reduction allocation for the gas alarm system will 10:1.

9.3.4   Protection measures

The conclusion from the safety allocations is that two safety-related instrument functions would be needed for the plant:

1. Each autoclave will need its own SIS function for the neutralizer injection and trip to vent. We have established that a SIL-1 system will be satisfactory.
2. A gas or vapor release alarming system will be needed to receive notice from any of the four autoclaves that it is close to operation of its bursting disk. This signal will be obtained from a pressure switch at the bursting disk. A problem here is to find an arrangement that gives early warning of release but avoids false alarms. However, an important point to note is that the gas alarming system must be fully independent of the neutralizer SIS. This is because it must be an independent layer of protection and must not carry any risk of common cause failure at the same time as the neutralizer fails. The safety integrity rating will need to be SIL-1 to satisfy the RRF of 10.

9.3.5   Safety requirements specification for the SIS + neutralizer

The details of SRS should be developed as soon as the safety allocations are considered to be practicable and realistic. The safety allocations and the risk reduction diagram would possible have been presented to the Hazop study as part of the design intent. Once the Hazop study team has approved the general intent of the safeguards, the SRS for the instrument system can be drawn up. We show overleaf an example of the SRS information that would be set down for the design basis of the SIS.

For simplicity we have left out the general requirements section of the SRS. The example shows the functional and the safety integrity sections only.

SRS No.:	Function No:	Title of Safety Function: Autoclave Neutralizer Trip			
AS-1	A1.1	Prepared by:	DMM	Date of first issue:	1 Nov 2002
Revision number:	Orig.	Revised by:		Date of rev:	
Approvals:		Approved by		Date of approval	

General description of the safety function for this data sheet

(Include here the method of achieving the safe state of the process in response to a demand on the SIS.)

The function causes a neutralizer solution to be injected into the autoclave under nitrogen pressure either when tripped by an operator pushbutton or when autoclave pressure exceeds the trip value. The injection is achieved by venting air supply pressure from valve XV-1, which is spring, loaded to open on loss of air. The contents of the chamber are forced into the autoclave by nitrogen from a high-pressure supply. At the same time a second output of the SIS will trip open a vent valve leading to the scrubber system that can absorb vapors blowing down from the autoclave. The capacity of the blow down is insufficient to handle the maximum delivery of the runaway reaction and the function depends on the neutralizer to stop the reaction.

The function shall have devices to confirm it has been charged with fluid and to prove that it is under pressure sufficient for the injection against maximum design pressure for the autoclave. The function shall indicate if it is not charged or pressurized and this condition shall prevent the filling of the autoclave with monomer by means of an interlock on the monomer feed valve.

Operating modes (Describe the modes of operation of the process (e.g. start up, recycle, normal, stand by shutdown), identify which modes of operation require this SIS function to be active.)

The trip device shall be 'armed' for all operating modes except during the 'charging' mode when it is being filled with neutralizing solution.

Define any requirements for overrides or bypasses of the SIS function.
The function does not require any bypasses. Testing will be done in between batches.

Process measurements or sensor functions as inputs to the SIS

Description	Normal Range	Trip Point
Variable 1: Autoclave pressure	0–25 bar g	12 bar g
Variable 2: N2 charge pressure	20–25 bar g	20 bar g low alarm via switch
Variable 3: Level in chamber	0–100%	50% low-level alarm point

SIS Output actions and actuators: Output1 to injection valve XV-1, output 2 to vent valve XV-2, output 3 to interlock the monomer feed valve XV-3

Notify any dangerous combination of output states: None found

Input/output and sequential trip events to be described here or by diagrams listed below. Include all logic, any math functions and any required permissive.

(1)   Output to injection valve and to autoclave vent valve will energize when trip reset button is operated provided no E-Stop condition is present and if autoclave pressure is at atmospheric.

(2)   Trip available status occurs if chamber pressure is above 20 bars and if level is above 50% and if reset has been successful.

(3)   Injection and vent valves will de-energize and valves will be vented to open when trip occurs. Valves will stay de-energized until reset occurs.

Reference to cause and effect diagram or trip logic diagram	Document Number & Rev No.	Title:
		Trip and reset logic diagram for NIS trip A1.1
Define the safe state of the process to result from the action of this SIS. Consider also any combination of safe states that could create a hazard when occurring concurrently	Safe state is when pressure has reduced to atmospheric after injection and venting to scrubber system Nitrogen will occur in autoclave after injection. Trip should be reset after injection to avoid contamination of chamber	
Assumed sources of demand on the SIS	Autoclave high pressure due to reaction runaway. E-stop action by operator in response to runaway alarms Runaway causes assumed: (1) Loss of coolant, (2) Cooling control fault due to instrument faults, valve or wrong operation of TIC function by operator, (3) Loss of stirrer motion due to power loss or seizure	
Assumed demand rate	0.5 per year arising from 2 process events per year with 25% failure rate for alarm and operator response.	
SIS mode of operation	Low Demand: YES	High demand/ continuous: NO
Intended proof test interval	3 months	
Required response time to bring the process to a safe state	(Not applicable to prevention interlocks) 10 s for injection. Pressure and temperature reduction indicated within 60 s	
Requirements for manual shutdown Define the requirement for resetting the SIS after a shutdown	(Is a manually operated trip button required? Local in the field or in the control room?) Yes. To be provided in the control room (Is auto-resetting allowed?) Manual reset. Auto reset not allowed due to need to ensure vent to scrubber remains open	
SIS Interface requirements for this application	(Record here any application-specific interfacing signals, e.g. list of variables, logical states, states of outputs, alarms and messages. Also record any signals to BPCS and to external control systems.) Detail sheet to be supplied. Critical signal is: Trip status to batch controller to prevent batch starting if trip is not reset and available	

SRS No.:	Function No.:	Title of Safety Function: Autoclave Neutralizer Trip			
AS-1	A1.1	Prepared by:	DMM	Date of 1st issue:	1 Nov. 2002

The required safety integrity level (SIL)	SIL-1
Target failure rate of the safety function (data only applicable to quantitative methods for SIL)	Target PFD$_{avg}$: 0.05

SIS mode of operation: (indicate by 'yes/no') (Low demand is used when demand rate is less than approximately 2 test intervals) (This definition based on IEC 51511-1 clause 3)	Low Demand: YES (IEC 61508-1 table 2)	High demand/ continuous: NO (IEC 61508-1 table 3)

Assumed demand rate: (used for design guidance only)	0.5 demands per year
Assumed proof test interval (to be finalized during detail design)	3 months. The chemical charge in the chamber to be replaced and checked every 3 months

Fail-safe design basis: (mark applicable box)	Energize to trip: NO	De-energize to trip: YES

Action needed to maintain a safe state in response to the detection of a fault in the SIS (e.g. Alarm only or auto-shutdown + alarm, revert to single channel if redundant design is used)	Alarm and cancel trip available status to batch controller
Action in response to power failure at the SIS	Allow outputs to de-energize and accept possible loss of batch
Reliability requirements if spurious trips may be hazardous	No break power unit to be provided. Essential to retain availability of trip during plant power outage
Is it expected that a fault tolerant (i.e. redundant) SIS will be used?	Fault tolerant SIS has been assumed
Maximum planned repair time for a faulty SIS channel. (This repair period is used in calculating the failure rate expected for a redundant channel SIS)	48 h
Requirements to enable proof testing to be performed (These include constraints on operation during proof testing and the facilities to be provided for online/offline testing)	Proof testing to be done after a batch has been completed and removed. No bypasses or overrides to be fitted Signal injection port required for pressure sensor

SRS No.:	Function No.:	Title of Safety Function: Autoclave Neutralizer Trip	
Electro magnetic immunity limits: (generally defined by IEC 61000-1) Add details here if EM noise environment is considered to be higher than typical for a process control environment or if a higher than normal immunity level is required		(Normally, the SIS will be required to have electromagnetic immunity performance that will comply with the testing conditions defined by IEC 61000-1-4. More severe immunity may be needed if the environment is exceptionally severe) No special electrical noise conditions notified	
Environmental conditions required for the SIS: state temperature, humidity, dust and vibration extremes expected: allow for storage, installation, transport, installation, commissioning, and operation		Sensors	0–50 °C No unusual extremes
		Logic solver	Control room-mounted ambient and dust protected conditions
		Actuators	0–50 °C No unusual extremes

This completes the safety requirements specification. The draft shown here is a first attempt and will need to be refined but it serves to indicate the baseline information that is required to enable the SIS design to proceed.

This example has taken us from the Hazop study output and has utilized the quantitative method of hazard analysis to arrive at a reasonable level of risk reduction to be provided by SIS and non-SIS protection measures.

9.4   Conclusion

The safety requirements (SRS) for any protection measure should be specified in a systematic and structured manner. The use of a fixed format document based on the contents list called for in IEC 61508 and 61511 will provide a reliable basis.

The information placed into the SRS should be consistent with the conclusions of the hazard studies and any hazard analysis studies linked to them. The most desirable approach is to arrange for the SRS to be compiled at the time of the hazard study and to ensure that all members of the hazard study team are able to verify its contents. The advantage of doing both jobs at the same time is clearly that the persons who understand the hazard will be available to confirm the performance needs of the protection measures.

Appendix A

References used in the manual

Ref. No.	Title/Subject	Origin
1	Out of Control: Why control systems go wrong and how to prevent failure. ISBN 0-7176-0847-6	UK Health and Safety Executive. HSE Books. www.hse.gov.uk
2	HB142-1999 A basic guide to managing risk using the Australian and New Zealand Risk Management Standard	Standards Association of Australia. PO Box 1055 Strathfield NSW 2135 www.Standards.com.au
3	Tolerable Risk Guidelines	Edward M. Marzal: Principal Engineer, Exida.com. www.exida.com
4	HAZOP and HAZAN by Trevor Kletz 2nd edn 1986	I Chem. Eng Rugby, UK
5	The design of new chemical plants using hazard analysis	I Chem. E Symposium series no. 47
6	Guidelines on a Major Accident Prevention Policy and Safety Management System, as Required by Council Directive 96/82/EC (Seveso II) ISBN 92-828-4664-4, N. Mitchison and S. Porter (Eds)	European Commission – Major Accident Hazards Bureau *It is available as a Free download from* Luxembourg: Office for Official Publications of the European Communities, 1998. www.mahbsrv.jrc.it
7	IEC 61882: Hazard and Operability Studies (Hazop studies) – Application Guide. 1st edn 2001–05	International Electro-Technical Commission, Geneva, Switzerland. Download/purchase from: www.iec.ch
8	Guidelines for Process Hazard Analysis:, Hazards Identification & Risk Analysis by Nigel Hyatt	Dyadem International Ltd Toronto Canada. www.dyadem.com
9	Hazard and Operability Study Manual. AECI Engineering Process Safety	Ishecon, Modderfontein, GP, RSA. D Rademeyer
10	Hazop Guide to Best Practice: by Frank Crawley, Malcom Preston and Brian Tyler. ISBN 0-85295-427-1	Published by: Inst of Chemical Engineers, Rugby, UK www.icheme.org.uk

Ref. No.	Title/Subject	Origin
11	Alarm systems: A guide to design, management and procurement. EEMUA Publication No. 191. 1999	Engineering Equipment and Materials Users Association
12	IEC 61508 Functional safety of E/E/PES systems. Parts 1 to 7	International Electro-Technical Commission, Geneva, Switzerland. Download/purchase from: www.iec.ch
13	IEC 61511 Safety instrumented systems for the process industry sector (available soon)	www.iec.ch when published.
14	ANSI/ISA –S84.01 Application of safety instrumented systems for the process industries	isa.org
15	DEF 00-55 Hazop studies on systems containing programmable electronics	UK Defence dept. Free download: www.dstan.mod
16	Computer Hazard and Operability Study or 'Chazop'. Benefits and Applications. Raghu Raman and Steve Sylvester	Halliburton (Australia) Pty Limited t/a Granherne Bridgeport House, Level 1, 5-7 Havilah Street, Chatswood NSW 2067, AUSTRALIA raghu.raman@halliburton.com

Suggested reading

Item No.	Title/Subject	Origin/Author
1	IEC 60812 Analysis techniques for system reliability-Procedure for failure modes and effects analysis (FMEA)	www.iec.ch
2	Reliability Maintainability and Risk, 6th edition 2001	Butterworth-Heinemann / Dr David J Smith
3	Safety Shutdown Systems: Design, Analysis and Justification, 1998	Paul Gruhn and Harry Cheddie /ISA
4	Guidelines for Safe Automation of Chemical Processes, AICE 1993	Centre for Chemical Process Safety of the American Institute of Chemical Engineers
5	Guidelines for Hazard Evaluation Procedures, AICE 1999	CCPS as above
6	MIL-STD-1629A Procedures for performing a failure modes, effects and criticality analysis	US Dept of Defense (1980)
7	Loss Prevention in the Process industries. Vols 1 and 2	F P Lees/ Butterworth-Heinemann London, 2nd edn

Appendix B

Some websites for safety systems information

The following websites may be of interest for those searching for more information or support for process safety-related projects. The list is only a sample of the range of facilities available.

Subject	Website	Comment
American Institute of Chemical. Engineers Center for Chemical Process Safety	http://www.aiche.org/ccps	Center for Chemical Process Safety. Information source for chemical safety data and practices. Training courses in plant safety
Asset Integrity Management	http://www.assetintegrity.co.uk	Located in Scotland, UK, this company provides software products and technical services for safety instrumentation in chemicals, oils and gas. An alarm management package is available to assess alarm criticality. See also Appendix D
Australian Standards	http//:www.standards.com.au	Comprehensive service with downloads of many national standards for safety. Includes AS 61508 parts 1 to 7 identical to IEC 61508 and less expensive
Conformity assessment (CASS)	http://www.siraservices.com	UK research and services organization for instrumentation. IEC 61508 conformity assessment training and services
Dyadem	http://www.dyadem.com	Canada-based PHA consulting. Software products for PHA and risk analysis including Hazops and FMEA

Subject	Website	Comment
European Process Safety Centre	http://www.epsc.org.uk	Information and publications on process safety. Source of EPSC Guide to Hazop Studies published by I Chem. E
EU Directive Information	http://europa.eu.int/comm/ environment/seveso/ index.htm#1	EU center for information on Chemical Accident Preparedness Prevention and Response. Provides details on the implementation and history of Seveso 2 directive
Exida	http://www.exida.com	US Consulting/engineering group. Manuals and study courses for Certified Functional Safety Expert. Provides SIL calculation software tools with reliability databases. Informative newsletter
Factory Mutual	http://www.fmglobal.com	Safety equipment testing laboratories and insurers. Certification body
Health and Safety Exec. (UK)	http://www.hse.gov.uk/sources/ index	Major contributor to occupational safety technologies and practices. Legally appointed body for safety supervision in UK industry. See index for vast range of data sheets, books and guides
HSE Power and Control Newsletter	http://www.hse.gov.uk/dst/ sctdir	HSE safety specialists provide informative newsletters
IEC. International Electro-technical Commission	http://www.iec.ch/home	Develops international standards in all areas of electronics, communication, consumer products, safety instrumentation, etc. Bookstore for IEC standards
Instrument Society of America	http://www.isa.org	Bookstore for ISA S84.01. Specialist section on safety systems
Jenbul Consultancy	http://www.jenbul.co.uk	Management and leadership of Hazop studies. Risk assessments, etc. Site safety management services
OHSA _USA	http://www.osha-slc.gov http://www.osha.gov/oshastats.gov	US Dept of Labour, Occupational Safety and Health Administration. Vast database of accident reports, regulatory information and accident statistics
Simmons Associates	http://www.tony-s.co.uk	UK consultancy for software management and instrumentation safety practices. Downloads of technical briefs on safety subjects. Support on IEC 61508 compliance and documentation

Subject	Website	Comment
TUV Services in Functional Safety	http://www.Tuv-global.com/sersfsafety	Based in Germany and USA with offices in many countries. Certification laboratories for safety system devices and PLCs. Find list of PES certifications Papers on certification, etc. Details of TUV Certified Functional Safety Expert qualification/exam
UK Defense Standards	http://www.dstan.mod.uk	Free standards on download. See computer hazops standard
UK Institute of Electrical Engineers	http://www.iee.org/Policy/Areas/SCC/index.cfm	Safety competency and commitment website. See also competency newsletter
UK Institute of Chemical Engineers	http://www.iche.co.uk	Publications on Hazop studies
Oil and Gas ESD	http://www.oilandgas.org	Safety code of practice
US Chemical safety board	http://www.csb.org	Accident reporting and evaluation service. Statistics and reports

Appendix C

Notes on national regulations relevant to hazard study and safety management

Key USA regulations for process safety

The USA has two major regulations for the control of process safety under the umbrella of the OHS Act. These are shown in Figure C.1.

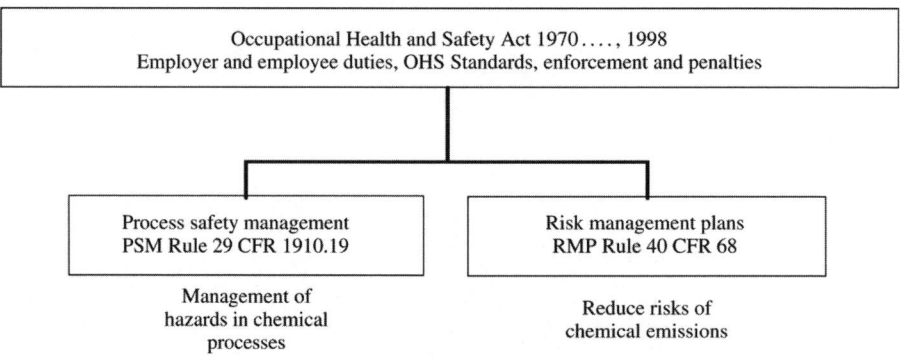

Figure C.1
USA Health and Safety Regulations

Process safety management (29 CFR 1910.19)

This regulation, known as 'the PSM rule' was introduced to deal specifically with the management of hazards in chemical processes. Some 30 000 process plants in the USA are subject to requirements of this rule.

Its purposes are:

- To prevent or minimize the consequences of catastrophic release of toxic, reactive, flammable or explosive chemicals
- Such releases may result in toxic, fire or explosion hazards.

With regard to process hazard studies (PHA) OHSA recognizes their value to achieve the following purposes:

- To identify and analyze the hazards of the process to ensure that safeguards are adequate.
- To provide documentation on the hazards of the process, so that it can later be accessed by any employee or contractor to make informed decisions on operating or changing the process.

According to OHSA requirements, PHA be conducted at three key stages:

For new facilities a PHA shall be conducted to ensure the hazards of the process are understood before the facility is operated.

For existing facilities, an initial hazard evaluation shall be conducted to provide a baseline understanding of the hazards. (It appears that a full study was not expected for existing plants because most employers had not previously been obliged to do so.)

As changes are made to the process, a PHA shall be conducted to ensure that the hazards of the proposed change are well recognized and analyzed before the change is implemented.

The scope of the PHA work is also spelt out by OHSA and we shall look at this some more in later modules.

The conclusion to be drawn at this stage is that employers in the USA are clearly required to perform systematic hazard studies on plants that are expected to be hazardous. Further to this is the need to comply with the measures to reduce harmful releases and emissions as described below.

Risk management plans (40 CFR 68)

The Clean Air Amendment Acts of 1990 have introduced a further requirement on all users of chemical substances to ensure they have adequate risk management plans to reduce the chances of chemical emissions. This regulation is administered by the US Environmental Protection Agency and is known as the 'RMP Rule' or Rule 40 CFR 68.

Its purposes are:

- To reduce risks of chemical accidents without command and control governmental regulations
- To encourage public knowledge which leads to public pressure motivating companies to operate plants more safely.

The RMP Rule requires that all affected companies or government agencies/facilities register their Risk Management Plans by June 1999. The scope of the rule applies to facilities handling more than defined amounts of 140 named substances.

Some 66 000 facilities ranging from major installations to small factories and offices are subject to this rule, but its level of application varies according to the potential for harm and the past history of each plant.

- *Program 1:* Worst-case scenario does not affect members of the public. Requires limited hazard assessment.
- *Program 2:* Hazard review to identify potential malfunctions and operational hazards. Requires steps to monitor and detect releases.
- *Program 3:* Requires a rigorous step-by-step hazard analysis of process equipment and procedures to identify potential accidental releases.

Key regulations for process safety in the UK

The European Union (EU) requirements for process safety are associated with the implementation of EU directives. The EU legislators issue directives requiring member states to implement the principles and requirements of a directive through their national legislation. Hence the laws in one member state will hopefully be very similar to those in any other EU member state. The main regulations for safety practices applicable to process industries are shown below.

The underlying principle in all safety-related legislation is that employers and users should conduct risk assessments and demonstrate that they have achieved adequate safeguards against unacceptable risks. In particular we should note the following.

The basis of British health and safety law is the Health and Safety at Work Act 1974. The act sets out the general duties that employers have towards employees and members of the public, and employees have to themselves and to each other. These duties are qualified in the Act by the principle of 'so far as reasonably practicable' (see Alarp in module 1).

The Management of Health and Safety at Work Regulations 1992 (MHSWR) generally makes more explicit what employers are required to do to manage health and safety under the Health and Safety at Work Act. Like the act they apply to every work activity. As is the case with most of the main UK safety regulations, the employer is required to carry out a risk assessment.

- The MHSWR requires all employers to carry out a basic risk assessment for activities at the places of work.
- The COMAH regulations are the UK enactment of the Seveso II directive (see Module 1). These require that major hazard plants provide a safety case report that includes the systematic identification of hazards and a risk management program to provide continued monitoring of the risk control measures.
- The Control of Substances Hazardous to Health Regulations 1994 (COSHH) requires employers to assess the risks from hazardous substances and take appropriate precautions.
- The Provison and Use of Work Equipment Regulations 1992 (PUWER) requires hazards associated with machinery and equipment in the workplace to be evaluated and that such equipment should be safe.

These types of regulation make specific demands on employers to carry out formal risk assessments on the operations in their plants and to develop accident prevention and mitigation plans.

For the administration of COMAH, the UK Health and Safety Executive and the Environment Agency have jointly been designated as the Competent Authority in England and Wales, which means that they have powers to monitor and enforce the regulations. In Scotland, a separate body called the Scottish Environment Protection Agency (SEPA) is authorized to perform this function.

For further information on UK regulations we suggest as first step the reader should obtain the leaflet: HSC13 'Health and Safety Regulations, A Short Guide' from HSE website www.hse.gov.uk.pubns/hsc13.htm.

Key regulations for process safety in South Africa

The Occupational Health and Safety (OHS) Act

This is a general act requiring employers to provide for the health and safety of persons at work and for the use of plant and machinery. It also covers the provision of safeguards for the protection of the public from risk caused by activities at a work place.

The general duties of employers under this act include:

- The provision and maintenance of systems of work, plant and machinery that as far as reasonably practicable are safe and without risks to health
- Taking steps to eliminate or mitigate any hazard or potential hazard before resorting to personal protective equipment
- Arranging for safety and absence of risks to health in connection with production, processing, use, handling, storage and transport of article or substances
- Establishing what hazards are present at work, what safeguards are to be taken and providing the means to apply the safeguards.

The OHS act has many specific requirements for general safety, workplace conditions, machinery safety regulation and electrical safety requirements. Sections of the act also include the pressure vessel regulations and those for registration and protection of boilers.

Key regulations within the OHS act applicable to process hazards are as follows.

Hazardous chemical substances regulations

These include:

- The need for an assessment of potential exposure of employees to hazardous chemical substances and the repeating of the assessment at least every 2 years
- Requirements for air monitoring and preventing exposure of employee to airborne substances
- The need for safeguards to control the exposure of an employee.

Major hazard installation regulations

These requirements include:

- Notification to local authorities of any installation that has potential for a 'major incident'. Notification of any change that may increase the risk of an incident
- Risk assessments to be carried out at intervals not exceeding 5 years and risk assessments to be submitted to the local authority. Risk assessments to be done by a person 'competent to express an opinion as to the risks associated with the installation'
- Risk assessment reports to include process description, potential consequences of an incident, copy of site emergency plan, potential effects of the incident on any adjacent major hazard installation and on the public
- Site emergency plan to be updated at least once every 3 years and tested at least once per year
- Reporting of incidents to inspector and local emergency services.

Explosives Act

It defines operational requirements, classification of explosive materials and specifies particular protection measures appropriate for explosives manufacture. Requires the appointment of a full-time employee as a safety manager.

Hazardous substances Act

Regulates manufacture, storage, transportation, etc.

Appendix D

Software tools for hazard studies

Descriptions of some software products and services offered by companies for application in hazard studies. This list is not comprehensive and is limited to products for hazard studies.

Company	Website	Product
Asset Integrity Management	http://www.assetintegrity.co.uk	Located in Scotland, UK, this company provides a software application for recording Hazops, (SILHazop) that is interactively linked to an IEC 61508 SIL determination and test/maintenance strategy application. An alarm management package is available to assess alarm criticality
Dyadem International Ltd	http://www.dyadem.com	A software development company in Toronto, Canada that also provides a full range of risk management consulting services. They offer software for Risk Analysis, Process Hazard Analysis (PHA), Hazop, Failure Modes and Effects Analysis (FMEA), and Risk Management Plans (RMP). Product for PHA/Hazop: 'PHA-Pro 5'
Isograph: UK	http://www.isographdirect.com	Reliability software includes a fault tree, event tree, weibull, simulation, reliability centered maintenance, life cycle cost, Hazop, Markov and a reliability model Product for Hazop: 'HAZOP *Plus* V1.1'
Risk Management Technologies	http://www.xlweb.com	A firm located in Chennai, India offering: Hazop training and software; air dispersion modeling; failure and reliability analysis; risk assessment; and accident investigation . . . all targeted for the chemical process industries. Product for Hazop: 'Haz-RMT v.2.1'
Det Norske Veritas	http://www.dnv.com	Consequence and process hazard analysis software for chemical, petrochemical industries, onshore and offshore industries
Jenbul Consultants	http://www.jenbul.co.uk	Offering a wide range of experience in leading Hazard and Operability (Hazop) studies. Located in the UK

Appendix E

EPA case study of phenol resin hazards

This case study deals with a recurring problem that was experienced in the USA concerning the production of phenol resins. It has been included here as a useful example of typical chemical plant hazards and the suggested safeguards. It refers to the application of hazard studies and the need to ensure they are carried out properly. Various protection methods and design safeguards are described.

A hazard study at level 2 (or preliminary PHA) would highlight all of the points raised in this study for due attention.

The document is a 'Chemical Safety Case Study' issued by United States Environmental Protection Agency. It can be obtained as a download file from: *www.epa.gov/ceppo*. The details are as follows:

Title: 'How to prevent runaway reactions. Case Study: Phenol-Formaldehyde Reaction Hazards.'

Published by: US Government, Chemical Emergency Preparedness and Prevention Office (CEPPO).

Reference number: EPA 550-F-99-004. It was issued in August 1999.

Appendix F

Expanded guideword table for continuous processes

| Guideword | Derived Word | Deviations |
		Activity/Condition
More of	High-	Flow, pressure, temperature, density, viscosity, concentration, speed, velocity, current, voltage, load, tension, intensity, counts, reaction rate, transfer, force
	Rapid-	Changes, action, rise, increase, tall, decrease
	Too much-	Added, drained, removed
	Over-	Filled, charged, heated, exposed
	Excessive-	Filling, charging, heating, movement
	Long-	Duration, time
	Later-	Starting, initiated
Less of	Low-	Flow, pressure, density, temperature, viscosity, concentration, speed, velocity, current, voltage, load, tension, intensity, counts, reaction rate, transfer, force
	Slow-	Changes, action, increase, rise, fall, decrease
	Too little-	Added, drained, removed
	Under-	Filled, charged, exposed, heated
	Little-	Filling, heating, movement
	Short-	Duration, time
	Sooner-	Starting, initiated
None	None-	Existing, compliance
Not	Not-	Complete, started, moved, executed, functioning
No	No-	Flow, action, current, reaction, entry, matter
	Zero-	Speed, anige
	Off-	Track, target, switch
Reverse	Opposite-	Flow, rotation, action, food, polarity, current, direction, travel
	Restore-	State, condition
Part of	Poor-	Quality, performance, strength, resistance
	Incomplete-	Reaction, action, procedure step
	Partial-	Failure
	Too-	Soon, late
	Inadequate-	Services, protection

Guideword	Deviations	
	Derived Word	**Activity/Condition**
As well as	By-	Products
	Side-	Reactions
	Generations-	Static, magnetic fields, leakage, emissions, pollution, fouling, sealing, corrosion, erosion, mixing, vibration, stresses, noise
	Present-	Ignition, heat, radiation
	Simultaneous-	Separation
Other than	Instead-	Start-up, maintenance, shutdown, testing, emergency
	Wrong-	Material, catalyst, change
	Unacceptable-	Impurities
	Different-	Route, action, step

Appendix G

Methods of reporting

This appendix summarizes the reporting options for Hazops based on advice given in IEC 61822 and in EPSC Guide.

Recording options:

- Manual recording on prepared forms. Suitable for small studies.
- Word processor versions of manually recorded sessions. Advantage of legibility and ease of circulation. Be careful to ensure correct original version is held on computer files and backed up securely.
- Computer-based recording systems. These are widely used and have several advantages
  - If the on-screen information can be displayed clearly to all persons present in the study the headers, parameters and deviations can be seen as they are being discussed
  - Displays can be customized to suit company standards or study needs
  - The team can view what is being recorded
  - Databanks of possible causes, effects and frequencies are available as dropdowns at all times
  - Draft records and action notes can be printed for distribution immediately.
  - Facilities often exist for using risk matrix insertions leading to risk ranking if this option is required
  - Reporting options allow full and summary reports to be generated. Full worksheet records are retained within the database.

Examples of recording format and style

A typical reporting table based on examples in IEC 61822 is shown in Figure G.1. Note that details must be sufficiently clear to allow anyone consulting the records in later times to understand the meaning.

Hazop Study Title:					Sheet:				
Drawing No.				Rev No.:	Date:				
Team Composition					Meeting Date:				
Part considered:									

Design intent Material: Activity: Source:
Destination:
Description:

No.	Guide-word	Element	Deviation	Possible Causes	Consequences	Safeguards	Comments	Actions Required	No.	Action by:

Figure G.1
Typical Hazop study worksheet: based on examples in IEC 61822

Appendix H

Design and calibration of a risk graph

Before a risk graph can be used the project team must establish the definition of the parameters being used and on the design of risk graph to be used. In practice in the process industries there will be separate versions for three categories of hazard, i.e.:

- Harm to persons
- Harm to environment
- Loss of assets (production and equipment losses/repair costs).

All three versions of the risk graph can have the same basic layout but for environment and asset loss the parameter F, for exposure, is considered to be permanent and can be left out of the diagram. Calibration of the risk parameters involves setting down a table of values for the risk parameters. The meaning of the parameters must first be clear. Table H.1, taken from IEC 61511, can be used for this. In particular it is important to note the interpretation of the term W.

Parameter		Description
Consequence	C	Average number of fatalities likely to result from the hazard. Determined by calculating the average numbers in the exposed area when the area is occupied taking into account the vulnerability to the hazardous event
Occupancy	F	Probability that the exposed area is occupied. Determined by calculating the fraction of time the area is occupied
Probability of avoiding the hazard	P	The probability that exposed persons are able to avoid the hazard if the protection system fails on demand. This depends on there being independent methods of alerting the exposed persons to the hazard and manual methods of preventing the hazard or methods of escape
Demand rate	W	The number of times per year that the hazardous event would occur if no SIS was fitted. This can be determined by considering all failures which can lead to one hazard and estimating the overall rate of occurrence

Table H.1
Parameter descriptions table

Example of a risk table

Risk tables must align with any existing risk profile specified by a user company operations and sites.   Table H.2 shows a suggested set of consequence and frequency parameters. For an individual company the values would be adjusted to align with any existing risk matrices or risk profile charts.

Risk Parameter		Classification	Comments
Consequence (C) Average number of fatalities	C_A	Minor injury	1. The classification system has been developed to deal with injury and death to people. 2. For the interpretation of C_A, C_B, C_C and C_D, the consequences of the accident and normal healing shall be taken into account
	C_B	Range 0.01 to <0.1	
	C_C	Range >0.1 to <1.0	
	C_D	Range >1.0	
Occupancy (F) This is calculated by determining the length of time the area exposed to the hazard is occupied during a normal working period. Note – If the time in the hazardous area is different depending on the shift being operated then the maximum should be selected. Note – It is only appropriate to use F_A where it can be shown that the demand rate is random and not related to when occupancy could be higher than normal. The latter is usually the case with demands which occur at equipment start-up	F_A	Rare to more often exposure in the hazardous zone. Occupancy less than 0.1	3. See comment 1 above
	F_B	Frequent to permanent exposure in the hazardous zone	
Probability of avoiding the hazardous event (P) if the protection system fails to operate	P_A	Adopted if all conditions in column 4 are satisfied	4. P_A should only be selected if all the following are true: facilities are provided to alert the operator that the SIS has failed; independent facilities are provided to shutdown such that the hazard can be avoided or which enable all persons to escape to a safe area; the time between the operator being alerted and a hazardous event occurring exceeds 1 h
	P_B	Adopted if all the conditions are not satisfied	

Risk Parameter		Classification	Comments
Demand rate (W) without protection system To determine the demand rate, it is necessary to consider all sources of failure that can lead to one hazardous event. In determining the demand rate, limited credit can be allowed for control system performance and intervention. This must not exceed an RRF of 10. Credit can be allowed for non-SIS risk reduction factors if these have not been included in the initial hazardous event rate	W_1	Demand rate less than 0.01 per year	5. The purpose of the W factor is to estimate the frequency of the hazard taking place without the addition of the SIS
	W_2	Demand rate between 0.01 and 0.1 per year	
	W_3	Demand rate between 0.1 and 10 per year For demand rates higher than 3 per year higher integrity shall be needed	6. If the demand rate is very high (e.g. 10 per year) the SIS operating mode is 'high demand mode' and the SIL has to be determined by another method. The quantitative method is advised in this case

Table H.2
Draft risk table for personnel risks based on IEC 61511

The risk graph parameter table can then be used in conjunction with the typical risk graph as shown in Table H.3.

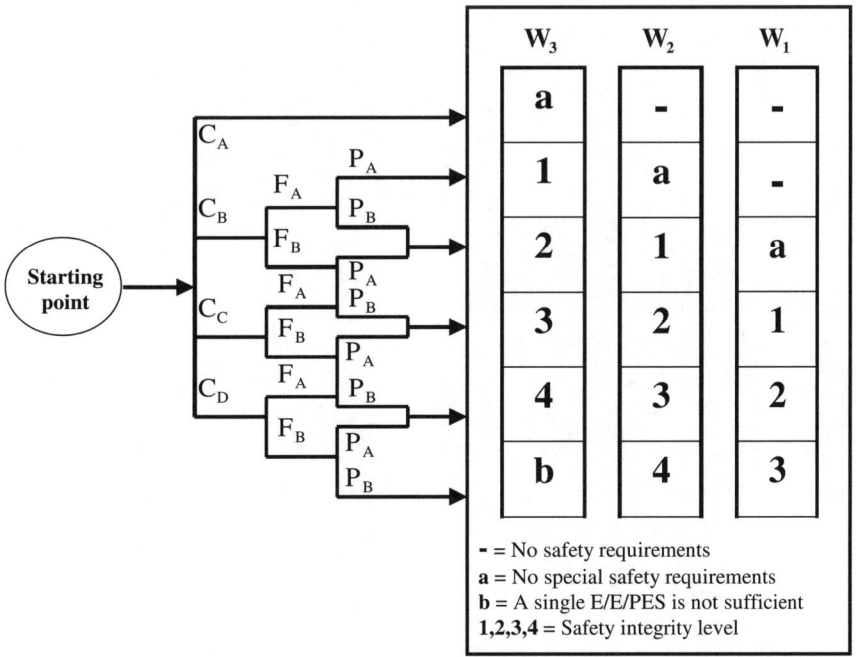

Table H.3
Parameter descriptions table

Calibration check

IEC 61511 requires that the parameters table and the use of the risk graph be validated by first testing its results against examples where the SIL rating is agreed with some confidence. This can be done conveniently by comparing the results of simple cases where a quantitative risk reduction model has been drawn. In practice several practical examples would serve to confirm the calibration of the risk graph.

Appendix I

Data capture sheet

Data capture sheet can be used in a preliminary hazard study or in a Hazop to check available information on a proposed trip solution.

Method

When a hazard study session identifies the possible need for a trip system the following three steps should be taken:

- Capture essential information (see data capture table below)
- Determine the required safety integrity level (SIL) of the trip system (see risk graph in Appendix H as one possible method)
- Decide the most likely configuration of the SIS loop (see Chapter 8 para. 8.1). This will lead to an estimation of cost and complexity.

(Data capture table to be used during hazard study for each item where an instrument trip function is requested.)

	Item	Data/Description
1	Hazop record item no.	
2	Equipment ref./description state number of units	
3	Hazardous event description and sequence of events leading to it	(Or see Hazop record sheet if adequate)
4	Estimated hazardous events per year due to all causes and without protection (i.e. demand rate) or probability description	
5	Consequence description	
7	Acceptable frequency for the hazardous event at the site	
8	Acceptable frequency per unit	(Divide above by no. of units)
9	Any additional non-instrument measures to reduce risk?	
10	Any additional instrumented measures to reduce risk? e.g Alarms, DCS-based interlocks?	

		Item	Data/Description
11		Is personnel exposure continuous or only a small time during the shift? For enviro. and financial losses use Fa	Continuous = Fa Small fraction = Fb
12		What is the probability that plant will be able to avoid the hazard if the trip were to fail on demand? State either: High or Low	High = Pa Low = Pb
13		Basic description of the trip function. Note any operating modes that require bypass of the trip	
14		What is the safe state of the process to be reached after trip action? What is the available time to reach the safe condition?	
15		What are the risks created by a spurious trip? What are the costs per trip?	
16		What will be an acceptable frequency for spurious trips (e.g. 1 per 10 years) per unit of equipment	

Table I.1
Data capture table

The data in Table I.1 is required for the SIL determination by the risk graph in Appendix H and for the preliminary design of the SIS to meet basic performance needs. Note that for a complete safety life cycle design task, more information will be required from the design team (see safety requirements specification example in Chapter 9), but this can be obtained at a later stage.

Appendix J

Glossary of terms used in hazard studies and safety-related systems

Availability   The probability that an item of equipment or a control system will perform its intended task. See safety availability.

BPCS (basic process control system)   Generic term used to describe any control system equipment provided for the normal operation of a plant or machine. A BPCS may or may not include safety functions.

CASS   Conformity assessment of safety-related systems. Refers to the developing methods for assessment of project execution, equipment design as well as functional safety management capabilities. In the UK, accredited certification bodies will be available at a future date to offer CASS assessment services to industry.

Cause and effect diagram   A matrix drawing showing the functional process safety interlocks between inputs and outputs of a safety system (see also FLD).

Common cause failure   Failure as a result of one or more events, originating from the same external or internal conditions, causing coincident failures of two or more separate channels in a multiple channel system (see also Systematic failures).

Coverage factor   See Diagnostic coverage.

Covert failure   A non-revealed defect in a system that is not detected by the incorporated test.

De-energized safe condition   In this context: the electrical or pneumatic valves, which can shutdown the guarded process, are energized during the normal (safe) process situation. If an unsafe condition arises, the (spring-loaded) valve will close, because the energy is cut off.

Diagnostic coverage   The efficacy of the self-diagnostics of an SIS, which makes it possible that a system successfully detects a specific type of component or software fault. This is defined in IEC 61508 as 'fractional decrease in the probability of dangerous hardware failures resulting from the operation of the automatic diagnostic tests'.

Diagnostic coverage factor (also known as C-factor)   The C-factor comprises the percentage of failures in modules, software, external wiring, internal wiring, cables, interconnections and other functions that are detected by the built-in test functions, or by a suitable test program. It can be expressed in a probability or in a factor that is always smaller than 1 (e.g. C = 0.95) or as a percentage (e.g. 95%).

DCS   Distributed (or digital) control system. A process control system based on computer intelligence and using a data highway to distribute the different functions to specialized controllers.

Dynamic logic circuit   In this context: the valid logic state can only exist and perform logic control, if the circuit is activated continuously, using alternating logic signals.

Emergency shutdown   Commonly used terminology to refer to the safeguarding systems intended to shutdown a plant in case of a process parameter limit-excess. See also SRS and SIS.

EMI   Electrical-magnetic interference.

EMC   Electrical-magnetic compatibility.

E/E/PES (electrical/electronic/programmable electronic system)   System for control, protection or monitoring based on one or more electrical/electronic programmable electronic (E/E/PE) devices, including all elements of the system such as power supplies, sensors and other input devices, data highways and other communication paths, and actuators and other output devices.

EUC (equipment under control)   Equipment, machinery, apparatus or plant used for manufacturing, process, transportation, medical or other activities.

EUC control system   System which responds to input signals from the process and/or from an operator and generates output signals causing the EUC to operate in the desired manner.

Note: The EUC control system includes input devices and final elements. See also BPCS.

Fail-safe   A control system that, after one or multiple failures, lapses into a predictable safe condition.

Fault tolerance   IEC definition: 'ability of a functional unit to perform a required function in presence of faults or errors'.

FLD   Functional logic diagram. A graphical representation of the system functions, showing the logic gates and timers as well as the logic signal interconnections.

FMEA   Failure mode and effect analysis. See Chapter 7. Also, FMECA: Failure mode and effect criticality analysis. Applies when FMEA is extended to identify dangerous or critical failure modes.

Also, FMEDA, Failure mode and effect diagnostic analysis – term used for identifying failure modes detected or not detected by diagnostics and for calculating how diagnostics reduce the incidence of dangerous failure conditions.

Functional safety   Part of the overall safety relating to the EUC and the EUC control system which depends on the correct functioning of the E/E/PE safety-related systems, other technology safety-related systems and external risk reduction facilities (definition from IEC 61508 part 4).

Hazard analysis   Term applied to the evaluation of probabilities and consequences of a recognized hazard.

Hazop Term applied to the structured and systematic examination of a process or system of parts to find possible hazards and operability problems. Hazop is one of a number of recognized hazard study methods.

HMI Human to machine interface or 'operator interface', usually a computer screen to present the actual process and system status.

IEC International electrotechnical commission. Based in Geneva. Develops a vast range of internationally supported standards. See website list.

Inherently fail-safe A particular designed dynamic logic principle that achieves the fail-safe property, from the principle itself and not from additional components or test circuits.

ISA (instrument society of America) Based in Research Triangle Park, North Carolina. Develops standards, technical reports and training material for the complete range of instrumentation with strong emphasis on process industries. See website list.

Logic solver Components or subsystems of E/E/PES that execute the application logic. Electronic and programmable electronics include input/output modules.

MTBF Mean time between failures. This term is normally applied to serviceable equipment, typically instrument sensors, valves or PLCs. Hence normally used in SIS reliability calculations.

MTTF Mean time to fail. This term is normally applied to disposable single life components such as relays or resistors which are replaced when they fail. Numerically the same as MTBF when calculating reliability of an SIS.

MTTR Mean time to repair. The mean time between the occurrence of a failure and the return to normal failure-free operation after a corrective action. This time also includes the time required for failure detection, failure search and restarting the system.

Nuisance failure See Spurious trip.

Overt faults Faults that are classified as announced, detected, revealed, etc. Opposite of 'covert fault'.

Probability of failure on demand (PFD) The probability of a system failing to respond to a demand for action arising from a potentially hazardous condition. This parameter degrades (increases) during the mission time or test interval time. Therefore the average figure, PFD_{avg} is used in calculating the reliability of a safety system over a given mission time. PFD equals 1 minus safety availability.

PES Programmable electronic system. The term includes PLCs, SCADAs, etc. as well as any instrument using a programmable device such as a 'smart' transmitter.

PLC Programmable logic controller.

Proof test A 100% functional system test. In practice, this is only possible when the SIS is disconnected from the process. Hence online proof testing may leave a small fraction of the SIS untested. Also termed 'Trip testing'.

Redundancy (identical and diverse) Identical redundancy involves the use of elements identical in design, construction and in function with the objective to make the system more robust for self-revealing failures. 'Diverse redundancy' uses non identical elements and provides a greater degree of protection against the potential for common cause faults. It can apply to hardware as well as to software.

Reliability   The probability that no functional failure has occurred in a system during a given period of time.

Reliability block diagram   The reliability block diagram can be thought of as a flow diagram from the input of the system to the output of the system. Each element of the system is a block in the reliability block diagram, and the blocks are placed in relation to the SIS architecture to indicate that a path from the input to the output is broken if one (or more) of the elements fails.

Revealed failure   A failure in a system that is detected by the system's self-diagnostics.

Safety availability   Probability that a SIS is able to perform its designated safety service when the process is operating. The average probability of failure on demand (PFD_{avg}) is the preferred term (PFD equals 1 minus safety availability).

Safety instrumented systems (SIS)   System composed of sensors, logic solvers, and final control elements for the purpose of taking the process to a safe state when predetermined conditions are violated. Other terms commonly used include emergency shutdown system (ESD, ESS), safety shutdown system (SSD), and safety interlock system.

Safety life cycle   Necessary activities involved in the implementation of safety-related systems, occurring during a period of time that starts at the concept phase of a project and finishes when all of the E/E/PE safety-related systems, other technology safety-related systems and external risk reduction facilities are no longer available for use.

SCADA   Supervisory control and data acquisition. This term is most commonly applied to PC-based equipment interfaced to plant via PLCs or input/output devices.

SER   Sequence of events recorder, based on real-time state changes of events in the system.

SIL   Safety integrity level defining a level of confidence in the risk reduction capabilities of a safety-related device or control system. For hardware aspects this is defined by failure probabilities arranged in order of magnitude. For software and design aspects the SIL relates to the measures taken to ensure quality and freedom from systematic errors in the design. In practice the SIL range is from 1 to 4.

Solid-state logic   A term used to describe circuits whose functionality depends upon the interconnection of electronic components as semiconductors, resistors, capacitors, magnetic cores, etc. and which does not depend on programmable electronics.

Spurious trip   A plant trip arising out of an overt or detected equipment failure in the SIS or an erroneous assessment of the situation (e.g. error in the logic functions). A shutdown is initiated, though no real impairment of safety exists. Also referred to as a 'false trip' or a 'nuisance failure'. Spurious trips can contribute to the hazard rate of the plant through the disturbances so caused.

Systematic failures   Failures occurring in identical parts of a (redundant) system due to similar circumstances. History shows that also errors in specification, engineering, software and environmental factors, such as electrical interference or maintenance errors must be considered. Such faults can only be eliminated by a modification of the design or of the manufacturing process, operational procedures, documentation or other relevant factors.

TMR (triple modular redundancy)   An architecture for SIS logic solvers to achieve fault-tolerance by a 2 out of 3 voting configuration using identical redundant modules.

Trip   A shutdown of the process or machinery by a safety system.

TÜV (Technische Üeberwachungs Verein)   A testing laboratory in Germany that certifies safety of equipment in terms of compliance with international standards or German national standards.

Unrevealed failure   A failure that impairs the system safety, but remains undetected (see also Covert failure). It is related to the risk (PFD) involved in various types of processes. These type of failures can accumulate in a safety system, causing a degradation of the safety performance (SIL), as a function of time.

Practical exercises

Exercise 1: Calculating risk parameters

This practical exercise supports module 1.

Subject	Calculation of individual risk (IR) and fatal accident rate (FAR) as theoretical risk factors. Interpretation of relative values for a decision on risk reduction
Objective	To assist participants to become familiar with risk criteria and their interpretation
Relevance	An important component of the quantitative methods for evaluation of risks. It is helpful to have the ability to estimate a given risk in terms that can be compared with industrial accident statistics
Starting information	A hazard study of a chemical plant producing a flammable gas has found that the possibility exists of an explosion within a gas mixing and conversion unit. The plant has two such units operating under the control of a shift team manned by an average of five persons per shift. The problem was referred for hazard analysis, which found that the likelihood of an explosion in a single gas unit was 1 event per 500 years and the consequence would possibly lead to the death of a person within 5 m of the unit. From the layout of the plant it was considered that on average there would be one shift worker within this 5-m range
Task detail	**Practical is for individual participants** **Task 1** *Step 1*: Evaluate the risk to workers in comparison with industry norms by calculating the figures for FAR for the workers at the plant. Assume three shifts are employed per 24 h and each worker logs 2000 h on site per year.

| | *Step 2*: Compare the result with the FAR table in the manual and suggest what improvements, if any, are required in the safety levels at the plant.
Reminder: FAR = fatalities per 10^8 exposed hours
Task 2
Step 1: Calculate the IR for a worker at the plant. This unit is the individual risk per year for the exposed worker.
Step 2: Compare the result with the Alarp range values noted in the manual for exposed workers and suggest what improvements, if any, are required in the safety levels at the plant.
Reminder: IR = chance of fatality due to the event- frequency of event |
| **Time allowed** | Thirty minutes including time for discussion of answers |

Exercise 2: Preparing a risk matrix

This practical exercise supports module 2.

Subject	Preparing a risk matrix
Objective	To assist participants to become familiar with risk matrix principles
Relevance	Risk matrices are generated at the start of a hazard study project to provide the team with an agreed basis for describing the level of risk in their application. This in turn allows ranking of risks and allows priorities of risk problems to be identified
Task detail	Practical is for individual participants 1. Use the template overleaf to create your version of a risk matrix for a plant or situation you may be familiar with. Arrange the scoring system so that the highest-ranking risk will be 1 and the lowest rank will be 25. 2. For each probability of occurrence code insert your own description and your own suggested rate per year. 3. For each severity of consequence insert your own description of harm to persons and possible damage or loss to the plant and production. 4. Now draw on the matrix where you wish to see the Alarp boundary lines for highest tolerable risk and highest negligible risk. 5. Complete the exercise by writing down one test case to validate the working of your risk matrix
Time allowed	Thirty minutes including time for discussions

Alarp diagram for convenience is shown here:

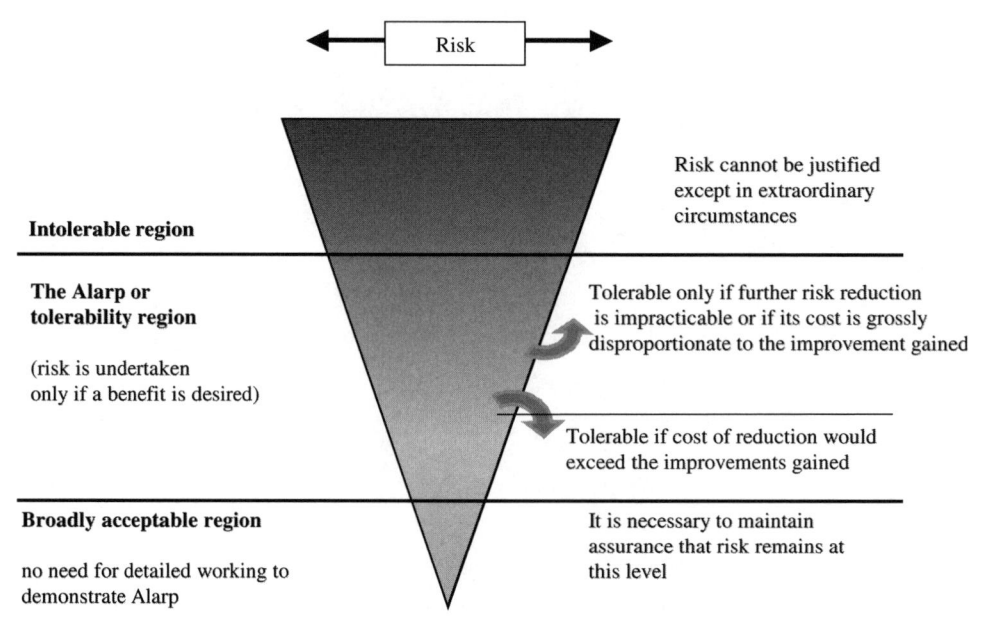

Draft risk matrix for: (Insert description of type of plant or activity)..
....
....

= rating number

Consequence Descriptors				Frequency of Occurrence				
Plant	Enviro.	Persons	Ratings	#	#	#	#	#
			#					
			#					
			#					
			#					
			#					

Frequency of Occurrence Event rate per year					
Occurrence description					

Describe test scenario here:
....
....
....

Exercise 3: Preliminary hazard study

This practical takes place after the basics of preliminary hazard studies in module 2 have been completed. Its objective is to involve participants in a typical preliminary hazard study using checklist procedures to identify risks. A report back session is used to compare the results.

Subject	Application of hazard study 2 and 3 methods to a raw gas holder
Objective	To identify potentially hazardous events in a proposed design and to propose safety measures to reduce risk
Relevance	The study forms part of a typical safety life cycle activity
Task detail	**Participants form into groups of three persons to form mini hazard study groups.** For each group: 1. Study the following 'starting information' which provides a description of a proposed gas holder system to be installed between the gas producer stage and the gas compression stage of an ammonia plant (typical hazard study 1 output). 2. Use the hazard study 2 guidelines in the workshop manual (module 3) to identify potential risks. Each team should apply the guide diagram 1 to identify possible hazards and basic causes. Use guide diagram 2 to identify possible consequences. 3. The team then considers the possible sequence of events that might be the cause of each of the hazardous events it has identified. 4. The hazards, causes and consequences are to be recorded in the attached reporting form. 5. The team should then discuss measures to reduce the likelihood and/or the consequences of the events. These measures are to be recorded against each event in the reporting form. Follow-up actions should be listed in the last column. 6. Finally the team should try marking on to the study reporting form (given at the end of Exercise 2) any suggested safety instrumented functions with sensors and actuators
Time allowed	One hour including time for discussions

Starting information

IEC safety life cycle Phases 1 and 2 or process hazard study 1 provide the following information.

IEC Ref.	Requirement	Information
7.2.2.1	EUC and required control functions	See Figure Ex. 1 and description of the process
7.2.2.1	Physical environment	Inside the chemical complex but 100 m from a canteen and recreational area which 500 persons used daily
7.2.2.2	Likely sources of hazard	Large volumes of raw gas that is not flammable until mixed with oxygen. Oxygen is used in the upstream gasification stage, hence there is a possible risk of producing flammable or explosive gas mixtures. Potential for internal explosions or fires, external explosions or toxic gas releases
7.2.2.3	Hazard info.	Raw gas consists of 25% H_2, 60% CO, 10% CO_2, 5% other gases inc H_2S (smelly) Raw gas is flammable in range 7.3–74.5% in air Toxic component is CO with a short-term exposure limit of 400 ppm and immediate escape level of 1500 ppm
7.2.2.4	Current safety regulations	OHS limits for toxic exposure
7.2.2.5	Interactions with other EUCs	Gas holder volume (level) rises and falls in response to mismatch between upstream and downstream plant rates. Gas holder contents must be isolated if either upstream gasification stage or downstream processing stage is shutdown
7.3.2.1	Physical equipment in the scope of the hazard and risk analysis	See Figure Ex. 2 and description of the process
7.3.2.2	External events to be taken into account	1. Tripping of 1 to 6 gasifiers 2. Tripping of 1or 2 gas blowers 3. Tripping of downstream plant 4. Tripping of 1 or 2 compressors 5. Flame out of gasifier or oxygen leak into gasifier output
7.3.2.3	Subsystems associated with hazards	Cooling water supply Raw gas flare system

Information to be developed by the hazard study team

In this practical the results required from the hazard study teams are shown in italics in the following table which shows their relevance to the safety life cycle.
IEC safety life cycle Phases 2 and 3 require the following information.

IEC Ref.	Requirement	How the Information is Obtained
7.3.2.4	*Type of accident-initiating events that need to be considered*	*Determined by hazard study 2 method*
7.4.2.1	Hazard and risk analysis to be done	Follows after the risks have been identified
7.4.2.2	Consideration to elimination of the hazards	Part of hazard study, refers back to designers
7.4.2.1	*The hazards and hazardous events of the EUC and EUC control system (in all modes of operation) to be determined*	*Determined by hazard study 2 method. In this practical ignore start-up modes*
7.4.2.4	*The event sequences leading to the hazardous events*	*Determined by hazard study 2 method*
7.4.2.5	*Likelihood of the hazardous event*	*Hazard study 2 to make an assessment.* Further studies assisted by fault tree analysis
7.4.2.6	*Potential consequences*	*Preliminary from the hazard study 2 consequence table.* Further details as part of risk analysis
7.4.2.7	EUC Risk for each hazardous event	Determined by hazard analysis

Participants should suggest risk reduction measures for any identified hazards. (In preparation for the specification of overall safety requirements: IEC safety life cycle Phase 4. The hazard study 2 is often used to propose the safety functions that will be incorporated into Phase 4 after analysis.)

Raw gas holder: scope and description of process plant (equipment under control)

A coal-based ammonia plant has been designed with two raw gas production streams, each stream consisting of three gasifiers supplying a large gas blower (see Figure Ex. 1). The gas from the blowers has to be cleaned by passing it through an electrostatic precipitator section before it is supplied to a pair of gas compressors. The compressors deliver the gas into a 'rectisol' stage in which the hydrogen is separated from the carbon monoxide and dioxide components.

The raw gas holder provides buffer storage of gas between the gas production and gas compression/rectisol stage to allow for short-term mismatches between the rate of gas production and rate of compression. This is essential for practical start up of the plant and for coping with changes in the rates of gas production and gas compression.

The gas holder provides a variable capacity storage of 1000–10 000 cubic metres (m^3) of raw gas at a virtually constant pressure of 35 mbar gage (i.e. above atmospheric pressure). This is achieved by allowing the top section of the gas holder to rise and fall as the net flow into the gasifier varies from positive to negative.

The rate of gas production is 20 000 m^3/hr per gasifier, hence with all six gasifiers running, the gas production rate will be 120 000 m^3/hr. The rate of the downstream gas compressor stage is to be controlled to match the inflow of gas so that the level of the gas holder is stabilized at typically 50%. The EUC control system comprises an automatic level control loop operating on the input control valves of the compressors supported by manual controls that allow the operator to decide if one or two compressors are to be run to balance the load.

One compressor can handle 1–3 gasifiers and two compressors can handle 3–6 gasifiers. Hence if one gasifier suddenly trips out the compression rate must be reduced by 16.6% or 20 000 m^3/hr to rebalance the flows. The gasholder capacity at 50% level provides a reserve of 5000 m^3, equivalent to 15 minutes of mismatched operation before the capacity falls to zero. In another scenario if one gas blower trips the compression rate may have to be reduced by 50% and the re-balancing time available may fall to 5 minutes.

Physical equipment in the scope of the hazard and risk analysis

The EUC scope for hazard study covers the gasholder and the precipitators stage starting at the discharge point from the gas blowers and ending at the inlet point to the gas compressors. The equipment includes a flare stack used to divert and burn off raw gas when starting up the gasifiers or in emergency conditions (see Figure Ex. 2, in which a simplified version of the equipment is shown).

The gasholder consists of a 30-m-diameter domed top section floating above a water-filled bottom section. As gas flows into the top section it will rise between spiral guide rails in response to the small rise in pressure. The internal gas space is sealed by water which is maintained at constant level in the bottom section and which is pushed up the sides of the bottom tank by the pressure in the dome. The sides are open to atmosphere and hence provide an ultimate pressure relief for the gasholder.

A drainage system is provided at the bottom of the inlet and outlet pipe u-sections or lutes to continuously remove condensate deposited by the incoming gas as it cools. Hence a pressure-sealing water trap or seal pot is provided. When the plant is shutdown these lutes are used as isolators to trap gas in the holder by flooding the lutes from a water tank.

The EUC control system comprises:

- The level control loop described above
- Remote manual controls to adjust the compressors
- High and low level alarms on the gasholder level taken from the level controller
- Remote manual controls for opening and closing the flare
- Remote manual controls for motorized valves used to isolate the gasholder by flooding the lutes.

The precipitators are large volume rigid chambers with electrostatic plates which attract the charged dust particles in the gas and hence clean the gas stream. The precipitator plates are likely to produce sparks.

Hazop. 2 Study Reporting Form	Project: Practical 2: Raw Gas Holder Hazard Study		Drg. No.: RGH Diagrams 1 and 2	Rev No.: 1
Team Members:			Date	Sheet No. of Meeting No.
Hazardous Event or Situation	Caused by / Sequence of Events	Consequences Immediate / Ultimate	Estimated Likelihood / Suggested Measures to Reduce Likelihood	Emergency Measures (Reduce Consequences)	Action Required

Hazop. 2 Study Reporting Form	Project: Practical 2: Raw Gas Holder Hazard Study			Drg. No.: RGH Diagrams 1 and 2	Rev No.: 1
Team Members:				Date	Sheet No. of Meeting No.
Hazardous Event or Situation	Caused by / Sequence of Events	Consequences Immediate / Ultimate	Estimated Likelihood / Suggested Measures to Reduce Likelihood	Emergency Measures (Reduce Consequences)	Action Required

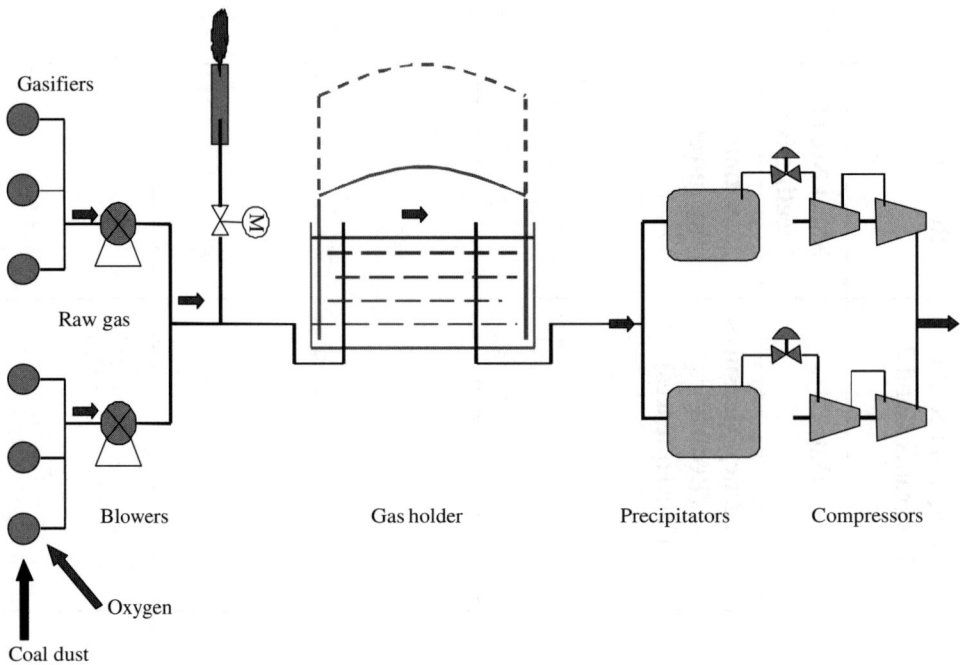

Figure Ex. 1
Flowsheet for raw gas transfer

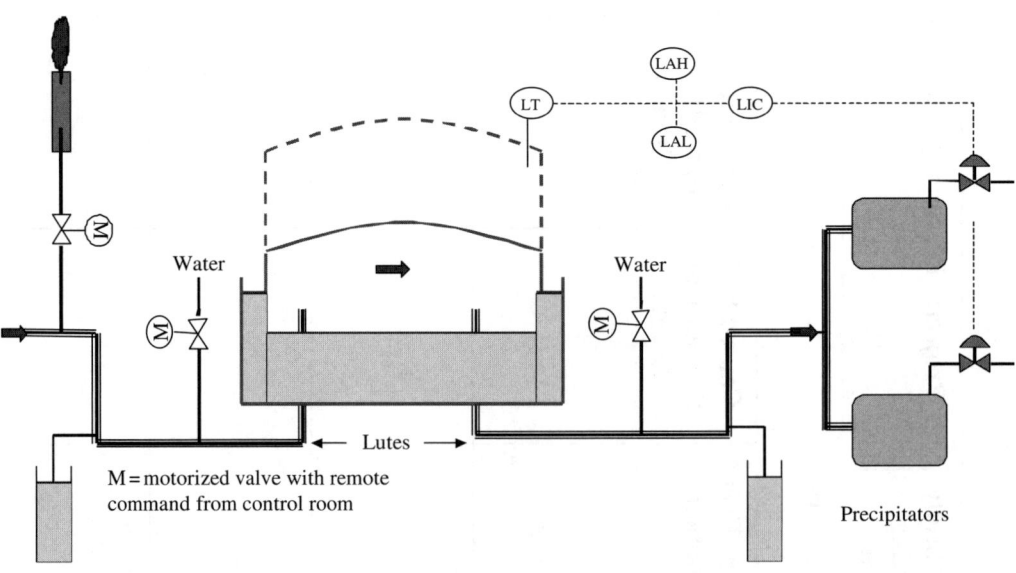

Figure Ex. 2
Detail of raw gas holder before Hazop

Exercise 4: Hazop method for a simple continuous process

This practical exercise supports module 4.

Subject	Hazop study on an oil vaporizer
Objective	To assist participants to become familiar with basic Hazop method and the application of guideword
Relevance	The example helps learning through participation in the Hazop procedure
Task detail	Practical is best done by forming study groups of 3 or 4 participants 1. Examine the diagram (Figure Ex. 3) and read through the overall description of the process 2. Select the parts for study. The suggested approach is to use two parts 3. For each part identify the elements and parameters to be studied 4. Examine for continuous operating mode. Ignore start-up mode for this exercise 5. Sketch out your own version of the guideword matrix and mark in the guideword/parameter combinations to be tested 6. Take the first part, mark it up on the diagram and apply the deviation questions to all participants in your team. You may use either the guideword-first or the element-first methods 7. Use your matrix of guidewords to tick off each one that has been tested 8. Use the attached worksheet forms to record only those deviations that require some action 9. Repeat for the 2nd part
Answers	We shall review the answers in the workshop by discussing the Hazop report sheets you have produced
Time allowed	One hour

Description of the process

The diagram shows an oil vaporizer unit intended to provide a continuous feed of vaporized oil to a downstream process unit. (This would normally be detailed as well but we are not going that far.)

The unit consists of a furnace containing a heating coil and natural gas burners. The oil enters the heating coil as a liquid, converts to vapor in the coil and is superheated as it passes through the coil and out to the process. Heating is provided by the natural gas combining with air in the burners and producing a hot flame in the firebox. The combustion gases exit by convection through the stack.

Oil is supplied to the coil from a constant pressure source and its flow is controlled by the flow loop FIC-1. The alarm FAL-1 has been provided to alert the operator to reduced flow conditions.

Natural gas is supplied to the unit at high pressure and regulated to a low-pressure supply by PCV-2. Gas is fed to the pilot burner through PV-3, which operates under control of an automatic ignition sequencer. Once a flame is established in the firebox, the main burner valve TV-1 can be opened and set to automatically control the exit temperature of the oil vapor.

The designer of the vaporizer has provided the following safeguards: If gas pressure becomes too high at the burner inlets the pressure switch PSH-1 will trip shut the main and pilot gas valves. The same valves will be tripped if the exit temperature of the oil vapor becomes too high or if the detector finds no flame in the firebox.

The design intent is that the oil flow set point should be adjusted on demand from the downstream process and that the vapor exit temperature should be reasonably constant under control from TIC-1. The unit will then become a load following service to the downstream use.

Guideword/deviations table

Element (Parameter)	Guideword									
	No	More	Less	Opposite	Part of	As well as	Early/ Late	Other than	Other	Other

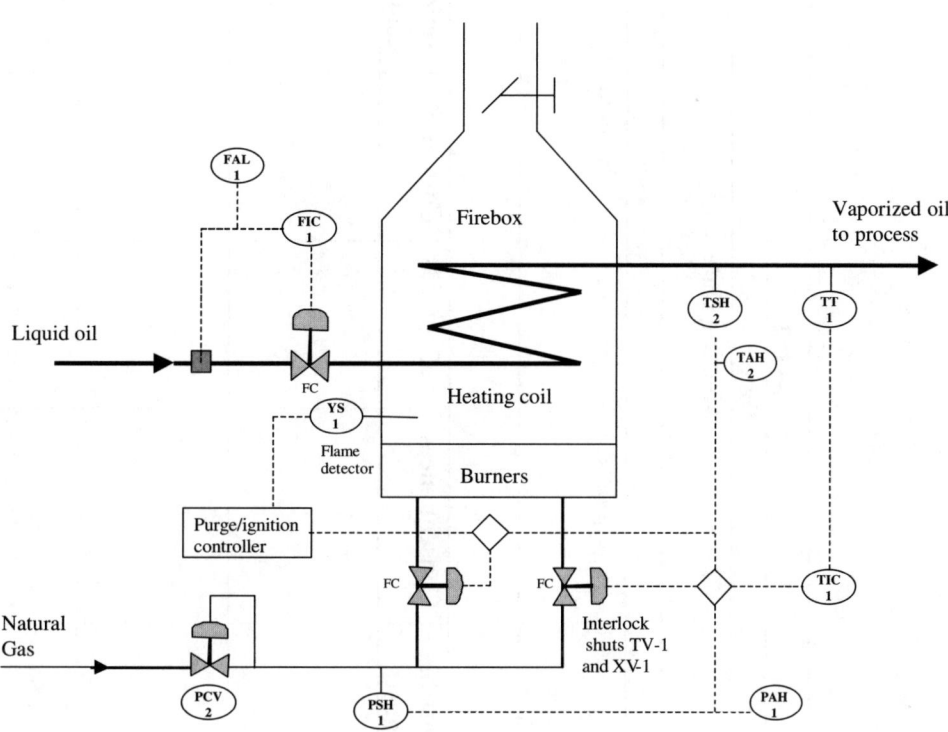

Figure Ex. 3
Diagram of oil vaporizer

IDC Hazop Study Worksheet

Hazop Study Title:		Sheet: 1
Drawing No.:	Rev No.:	Date:
Team Composition		Meeting Date:

Part considered:			
Design Intent	Material Description:	Source:	Destination:

No.	Guide word	Element	Deviation	Possible Causes	Consequences	Safeguards	Comments	Actions Required	Act. No.	Action by

IDC Hazop study worksheet (Continuation sheet)

Hazop Study Title: **Sheet:**

Part considered   **Date: 18 Nov. 2002**

No.	Guide word	Element	Deviation	Possible Causes	Consequences	Safeguards	Comments	Actions Required	Act. No.	Action by

Exercise 5: Hazop method for a batch process (procedural Hazop)

This practical exercise supports module 4.

Subject	Hazop study on an ethylene oxide sterilizer
Objective	To assist participants to become familiar with the procedural Hazop method
Relevance	The example helps learning through participation in the Hazop procedure. The ETO sterilizer was studied in modules 2 for preliminary hazard analysis. This practical extends the level 2 study into the level 3 or Hazop stage
Task detail	Practical is best done by forming study groups of 3 or 4 participants 1. Examine Figure Ex. 4 and read through the overall description of the process 2. Examine the operating cycle and define the phases to be studied individually 3. For each phase identify the elements and parameters to be studied. 4. Examine for each procedural step in each phase 5. Sketch out your own version of the guideword matrix and mark in the guideword/parameter combinations to be tested 6. Take the first phase and apply the deviation questions to all participants in your team. You may use the guideword-first method 7. Use the matrix of guidewords to tick off each one that has been tested 8. Use the attached worksheet forms to record only those deviations that require some action 9. Repeat for Phases 2 and 3. The study will stop after Phase 3 to save time for review
Answers	We shall review the answers in the workshop by discussing the Hazop report sheets you have produced
Time allowed	One hour

Description of the process

The proposed design of the sterilizer system is represented in the block flow diagram shown in Figure Ex. 4.

The essential features of the design are:

- The sterilizer chamber doors are closed and sealed gas-tight for vacuum by means of inflatable silicon rubber seals. Doors are operated manually.
- A PLC control system sequences valves and drives to execute the operating cycle without operator intervention.

- Evacuation of the chamber is done by a vacuum pump delivering its discharge through a water column absorber and exhausting to atmosphere.
- Ethylene oxide (ETO) is supplied from an individual gas bottle with enough capacity for one or more cycles. The bottle has a dip pipe and the liquid gas is passed through a water-heated evaporator to supply the gas into the chamber. The gas bottles and evaporator are sited in a separate room from the operating areas. The dosage of ETO is measured by the pressure rise achieved as ETO is added to the chamber. The ETO bottle is checked for loss in weight by an electronic weigh scale connected to the PLC.
- The proposed operating cycle, see Figure Ex. 4, replaces air in the chamber with nitrogen before admitting ETO. The operating point and the transient gas mixture values can be seen on the flammability diagram in Figure Ex. 5.

Operating cycle: all sequencing and valve operations are by PLC automatic sequence.

Phase 1: Prepare for entry

All valves closed. Both doors must be closed. Door controls are disabled. Press 'entry'

- *Step 1:* Door seals inflate
- *Step 2:* Start vac pump, open vac valve and reduce pressure to 50 mbar
- *Step 3:* Close vac valve, open vent valve, allow pressure to rise to atmospheric
- *Step 4:* Door seals deflate and door controls are enabled after purge cycle complete.

Phase 2: Load material into chamber

- *Step 1:* Open inlet door, keep outlet shut. Ventilator valve opens and fan starts automatically from door open limit switch. This draws clean air into chamber
- *Step 2:* Operator enters chamber with pallet trolley and loads pallets onto floor. Several journeys are executed as needed
- *Step 3:* Close inlet door. Ventilator valve closes, fan stops, recirc fan starts.

Phase 3: Fill chamber with gas mixture

Preparation

Operator connects ETO bottle to hose and opens bottle valve to evaporator. Opens hot water valve to ETO evaporator unit. Presses 'start'. PLC automatic sequence is as follows:

- *Step 1:* Door seals inflate
- *Step 2:* Start vac pump, open vac valve and reduce pressure to 50 mbar
- *Step 3:* Open N2 valve to raise pressure, to 300 mbar, close N2 valve
- *Step 4:* Close vac valve, Open ETO valves XV-1 and XV-2 and admit ETO as vapor to chamber. Pressure rises to 700 mbar, close ETO valves
- *Step 5:* Keep all vales closed, monitor pressure to be constant, hold for 4 h to allow full sterilization then go to Phase 4. If pressure rises by more than 100 mbar, sound alarm and go directly to Phase 4.

Phase 4: Purge cycles

- *Step 1:* Start vac pump, open vac valves and remove gas mixture from chamber, pulling pressure down to 20 mbar
- *Step 2:* Open air valve to allow pressure to rise to atmospheric
- *Step 3:* Repeat steps 1 and 2 for 4 cycles
- *Step 4:* Door seals deflate and door controls enabled after purge cycle complete.

Phase 5: Remove sterilized goods from chamber

- *Step 1:* Open outlet door, keep inlet shut. Ventilator valve opens and fan starts
- *Step 2:* Operator enters chamber and removes sterilized goods
- *Step 3:* Close outlet door. Return to Phase 2.

Design intent

The operating cycle incorporates a number of safeguards that have been hardwired:

- Only one door can be opened. If one door is open, the controls for the other are disabled. This reduces the risk of passing unsterilized goods through the chamber to the sterile goods area. All bypass routes are blocked to pallet trucks
- Door seals cannot inflate until doors are shut
- Ventilation valve and fan start automatically when a door is opened and stop when both are closed
- N2 valve and ETO valves cannot be opened unless pressure in the chamber is below atmospheric. This reduces risk of injecting N2 or ETO whilst chamber doors are open
- Ventilation valve opens automatically if pressure in the chamber becomes positive. This reduces any chance of overpressure in the chamber
- The purge cycle at Phase 1 is to reduce risk of ETO being in the chamber when the doors are opened for entry.

Guideword/deviation matrix

The following table can be used to construct the deviations to be checked in the study. The first phase has been set up as guide. Please continue entries up to the end of Phase 3.

Element (Parameter)	No.	More	Less	Opposite	Part of	As well as	Early/ Late/ Before/ After	Other than	Where else
Power on	X						X		
Phase 1									
Inflate door seals	X	X	X		X				
Vac to 50 mbar	X	X	X	X		X	X	X	X
Vent	X		X			X	X	X	

Element (Parameter)	No.	More	Less	Opposite	Part of	As well as	Early/ Late/ Before/ After	Other than	Where else
Deflate seals	X			X			X		
Phase 2									
Open Inlet door									
Ventilation									
Loading action									
Close door									
Chamber contents									
Phase 3									
Connect ETO									
Press start PB									
Inflate door seals									
Vac to 50 mbar									
N2 addition step									
N2 flow and quantity to 300 mbar									
N2 Pressure									
N2 composition									
ETO addition step									
ETO Flow (via vaporizer)									
Quantity									
Pressure									
Temperature									
Composition									

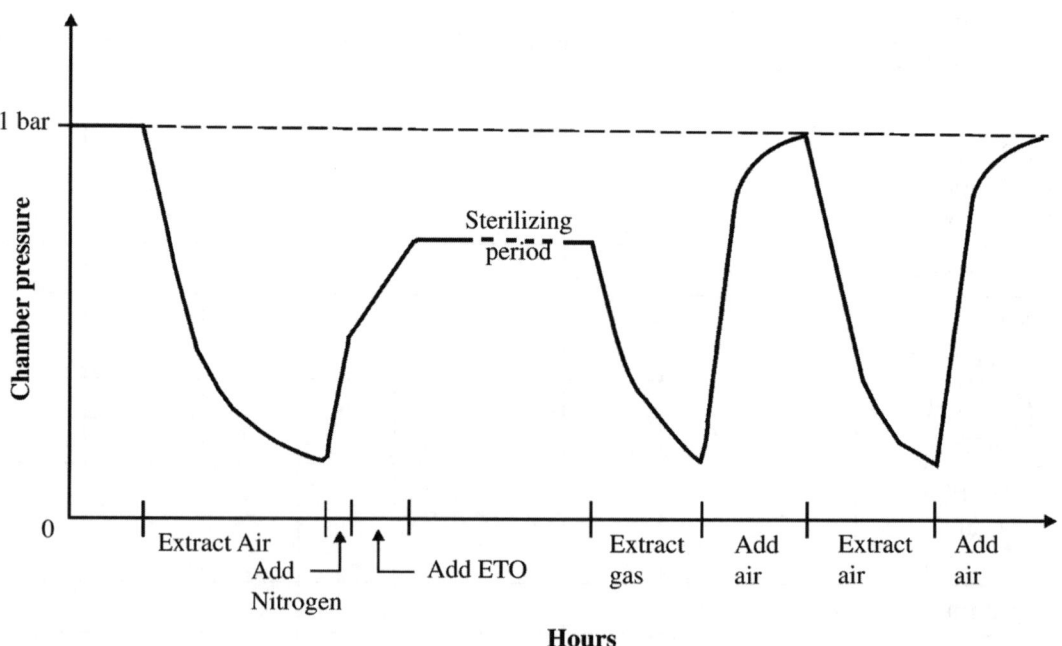

Figure Ex. 4
ETO sterilizer flow diagram

Figure Ex. 5
ETO sterilizer operating cycle.

Report form/worksheet for procedural Hazop

Hazop Study Title: **Sheet:**

Part considered: **Date:**

No.	Element	Deviation	Possible Causes	Consequences	Safeguards	Comments	Actions Required	Act. No.	Action by

Exercise 6: Fault trees

This practical exercise requires participants to construct a fault tree diagram using the basic principles introduced in module 7. It uses an example of a simple reactor with automatically controlled feeds that has the potential to cause a serious risk to plant personnel. Once the basic fault tree has been drawn, the model is to be adjusted to incorporate a safety instrumented system and to demonstrate the resulting risk reduction.

Subject	Application of a fault tree analysis to a chemical reactor	
Objective	To model the relationship between possible faults and hazards in a proposed design and to show the effects of risk reduction measures	
Relevance	The study forms part of the detailed risk analysis that follows from a hazard study. Provides exercise in using the guidelines set out in Appendix 1 to mod 3. See also additional guidance on page 2 of this practical	
Task detail	1. Draw a fault tree for the example of a reactor hazard given in the starting information below 2. Calculate the explosion rate for EUC without protection 3. Add into the fault tree a simple trip protection against high fuel flow or low oxidant flow 4. Calculate the new explosion rate 5. Review the results in the workshop Later in the workshop we can evaluate failure rate of the trip system and decide the proof test interval. Then add in hazard rate calculations	
Time allowed	Thirty minutes	

Starting information

Ref.	Requirement	Information
Task 1	Fault tree for reactor explosion based on 'explosion' as the top event	Figure Ex. 6 shows a reactor with a continuous feed of fuel and oxidant. An explosion can occur inside the reactor if the mixture becomes explosive and a source of ignition is found. In this case we might suppose the source is a hot catalyst inside the reactor presenting a 75% probability of igniting the explosive gas mixture when it occurs The mixture can become explosive if the fuel flow becomes too high relative to the oxidant flow. Include for the possibility of sudden loss of feed in either stream The failure modes of the control loops are to be shown as events based on faults in the sensors and the controllers. Assume controllers are kept on auto

Ref.	Requirement	Information
Task 2	Calculate the explosion rate	Assume all instrument failure rates are 0.1 per year Assume the feeds fail at 0.2 per year
Task 3	Model the effect of fuel flow trips	On the fault tree, add the logic for a high fuel/oxidant flow ratio trip to shutdown the fuel supply. This arrangement is shown in Figure Ex. 7
Task 4	Calculate the new explosion rate	Assume a PFD of 0.02 for the trip. Do not decompose the elements of the trip at this stage
Task 5	Review the results	Group exercise: specimen answer will be reviewed
Task 6	Evaluate the fault tree for the trip system and calculate the failure rate without trip testing	Assume 0.1 per year rates for instruments. We will then demonstrate trip testing to improve the PFD_{avg} to 0.02

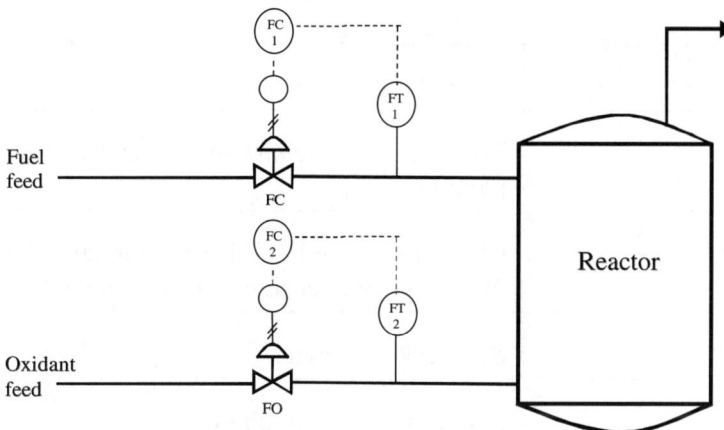

Figure Ex. 6
Reactor for fault tree practical

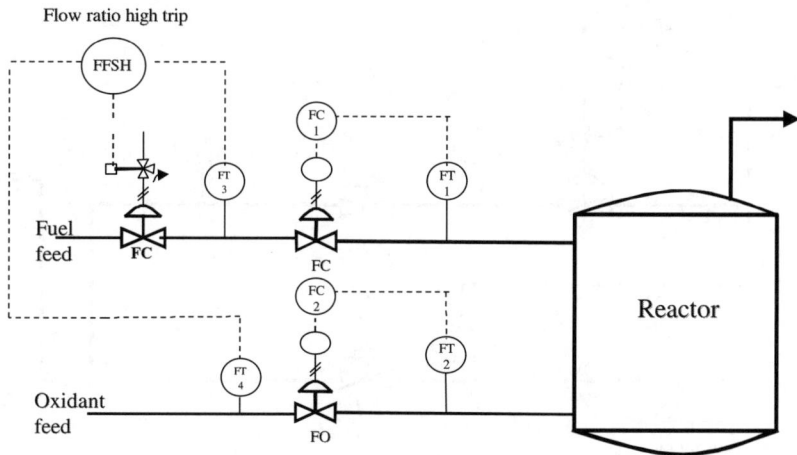

Figure Ex. 7
Reactor with flow ratio trip protection

Exericise 7: Determination of SIL by risk parameter chart

This practical exercise requires participants to determine the required SIL of a proposed safety instrumented system using the basic principles described in Chapter 6. It uses an example of a simple reactor with automatically controlled feeds that has the potential to cause a serious risk to plant personnel (see Figure Ex. 8). This practical explores a conventional paper-based method of SIL determination using one of the methods described in IEC 61511 part 2.

The practical can be extended to gain experience with using software tools for:

(a) The systematic determination of SIL requirements
(b) The recording of design details and the evaluation of the SIL that can be achieved by the design.

For the software exercise you will need a PC running MS Access 2000 and a copy of the demonstration version of 'SILclass'. For these items please refer to the document: 'Introduction to the IDC/AIM software safety tools demonstration packages' and to IDC Exercise No. 8.

Subject	SIL determination
Objective	To experience the task of SIL determination by using a risk parameter chart
Relevance	Chapter 6 of the IDC Hazop Manual, Para 6.5. The study forms part of the essential design steps for any SIS.
Task detail	See Task 1 overleaf
Time allowed	Approximately 30 minutes

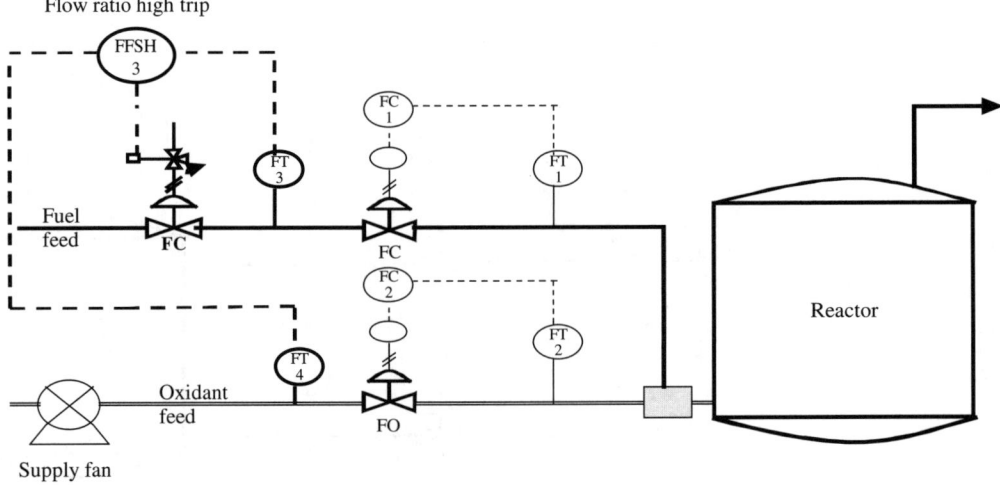

Figure Ex. 8
Reactor with two flow controls and a flow ratio trip function

Task 1: SIL determination by risk graph

Ref.	Requirement	Information
Step 1	Examine the process described here and as in Figure Ex. 8 Read the description of the problem given here and note the risk parameters	Figure Ex. 8 shows a reactor with a continuous feed of fuel and oxidant. Two independent flow control loops are set by the operator to provide matching flows of fuel and oxidant to the reactor An explosion can occur inside the reactor if the mixture becomes explosive and a source of ignition is found. In this case we might suppose the source is a hot catalyst inside the reactor presenting a 75% probability of igniting the explosive gas mixture when it occurs. The mixture can become explosive if the fuel flow becomes too high relative to the oxidant flow. The reasons for this will be: 　Failures of the basic control system 　Operator error in manipulating the controls 　Sudden loss of oxidant feed A safety instrumented system is proposed with a separate set of flow meters connected to a flow ratio measuring function that is designed to trip the process to safe condition if the fuel flow exceeds the oxidant flow by a significant amount. The tag number for this function is FFSH-03 Assume that the following information has been decided for the reactor. 　The total frequency of the events leading to an explosive mixture is approximately once every 10 years (see fault tree diagram) 　The consequence of the explosion has been determined by a study to be a vessel rupture with a 1 in 5 chance of death or serious injury to one person 　The occupancy in the exposed area is less than 10% of the time and is not related to the condition of the process 　The onset of the event is likely to be fast with a worst-case time of 10 minutes between loss of oxidant and the possible explosion 　The material released from an explosion is not harmful to the environment 　The reactor is critical to production and will take 2 months to replace
Step 2	Decide on the risk parameters according to Table Ex. 1	Consequence parameter: C......................

Ref.	Requirement	Information
	Please note that this table is based on a version originally published in IEC 61511 and is also shown in Appendix H of the IDC manual	Avoidance parameter: F............................
		Hazard avoidance parameter: P.................
		Frequency parameter: W...........................
Step 3	Read off the required SIL for the protection system FFSH-03	Apply the designated parameter codes to the risk parameter graph shown in Figure Ex. 9 to determine the SIL required for the safety instrumented function
Step 4	Repeat the procedure from step 1 for a new case where the reactor design has been fitted with a blowout disk that reduces the chances of vessel rupture to less than 5%	Enter the revised Consequence parameter C........... Record the revised SIL target for the safety system ...

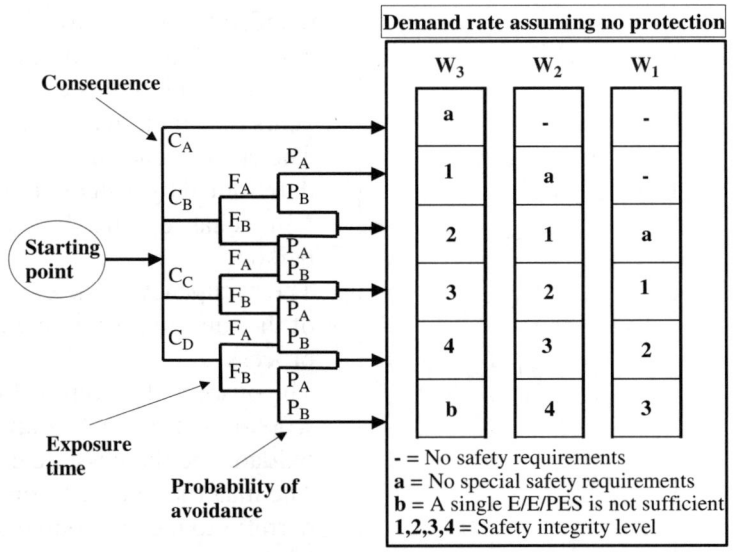

Figure Ex. 9
Risk graph for general use based on IEC 61508/61511

Example risk table for personnel risks based approximately on IEC 61511 draft. This table is not intended for project use, it is only suitable for training exercise purposes. Refer to standards for guidance.

Risk Parameter		Classifications	Comments
Consequence (C) Number of fatalities or likelihood of fataility	C_A	Minor injury	1. The classification system has been developed to deal with injury and death to people 2. For the interpretation of C_A, C_B, C_C and C_D, the consequences of the accident and normal healings shall be taken into account
	C_B	Range 0.01 to <0.1	
	C_C	Range >0.1 to <1.0	
	C_D	Range >1.0	
Occupancy (F) Basis: the time during which the area exposed to the hazard is occupied during a normal working period Note – If the time in the hazardous area is different depending on the shift being operated then the maximum should be selected Note – It is only appropriate to use F_A where it can be shown that the demand rate is random and not related to when occupancy could be higher than normal. The latter is usually the case with demands which occur at equipment start-up	F_A	Rare to more often exposure in the hazardous zone. Occupancy less than 0.1	3. See comment 1 above
	F_B	Frequent to permanent exposure in the hazardous zone	
Probability of avoiding the hazardous event (P) if the protection system fails to operate	P_A	Adopted if all conditions in column 4 are satisfied	4. P_A should only be selected if all the following are true: – Facilities are provided to alert the operator that the SIS has failed – Independent facilities are provided to shutdown such that the hazard can be avoided or which enable all persons to escape to a safe area
	P_B	Adopted if all the conditions are not satisfied	

Risk Parameter		Classifications	Comments
			– The time between the operator being alerted and a hazardous event occurring exceeds 1 h
Demand rate (W) without protection system To determine the demand rate it is necessary to consider all sources of failure that can lead to one hazardous event. In determining the demand rate, limited credit can be allowed for control system performance and intervention. This must not exceed a risk reduction factor of 10. Credit can be allowed for non-SIS risk reduction factors if these have not been included in the initial hazardous event rate	W_1	Demand rate less than 0.03 per year	5. The purpose of the W factor is to estimate the frequency of the hazard taking place without the addition of the SIS
	W_2	Demand rate between 0.3 and 0.03 per year	
	W_3	Demand rate between 3 and 0.3 per year For demand rates higher than 3 per year higher integrity shall be needed	6. If the demand rate is very high (e.g. 10 per year) the SIS operating mode is 'high demand mode' and the SIL has to be determined by another method. The quantitative method is advised in this case

Table Ex. 1
Example risk table for personnel risks

Note: This is not a definitive table, but it represents an example of a typical calibration of a risk graph that is appropriate for the process industries. Many alternative versions are in use throughout industry and companies are expected to adopt their own calibrations to suit their local circumstances.

Suggested results for task 1:

- The frequency parameter is: W_2
  Reason: the demand rate is estimated at 0.1 per year (range >0.03 to <0.3)
- The consequence parameter is C_B
  Reason: the chance of death is 0.2 (range >0.1 to <1.0)
- The exposure parameter is: F_A
  Reason: occupancy is less 0.1
- The probability of avoidance is: P_B
  Reason: the explosion has a rapid onset (<10 minutes) (range >0.1 to <1.0)

The risk graph indicates that the overall protection scheme will need to be SIL-2.

Task 2: SIL rating for environmental risk and asset loss

The draft of IEC 51511 suggests consequence codes for environmental damage in the range C_A to C_D. Try writing down your own consequence scale interpretations for environment and apply them to this example.

Repeat the exercise with a scale of asset loss that you believe would match up to the personal risk scale we have shown in Table Ex. 1.

When selecting the exposure or occupancy parameter F, note that for environment and asset loss the object of harm is always present, hence the parameter always defaults to F_B. Suggested results for task 2:

- With a low environmental impact except for the shock of an explosion the environmental rating is likely to be SIL-1.
- With a 2-month production loss the asset loss rating is likely to be SIL-2.

This illustrates that the SIL rating should be considered for all risks and that the cost of the safety system may be justifiable on the grounds of insurance value as well as for personal risk.

Conclusions

This completes the scope of Exercise No. 7. You will have seen how the risk graph calibration or scales are used to support the classification of the SIL required for a protection system that is to be added to the basic unprotected design.

Each safety function that you require for a given project must be recorded and must remain available for future reference. IEC 61511 shows ideas for typical recording forms that can be used for this purpose (see figure D3 in IEC 61511 part 3).

An alternative approach for recording the SIL determination reasoning is to use a software tool such as the AIM SILclass package. This is the subject of IDC Hazop Exercise No. 8, which repeats this exercise but uses SILclass to implement the recording and risk graph selection functions.

Exercise 8: Determination of SIL by using SILclass software tool with risk parameter chart

This practical exercise requires participants to determine the required SIL of a proposed safety instrumented system using a demonstration version of the SILclass software tool provided with kind permission by Asset Integrity Management Ltd (abbreviated hereafter as AIM). It uses an example of a simple reactor with automatically controlled feeds that has the potential to cause a serious risk to plant personnel.

The SILclass demonstration package has been supplied with a restricted set of options for data entry. This has been arranged to prevent the demonstration version being used for commercial and project purposes. However, once the data has been loaded the SILclass package will retain it for continuation without requiring a 'save' command.

The practical can be used to gain experience with using software tools for:

    (a) The systematic determination of SIL requirements
    (b) The recording of design details and the evaluation of the SIL that can be achieved by the design.

See Exercise No. 7 for a similar version of this practical intended for a paper-based risk graph procedure.

For this exercise the basic demonstration package must be loaded with sufficient data to cover the needs of this exercise. Begin by loading the package and entering the password 'SIL' as described in the table below.

Subject	SIL determination
Objectives	1. To experience the application of the SILclass software tool for SIL determination 2. To illustrate the application of a semi-quantitative risk graph (IEC 61511 calibration)
Relevance	The study forms part of the essential design steps for any SIS and shows how a software package can assist in presenting design options and in the permanent recording of the design basis
Task detail	See tasks 1, 2 below
Preparation	Before starting this exercise ensure that the SILclass demonstration software (ver. 4.1) has been correctly installed as described in the file: IDC SIL demo Introduce. Set your screen display resolution to 1024×768 pixels to ensure all SILclass screens are shown in full 1. SILclass opens at the database screen with a request for the password. Enter the password: 'SIL' and press enter. Double click the dropdown label 'Demonstration system' to open the 'Function Selection' screen

2. This exercise will utilize the first function shown in the list:
   Function Tag: Demo-Tag-1. Click on the function tag to
   open the 'function' screen that will be used to detail the
   exercise overleaf. You are ready to proceed with the
   practical

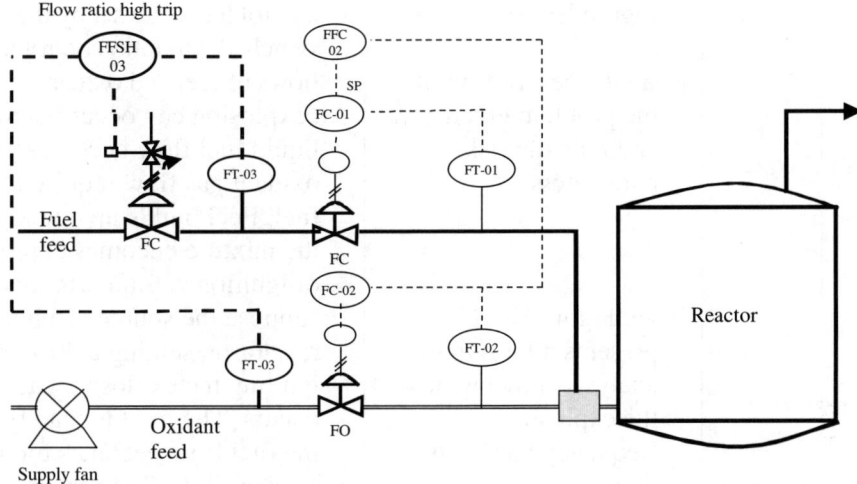

Figure Ex. 10
Reactor with two flow controls and a flow ratio trip function

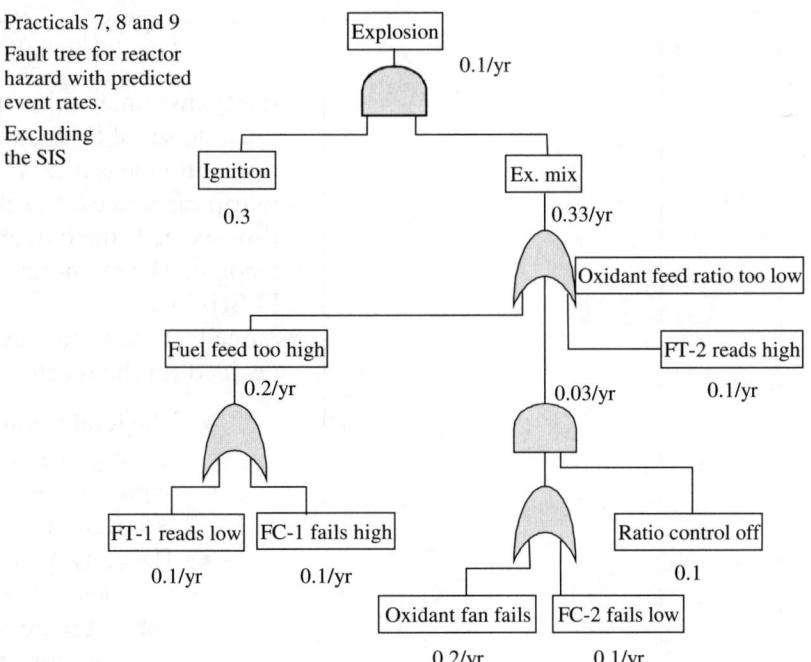

Figure Ex. 11
Fault tree analysis for reactor example shown without the SIS

Task 1: SIL determination by risk graph using the SILclass software tool

Ref.	Requirement	Information
Step 1	Examine the process defined here and as in Figure Ex. 10 Read the description of the problem given here and note the risk parameters Note: Figure Ex. 11 presents a fault tree analysis showing how the explosion frequency has been derived	Figure Ex. 10 shows a reactor with a continuous feed of fuel gas and an oxidant gas. Two flow control loops are set by the operator and switched into ratio control to provide matching flows of fuel and oxidant to the reactor An explosion can occur inside the reactor if the liquid fuel flow becomes greater than the oxidant gas flow required to react with the fuel. Fuel builds up quickly in the reactor and the mixture becomes explosive when a source of ignition is found. In this case we might suppose the source is a hot catalyst inside the reactor presenting a 30% probability of igniting the explosive gas mixture when it occurs. The mixture can become explosive if the fuel flow becomes too high relative to the oxidant flow. The reasons for this will be: • Failures of the basic control system • Operator error in manipulating the controls • Sudden loss of oxidant feed from the supply fan if the ratio control has not been switched on A safety instrumented system is proposed with a separate set of flow meters connected to a flow ratio measuring function that is designed to trip the process to safe condition if the fuel flow exceeds the oxidant flow by a significant amount. The tag number for this function is FFSH-03 Assume that the following information has been decided for the reactor. • The total frequency of the events leading to an explosive mixture is approximately once every 10 years (see fault tree diagram) • The consequence of the explosion has been determined by a study to be a vessel rupture with a 1 in 5 chance of death or serious injury to one person • The occupancy in the exposed area is less than 10% of the time and is not

Ref.	Requirement	Information
		related to the condition of the process • The onset of the event is likely to be fast with a worst-case time of 10 minutes between loss of oxidant and the possible explosion • The material released from an explosion is not harmful to the environment
		The reactor is critical to production and will take 2 months to replace
Step 2	Set up the Function details in SILclass for Demo-Tag-1 as described opposite or use your own descriptions	In the FUNCTION box type in the description for this function. IDC Exercise No. 8: Reactor trip on high fuel/oxidant ratio. Demo-tag-1 alias FFSH-03
		The function tag cannot be changed: leave as Demo-Tag-1
		In Design intent: The intent is to avoid the condition where excess fuel flows into the reactor. Whenever the flow of liquid fuel (measured in m^3/hr) exceeds the flow of oxidant gas (measured in NM^3/hr) by more than 3% the fuel supply valve FV-03 is to be tripped shut. It must remain shut until the trip is reset by authorized operator action. Click on the + sign and add this entry to the glossary
		In the box for consequence of failure on demand: Failure on demand will lead to a build up of excess fuel in the reactor leading to a possible explosion. The consequence of the explosion is expected to be a vessel rupture with 1 in 5 chance of death or serious injury to 1 person. Click on the + sign and add this entry to the glossary
Step 2	Insert details of the proposed SIS	Now select the risk graph type you want to use. In this practical the selection is for IEC 61508 (IEC 61511 calibration) In the Function screen find the command Life cycle in the toolbar at the top. Open the 'Life cycle' screen

Ref.	Requirement	Information
	Note: This step can be omitted for the SIL determination task but it is useful for making effective use of the package in the next design stages (Exercise No. 9)	Enter the data for the following: • Initiator tag description; Initiator stage comprises FT-03 and FT-04. FT-03 is a vortex meter. FT-04 is an orifice plate and dp transmitter. • Logic solver tag: FFSH-03, description: A hardwired solid-state logic solver is expected but a PES will be considered. • Final element tag: FV-03, description: FV-03 is an air-to-open, fire-safe ball valve with de-energies to trip solenoid valve Click on each of the + buttons to load these descriptors into the glossary Click on the button 'Associated tags'. This will open the 'Associated Tags' screen with a devices list and you can then itemize each tagged instrument in the proposed design. Here is the suggested data:
		FFSH-03: Flow fraction high trip function in logic solver. Includes conversion of flows to engineering units for ratio measurement. FT-03: Fuel flow to reactor FT-04: Oxidant flow to reactor FXV-03: Solenoid valve, de-energized to vent FV-03: Fire safe shut off valve for fuel to reactor To finish: close off the screen by using the icon on top right or top left and 'close'
Step 3	Classify the function for required SIL based on personal risk	At the top of the 'Function' screen select: 'CLASSIFY' from the command bar. This will open the risk graph screen. Check that the label on this screen is for 'Demo-Tag-1'. Otherwise go back and select the correct function tag Note the tabs for personal risk, environment and asset loss. Select personal risk: This screen will help you decide the SIL rating for the protection function based on personal risk

Ref.	Requirement	Information
	Note: The risk graph we are using has a scale of frequencies available. The scale is based on the values suggested in IEC 61511 for the process industries. Other risk graphs available in this package have alternative scales for estimating the frequency or likelihood of the event	1. Begin with 'Demand rate' and respond to the choices presented by referring back to the data in step 1. In the narrative box enter your reasons for the choice you have made 2. Now enter the parameter for the consequences box and provide a narrative description in the box personal safety narrative (see step 1). Note that the selection path has become highlighted on the risk chart 3. Now select the parameter for the exposure time parameter, also known as the occupancy (see step 1) 4. Finally select the parameter for the avoidance possibilities (see step 1) The SIL classification for this function will now be highlighted on the risk chart. The SIL classification for personal risk should now be seen as SIL-2. You can also clearly see the choices that have lead to the SIL rating At this point it is useful to experiment with changing the parameter choices to see the effect on the SIL classification. Finish this step by leaving your final data in place
Step 4	Classify the function for required SIL based on environmental risk	Select environmental risk: this screen will help you decide the SIL rating for the protection function based on environmental risk Complete the narrative box for environmental damage (see step 1 info.). Select the relevant parameters for consequence and avoidance. Note that occupancy does not feature in this graph since it is 100% The SIL classification for the environmental risk should now be seen as SIL-1
Step 5	Classify the function for required SIL based on asset loss	Note that the consequence scale has not been calibrated in the SILclass tool. Calibration of the scale is left to individual companies to decide based on the circumstances of the plant and the business. In this example where the loss is significant but not disastrous it is likely to be classed as 'Cb'. Note the result is a SIL-2 rating

Ref.	Requirement	Information
Step 6	Examine the overall SIL rating This completes the classification exercise for the first function	Assuming you now have SIL-2 for personal and asset loss and have SIL-1 for environmental damage you can see that the screen declares overall SIL for this function to be SIL-2 Close the classify screen and return to the Function screen. See that the SIL-2 rating is now recorded here against the Demo-tag-1 function
Step 7	Create ver. 2 of the above case with a modified reactor and reduced risk of injury/death Determine the SIL classification for this version	We suggest that you create another version of the safety function with a modified design of reactor. In this case the reactor has a bursting disk panel such that it will tolerate an explosion without rupturing the vessel. The probability of an injury or death is then assumed in this example to be reduced to 1 in 50 events Select function 2 and describe it as follows: IDC Exercise No. 8: Ver. 2 of Reactor trip on high fuel/oxidant ratio. Demo-tag-2 alias FFSH-03 for ver. 2
		Enter the same descriptions as before but use the Glossary Arrow button to find the text you saved during the original entries. Also repeat step 2 as above Classify this function as per steps 3, 4 and 5 but change the parameters to reflect the reduced consequence scale. Note that the result is a SIL-1 classification for personal, environment and asset loss
Step 8	Generate a function summary report	A printed summary or screen preview can be obtained for the two function options you have loaded Return to the 'Function Selection' screen. Select Demo-tag-1 at the top to filter out other choices. Now find the 'report'
		command on the tool bar. Go to 'print selection function summary' and select 'print' then select 'preview' You should see that you have a report with your records arranged logically. This would be suitable for hard copy records

Ref.	Requirement	Information
Step 9	Options when you have the time	Experiment with changing to another choice of risk graph by selection in the dropdown box of the Function screen. Note that you will have to re-enter your choices every time you make this change

Conclusions

This completes the scope of Exercise No. 8. You will have seen how the SILclass package ensures that all the relevant data and narratives are recorded systematically to support the classification. Clearly each function that you require to be recorded for a given project can be easily recorded and remains available for future reference.

The next step in the design process is to evaluate the feasibility of your intended design. This can be done by selecting the 'test interval' command from the 'Function Selection' screen. The procedures are documented in IDC Exercise No. 9.

Exercise 9: Evaluation of a preliminary SIS design using SILclass software tool

This practical exercise requires participants to examine the performance of a safety instrumented system design in terms of its ability to meet a required SIL rating. The practical uses a demonstration version of the SILclass software tool provided with kind permission by AIM. It uses an example of a simple reactor with automatically controlled feeds that has the potential to cause a serious risk to plant personnel.

The SILclass demonstration package has been supplied with a restricted set of options for data entry. This has been arranged to prevent the demonstration version being used for commercial and project purposes. However, once the data has been loaded the SILclass package will retain it for continuation without requiring a 'save' command.

The practical can be used to gain experience with using software tools for:

(a) The recording of design details and the evaluation of the SIL that can be achieved by the design
(b) Observing the effects of test intervals and system architectures on safety integrity.

This practical follows directly from Exercise No. 8 and it is therefore essential that at least the data entry and classification stages of Exercise No. 8 have been completed on the SILclass database before working through this practical.

For this exercise the basic demonstration package must be loaded with sufficient data to cover the needs of this exercise. Begin by loading the package and entering the password 'SIL' as described in the table below.

Subject	Evaluation of test intervals and safety integrity of an SIS
Objectives	1. To experience the application of the SILclass software tool for preliminary evaluations 2. To illustrate the factors influencing safety integrity in the basic design
Relevance	The study forms part of the essential design steps for any SIS and shows how a software package can assist in presenting design options and in the permanent recording of the design basis
Task outline	The exercise begins in SILclass by taking the proposed SIS for demo-tag-1 and setting up its basic architecture in terms of *initiator, logic solver and final element.* Each stage is assigned a configuration and the SILclass demo package supplies generic reliability data for failure to danger rates of the devices. SILclass already knows the required SIL for the

	design for the display indicates the range of test interval options available for the selected design to fall within an acceptable range of values for probability of failure on demand (PFD). It will be seen that the SIL-2 performance required for demo-tag-1 is difficult to achieve without using dual redundant (1oo2) configurations for the initiator and final element. The exercise moves on to use demo-tag-2, which has been classified as SIL-1. Test interval selection screen then shows that there is a wide range of options available for test intervals even using 1oo1 configurations
Preparation	1. Follow the preparation instructions for setting up SILclass ver. 4.1 or higher as given in Exercise No. 8 2. Ensure that the classification of demo-tag-1 and (optionally) demo-tag-2 has been completed as described in Exercise No. 8 3. Ensure that the 'life cycle' screen data fields have been completed as described in step 2 of Exercise No.8

Task detail: evaluation of the SIS test intervals and its safety integrity

Ref.	Requirement	Information
Step 1	Set up the proposed configuration for the SIS for demo-tag-1 (The process shown in Figure Ex. 10 is as described in Exercise No. 8)	Open SILclass and select the function Demo-tag-1. Check that this has been classified SIL-2. If not, see Exercise No. 8 Select 'Test Interval' on the toolbar and find the table with Reliability data. Set up the devices as follows; Initiator: transmitter: 1 out of 2, beta factor 0.1 Logic solver: Solid state; 1 out of 1 Final element: Valve type 2 : 1 out of 1 In this application the initiator stage comprises the combined transmitters FT-03 and FT-04 since both are needed for the measurement of flow ratio. The 1 out of 2 configuration means that both FT-03 and FT-04 would have to be duplicated. Clearly this is not desirable but it may be needed if SIL-2 is required

Ref.	Requirement	Information
		The beta factor of 0.1 is a typical value for common cause failures possible in both transmitters at the same time
Step 2	Set up the test intervals and try design choices. 1. Enter the intervals box, begin by setting the logic solver test interval to 2 years and set the test durations for all stages to 4 h and the maintenance intervals to 48 h. Set the test coverage percentages for the transmitters and valves at 95% 2. Now observe the colored grid: it should have a green area in the top left Experiment with clicking on the green or orange squares to see the combination of test intervals that are feasible for the transmitters and valves. Choose realistic values for proof test intervals such as 6 or 12 months 3. Experiment with changing the test coverage factors and test durations to see the effect on the available test interval range. Restore values to 95% 4. Try the design using 1 out of 1 configuration for the initiator. Note that the available test range is minimal. Unless the reliability figures can be improved this arrangement does not look very attractive 5. Try changing the valve to type 1; this has lower reliability but then try a 1 out of 2 configuration 6. Finally, restore the values used in step 4	
Step 3	Generate a Function summary report	A printed summary or screen preview can be obtained for the two function options you have loaded Return to the 'Function Selection' screen. Select Demo-tag-1 at the top to filter out other choices. Now find the 'report' command on the tool bar. Go to 'print selection function summary' and select 'print' then select 'preview' You should see that you have a report with your records arranged logically. This would be suitable for hard copy records of the design
Step 4	Repeat the design exercise using Demo-tag-2, which has been classified as a SIL-1 requirement under Exercise No. 8	Repeat step 2 activities after selecting Demo-tag-2 in the 'Function' screen. First check that this function is confirmed for SIL-1 as completed in Exercise No. 8 Note how the SIL-1 design can be comfortably met by single channel architectures for all stages of the safety instrumented system

Conclusions

This completes the scope of Exercise No. 9. You will have seen how the SILclass package links the design stage to the SIL requirement for each function.

Remember that for any logic solver that is shared between two or more functions you would expect to use a consistent test interval. Hence the design procedure begins by defining the logic solver tag and its test interval. The test intervals for the field devices for initiator and final element are then adjusted to match the target performance in the light of the reliability data.

This practical highlights the key steps in the preliminary design of any safety instrumented system. The database package provided by SILclass ensures that the design decisions and the maintenance assumptions used in the design are documented and remain easily available for the life of the plant.

Where the SIS design involves a complex arrangement of sensors or has complex functions the AIM package 'SILcalc' offers a more comprehensive calculation facility. A demonstration version of 'SILcalc' is available with the IDC/AIM demo package. It is instructive to experiment with the same proposed SIS configurations that have been used in this practical.

Answers to practical exercises

Exercise 1

Task 1

Evaluate the risk to workers in comparison with industry norms by calculating the figures for FAR for the workers at the plant. Assume three shifts are employed per 24 h and each worker logs 2000 h on site per year.

Compare the result with the FAR table in the manual and suggest what improvements, if, any, are required in the safety levels at the plant.

Reminder: FAR = Fatalities per 10^8 exposed hours

Answer

Exposed hours per plant = $2000 \times 5 \times 3 = 30\,000$ hrs per year
No. of fatalities per year at the plant = $2 \times 1/500 = 0.004$ or 4×10^{-3} (there are 2 units)
No. of fatalities per hour = $4 \times 10^{-3}/(3 \times 10^4) = 1.3 \times 10^{-7}$
FAR = no. of fatalities per 10^8 exposure hrs = $1.3 \times 10^{-7} \times 10^8 = 13$

The industry norm is 4 according to the table in the manual. Therefore the plant should improve the potential accident rate by a factor of approximately (13/4) 3.25. A target frequency for the explosion per unit should be changed from 1 per 500 years to less than 1 per 1625 years.

Note: In practice the background FAR for a normal industrial plant is approximately 2. The addition of this hazard should not raise the FAR above 4 so the target for this hazard should be 2. This would indicate that a risk improvement factor of at least (13/2) 6.5 is required. The target frequency for each reaction unit then becomes 1 event per 3250 years or 3.8×10^{-4} events per year.

Task 2

Step 1: Calculate the IR for a worker at the plant. This unit is the individual probability of death per year for the exposed worker.

Step 2: Compare the result with the Alarp range values noted in the manual for exposed workers and suggest what improvements, if any, are required in the safety levels at the plant.

Reminder: IR = chance of fatality × frequency of event

Answer

IR = (chance of fatality per event) × (no. of events per year)

= (1 in 15 workers) × (1 in 250 years) = $(1/15) \times (1/250)$

= 2.6×10^{-4}

Comparison with the Alarp region values shown in the manual suggests that the mid-range for exposed workers is 1.0×10^{-4}. So the IR should be reduced by 2.6. This would be done by reducing the frequency of the event through a protection measure or through redesign. If the consequence can be mitigated by design or protection this will be equally effective as it reduces the chance of fatality.

If we apply the same rule of background risk to this answer the reduction factor should be doubled. This will give a factor of 5.2, which aligns quite well with the answer using FAR. Obviously all results are dependent on the tolerable risk target that we choose.

Exercise 2

Exercise in creating a risk matrix

Draft risk matrix for: (Insert description of type of plant or activity)………
…Typical chemical process
....
....
....
5 = low rating number, 1 = high rating.

Consequence Descriptors				Frequency of Occurrence				
Plant	Enviro	Persons	Ratings	5	4	3	2	1
>25% destroyed	Major long term	Several fatalities	1 Catastrophic	5	4	3	2	1
> 5% destroyed or 20% capacity lost per year	Major Short term	One or two fatalities, several injuries	2 Severe	10	8	6	4	2
Repairs needed. < 20% capacity lost in years	Moderate but causing concern to authorities	10% chance of 1 fatality. Severe injury	3 Serious	15	12	9	6	3
Short-term production loss	Short-term excursion within annual limits	Several minor injuries	4 Medium	20	16	12	8	4
Production disturbance only	No impact	No injuries	5 Minor	25	20	15	10	5

ALARP High Line

ALARP Low Line

Event rate per year	10^{-4}	10^{-3}	10^{-2}	10^{-1}	1
Occurrence description	Very unlikely	Possible	Probable	Likely	Very likely

Describe test scenario here:
Example: Severe event expected to occur at average 1 per 100 years scores 6. Risk reduction essential to get well into Alarp region where score will be 10 or higher. Occurrence must be reduced to approximately 10^{-4} per year.

Result appears to be logical and reasonable for first analysis.

Exercise 3

Conclusion for the safety requirements spec.

The following functional safety requirements have been identified:

- High-level approach requires an urgent alarm
- High-level limit requires a trip to flare
- Low-level approach requires an urgent alarm for operator to reduce compressor load
- Low-level limit requires immediate trip of compressors. Probably followed by trip of in feed to flare
- Oxygen in feed to RGH requires a trip to flare, trip of compressors to stop oxygen reaching the next stage and then requires isolation of the RGH to prevent back flow and mixing with the in feed
- The table overleaf has been marked up to show the suggested SIS protection loops.

The following notes may be of help.

The level sensors used to initiate trips are separate from the process controller to avoid common cause failures:

- Continuous transmitters are used so that their operating conditions can be diagnosed at all times.
- Redundancy is suggested despite the cost due to the difficulty of the measurement.
- Two different types of level measurement might be a good idea to include diversity and avoid common cause errors. A reliability analysis should be carried out to find the best configuration.
- In the original implementation of this function the designers decided to use switched level detectors and install them as 2oo3 voting units. The benefits of continuous measurement were lost but the practicality of installing a reliable switching device outweighed this benefit. Many level switch sensors include good diagnostics which improve their fail to safety performance and reduce their fail to danger rate.

There is no requirement for additional actuators for example at the flare because the cause of the problem was not expected to be anything to do with the existing actuators, hence no risk of a common cause problem.

The oxygen detectors are arranged in a 2oo3 voting system since the risk of spurious trips from an analyzer device is high and their overall reliability is not so good. The cost of a trip is very high so the returns for having a high safety availability combined with a low spurious trip rate are attractive.

Hazop. 2 Study Reporting Form	Project: Practical 2: Raw Gas Holder Hazard Study				Drg. No.: RGH Diagrams 1 and 2	Rev No.: 1
	Team Members:				Date	Sheet no. 1 of Meeting No. 1
Hazardous Event or Situation	Caused by / Sequence of Events	Consequences Immediate/ Ultimate	Estimated Likelihood / Suggested Measures to Reduce Likelihood		Emergency Measures (Reduce Conseq uences)	Action Required
External fire	Ignition of gas escapes from RGH due to: 1. Loss of level control • Due to plant disturbances exceeding the range of compressors • High-level stops reached in RGH leads to gas pressure rise, leads to gas escape via seals. Gas ignited by sources of ignition. Or 2. Due to loss of water in seals due to evaporation or leakage. Gas escaping from seal is Ignited.	Limited plant damage and loss of production	Probable 1. Loss of level control is probable. Ignition is possible if sources of ignition are not prevented. Reduce likelihood by: 1. Prevent seals running dry by continuous feed of water and low flow alarm 2. Protection against high-level stops being reached: trip to divert gas feed to flare on approach to high level in RGH 3. Remote locating of flare 4. Haz. area classification.		1. Fire fighting equipment 2. Personnel prevented from access to lute areas.	1. Haz. analysis study 2. Classify area 3. Request design of highly reliable water supply to lutes 4. Manual monitoring or alarms on lute levels 5. Low flow alarm on water to lutes.
Internal fire	Flammable mixture formed by residual air at start-up or breakthrough of oxygen from gasifier. Mixture ignites at precips		Moderate risk due to possible problems with gasifiers, high risk at start-up. Reduce likelihood by: 1. Design gasifiers control to minimize chances of Oxygen breakthrough		Consider heat sensors and alarms at precips. Shut off gas to precips and trip gasifiers	Hazard analysis required Assessment of risk and RRF Specify SIS for protection against

		2. Install oxygen detectors at Inlet to RGH. Trip the in-feed to flare on detection of oxygen 3. Trip compressors as soon as trip to flare confirmed. Then isolate RGH by opening flood valves to lutes 4. Ensure full Nitrogen purge before admitting raw gas. Test for Oxygen at RGH exit during start up		Oxygen; Note this is a very expensive trip event and spurious tripping is to be avoided
Internal fire 2	Flammable mixture formed by entry of air to RGH. Sequence as follows: 1. One or more Gasifierro trip or a blower trips and compressor rates are not reduced to balance the flow 2. Negative pressure occurs if RGH reaches low-level stops and compressors continue to extract gas faster than inlet rate from gasifiers 3. Air entry occurs if excessive negative pressure condition arises in RGH 4. Mixture Ignites at precips	Potential for explosion in RGH as below	Probable due to frequent tripping of gasifiers and the need for operator intervention to balance the compressors. Reduce likelihood by: 1. High priority alarm on loss of gasifter or blower, urgent operator action to reduce rate of compressors 2. Tripping of both compressors if all gasifiers or blowers trip 3. Tripping of one compressor if low-level approach occurs in RGH 4. Tripping of one compressor if one blower stops 5. Tripping of both compressors If low-level limits are reached on RGH,	As above	Hazard analysis required Assessment of risk and RRF Specify SIS

Hazardous Event or Situation	Caused by / Sequence of Events	Consequences Immediate/ Ultimate	Estimated Likelihood / Suggested Measures to Reduce Likelihood	Emergency Measures (Reduce Conseq uences)	Action Required
Internal explosion	As above but fire burns back to gas holder where flammable mixture exists in large volume	Severe damage, rupture of gas holder, fragments Fatalities, injuries	Probable as for internal fire. Reduce likelihood by measures as above. Also • Design to prevent flashbacks from precips • Isolation valves to be operable from control room • Ensure emergency power always available to Isolation valves	Design RGH to blow out upwards, i.e. explosion panels in roof	Hazard analysis required Hazard analysis required Assessment of Risk and RRF Specify SIS
Unconfined explosion	Flammable gas escapes and forms a vapor cloud in air. Reason same as above for external fire	Loud noise, Missiles, blast damage, fatalities and injuries, to employees, not to public? (investigate) Large number of people close by	Rare event due to light gas rising and diluting in air. Possibility cannot be ruled out Protection as for external fire	Training of emergency services	Design review to improve equipment. Hazard study and risk assessment Evaluate RRF Specify SIS
Harmful exposure	1. Toxic gas escape from RGH. Events as for external fire but without ignition 2. Gas cloud blows across to adjacent plant area where high density of people exists 3. Due to maintenance personnel entry to vessels and lutes	Acute effect on employees. Not on public. CO exposure from a large gas release, e.g. at 2000 m^3/hr Large number of people close by likely to be injured employees/ Ill health Bad publicity	More probable than external explosion. Large population of adjacent plant is at risk 1. fencing to keep people further away 2. trip systems as for external fire	Area alarm system to alert staff to evacuate to safe areas or gas proof room	Note that this event has great potential for harm. Hence the combined risks of gas leakage amount to a serious hazard

				SIS must be designed to minimize the chances of an overpressure at the RGH
Chronic exposure	Toxic gas, low-level leakage, exposure mechanism, seals on valves maintenance	Chronic effect on employees /ill health	Probable 1. design for low leakage equipment, avoid valve sterns, etc. 2. provide personnel with gas detectors	
Pollution	Smells from sulphur in gas low level leakage, exposure mechanism, seals on valves		Probable Design for low leakage	

Exercise 4

Step 1

Select the parts for study. The suggested approach is to use two parts.
   The parts are:

- The oil pipe from liquid oil inlet to vaporized oil outlet. Includes the instrument/control functions
- The natural gas supply, burners and firebox. Includes the controls and instruments including furnace flame detector.

These are also known as 'change paths'. Mark up the flow diagram (P&ID) to show the paths.

Step 2

Set up the deviations table by first defining the components or parameters of the part under study. Then mark in the deviations that may be possible and are to be considered by the team.

Guideword/Deviations table for part 1: Oil Pipe and heating coil

Element (Parameter)	Guideword									
	No.	More	Less	Opposite	Part of	As well as	Early/ Late	Other than	Other	Other
Oil flow	x	x	x	x		x		(where else)		
Oil pressure	x	x	x							
Oil inlet temperature		x	x							
Oil outlet temperature		x	x							
Oil vaporize	x	x	x					x		
Heat transfer		x	x							
Pipe material						x		x		
Control	x							x		

Element (Parameter)	Guideword									
Table for part 2: Gas supply to burners and combustion through furnace										
	No.	More	Less	Opposite	Part of	As well as	Early/ Late	Other than	Other	Other
Supply-Gas flow	x	x	x				x			
Supply-Gas pressure	x	x	x							
Supply-gas composition						x		x		
Air flow	x	x	x							
Firebox combustion	x		x							
Flue gas flow			x							

Note: A prompting list is useful for creating the above deviations matrix. Team leader or designers decide on realistic level of detail for deviations. Choosing more detail at the early stages is probably safest. Unrealistic detail is easily disposed of in the Hazop. Applying each of the marked deviations leads to the following items being noted by exception. Items without any hazard or operability problem are omitted from the summary. Detail worksheets may optionally record the decisions on every marked deviation.

Hazop Study Title: Oil Vaporizer example: IDC Workshop								Sheet: 1		
Drawing No.: As shown in Exercise No. 4				Rev No.: 1				Date: 18/11/02		
Team Composition: D M Macdonald + Workshop team								Meeting Date: 18/11/02		

Part considered: The oil pipe from liquid oil inlet to vaporized oil outlet

Design Intent — Material: Furnace grade oil   Source: Oil stock tank via feed pump   Destination: Furnace oil burners
Description: See notes for detail. Oil is to be heated in the coils and vaporized in exit zone of the coil. Vapor condition presents characteristic temperature of 300 to 350 °C at exit.

No.	Guide Word	Element	Deviation	Possible Causes	Consequences	Safeguards	Comments	Actions Required	Act. No.	Action by
1.1	No	Oil flow	Oil flow stops	1. Supply fails 2. Flow control valve closed	Coil overheats, material may fail possible coking of coil and overheating	Low flow alarm, high temp., trip to shut off heating	Alarm may be too slow, high temp. may not be detected due to no flow	Consider trip to shut off gas on detection of low oil flow	1.1	Mech. Eng. 1
				3. Plugging of coil 4. Blockage downstream See 3 below	Oil will boil causing reverse flow, Possible coking of coil	As above		As above and consider pressure relief for trapped liquid	1.2	Mech. Eng. 1
1.2	More	Oil flow	More oil flow	1. Supply pressure high 2. Flow control fails high 3. Wrong set point	Fails to vaporize or superheat oil to process	None	Possible flooding of process, Lost production	Check process effects, Decide on low oil outlet temp. alarm and responses	1.3	Process designer .

1.3	As well as	Oil flow	Impurities in the oil Air in oil See under vaporize	Dirt in supply tank, dirty oil	1. Coking of coil 2. Blockages downstream	None shown	Check de coking abilities of design	Provide filter at supply point	1.4	Mech. eng.
1.4	Other (where else)	Oil flow	Oil flows into furnace	Break in coil due to over heating, corrosion, stress	Oil feeds fire in furnace, capacity to destroy furnace	None	Low likelihood event but needs improving	Provide fire shut off valve for oil, Hazan the need for oil flow trip	1.5	Process eng.
1.5	More	Oil outlet temp.		Control fault or wrong setting. Excess heating	Coking of coil	High temp., trip to shut off heating	Needs a pre-alarm to avoid trips	Provide high temp. alarm at TIC-1	1.6	Inst. eng.
1.6	No	Vapor- ize	Oil does not vaporize	No heating, Coking Too much flow of oil	Liquid oil to process, Process losses	None	Not acceptable	Low exit oil temperature alarm (process alarm)	1.7	Inst. eng.
1.7	Less	Heat tfr	Under heating of oil	Coking,	Inefficiencies	None	Regular checks on heat balance required	Update operating procedures to do efficiency check	1.8	Process
1.8	More	Heat tfr	Overheating of oil in coil	Loss of gas control. See part 2	As low oil flow see point 3		See point 3			
1.9	As well as	Pipe material	Pipe material addition = build up on outside or inside	Coking internally, sooting externally	Low heat tfr, as for point 6		See point 6	Ensure pipe is accessible for cleaning	1.9	Mech. eng.

Hazop Study Title: Oil Vaporizer example: IDC Workshop						Sheet: 1		
Drawing No.: As shown in Exercie no. 4				Rev No.: 1		Date: 18/11/02		
Team Composition: D M Macdonald + Workshop team						Meeting Date: 18/11/02		

Part considered: The gas supply to burners, firebox and flue gas exit Part No. 2

Design Intent

Material: Natural gas   Source: Site supply point take off branch   Destination: Furnace gas burners to flue

Description: See notes for detail. Gas let down pressure controlled by PCV. Flow of gas is controlled to maintain oil exit temp. set point. Gas burners have fixed air intake apertures to maintain fuel air ratio. Hot gases pass over heating coil, exit with natural draft via flue with manually set damper.

No.	Guide Word	Element	Deviation	Possible Causes	Consequences	Safeguards	Comments	Actions Required	Act. No.	Action By
2.1	More	Gas flow	High flow of gas	1. Loss of control 2. Failed control valve TV-1 3. High pressure	High firing rate, possible damage to furnace Unburned gas in furnace, possible explosion. See also 2.2	1. PSH-1 trips gas flow valves shut 2. Oil overheat detection TSH-2 trips gas flow valves shut	Design of burners should limit this possibility. Need to improve gas shut off	1. Check sizing of flow valve to limit overflow of gas 2. Hazan to consider independent shut off valve for gas.	2.1	1. Process eng. 2. Inst. eng.
2.2	No	Combustion	No flame in furnace	Flame out or failed ignition	Explosive mixture in furnace. Possible explosion and damage to oil line. Broken oil line will feed fire	Flame failure detector. Locks out pilot and control valves. Ignition control allows for purge before release of pilot valve interlock	Safeguards may not be adequate	Consider fire shut off valve on oil supply. Hazard analysis to be done on ignition controls	2.2	Project eng.
2.3	More	Combustion	Too much heat in furnace	Loss of gas flow control, wrong setting	Furnace temp. too high. See under heat to coil, part 1					

2.4	Flue Gas flow	Flue gas flow restricted	Damper left closed or blocked flue	Flue gas leaks to plant area. Injury to persons	Design prevents full closure of flues	Consider gas monitor and alarm	Hazard analysis to be done	2.3	Project eng.
Less	Oil flow	Less oil flow than intended	Low supply pressure, Control valve setting wrong	Same as point 1	Low flow alarm, High temp. trip to shut off heating	Safeguards are adequate. In this case the high temp. will be detected	None		
Oppo-site	Oil flow	Reverse flow of oil	Not possible except as under point 1						
No	Oil pressure	No oil pressure	Supply fails, Pump fails	No flow or low flow, same as point 1	Low flow alarm, High temp. trip to shut off heating	Same as point 1	As point 1		
More	Oil pressure	High oil pressure	Blocked line, Downstream shut off	Same as point 1		Same as point 1	As point 1		
Less	Oil pressure	Low oil supply pressure	Worn pump, leaks in joints	Production loss	Low flow alarm, High temp. trip to shut off heating	Acceptable	None		
More	Oil inlet temp.	High temp. in supply oil	None likely	None		Not feasible	None		
Less	Oil inlet temp.	Cold oil	Cold storage conditions	High viscosity, difficulty in achieving design flow	None	Acceptable	None		

Exercise 5

The element/deviations table is a useful way to set out the procedural steps. The crosses mark the deviations that should be considered. Some duplication inevitably occurs but many deviations will be trivial. However these are quickly processed in the study. A separate node should be defined for the vacuum and scrubber system as a continuous process.

During the study the team leader marks off the deviation crosses that have been considered and retains the copy for records.

The deviations table exercise covers up to the completion of ETO addition. The rest of the study would proceed in the same manner.

A sample of the worksheet results is shown after the guideword table.

Element (Parameter)	No	More	Less	Opposite	Part of	As well as	Early/ Late/ Before/ After	Other than	Where else
Power on	X						X		
Phase 1									
Inflate door seals	X	X	X		X		X		
Vac to 50 mbar	X	X	X				X		
Vent	X		X			X	X	X	
Deflate seals	X						X		
Phase 2									
Open Inlet door		X				X	X	X	
Ventilation	X					X	X		
Loading action	X	X	X	X	X	X		X	X
Close door	X								
Chamber contents	X					X			
Phase 3									
Connect ETO	X		X				X	X	X
Press start. PB	X						X		
Inflate door seals	X	X	X		X				
Vac to 50 mbar	X	X	X	X		X	X	X	X
N2 addition step	X	X	X				X		
N2 flow and quantity to 300 mbar	X	X	X						X
N2 pressure	X								
N2 composition						X			

Element (Parameter)	No	More	Less	Opposite	Part of	As well as	Early/ Late/ Before/ After	Other than	Where else
ETO addition step	X	X	X			X	X	X	X
Eto Flow (via vaporizer)	X	X	X	X					
Quantity		X	X						
Pressure	X	X	X						
Temperature									
Composition		X	X		X	X			

Resulting from the Hazop at Phase 3 the following modification was proposed to ensure that the chamber was leak-tested at the point just before injection of ETO.

Phase 3: Insert after step 2

Step 2A: Hold all valves shut and wait 5 minutes. Check pressure in chamber. If pressure rise is less than 10 mbar proceed to step 4. If rise is higher than 10 mbar sound alarm and hold for operator to reset the cycle at Phase 1.

Hazop Study Title: ETO Sterilizer Operating Cycle **Sheet: 1**

Part considered: ETO Sterilizer Cycle Phase 1: Loading **Date: 18 Nov 2002**

No.	Element	Deviation	Possible Causes	Consequences	Safeguards	Comments	Actions Required	Act. No.	Action by
1.0	Step1 Inflate door seals	No inflation, Under inflation, Step omitted	Loss of air supply or Equipment fault	Seals will leak, Vac will not be achieved	None	Acceptable for this phase but not for later phases	See below		
1.1	Inflate door seals	Before, door closed	Override on limit switch	Seals will split	Interlock	Acceptable if tamper proof	Fit tamper proof limit sw	1	Inst. eng
1.2	Step 1 inflate door seals	Overinflation	Regulator fault	Seals will split	None	Expensive damage	Fit pressure relief valve	2	Inst. eng.
1.3	Step 2 Vac	None, step omitted	Sequencer fault	Risk of residual ETO in chamber	Gas detector check by operator before entry	Acceptable			
1.4	Step 2 Vac	Less, vac target not achieved	Leak or faulty vac pump, or stuck valve	Indicates fault in chamber system	Use gas monitor	Acceptable			
1.5	Step 3 Vent	As well as, ETO or N2 valves opened in error	PLC output fault	Toxic conditions Nitrogen fills chamber, danger to person entering	ETO supposed to be disconnected, ETO monitor, no safeguard for Nitrogen	Acceptable / Not acceptable	Investigate Nitrogen monitor	2	Proj. eng.

Exercise 6

Figures An. 1, An. 2, An. 3, and An. 4, describe the fault tree analysis before and after the proposed SIS is fitted to the reactor hazard problem.

The failure modes described for the instruments can be shown in greater detail but have been kept to basics in this example.

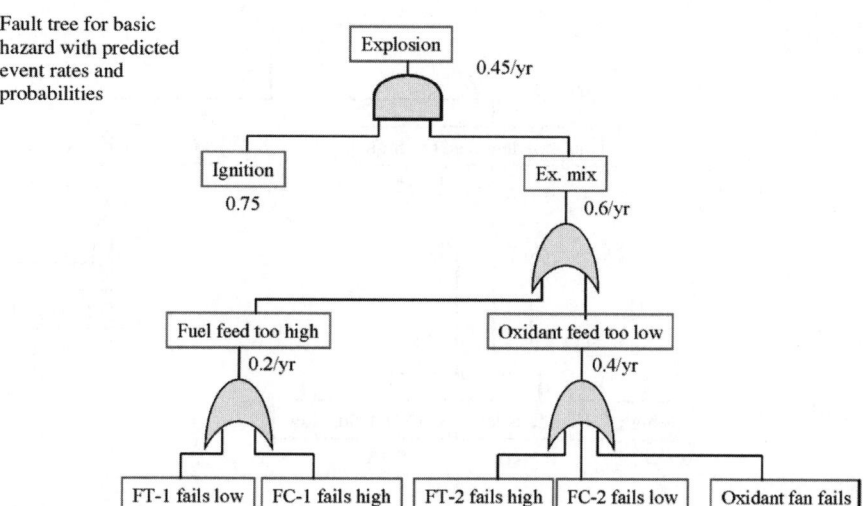

Figure An. 1:
Fault tree before protection

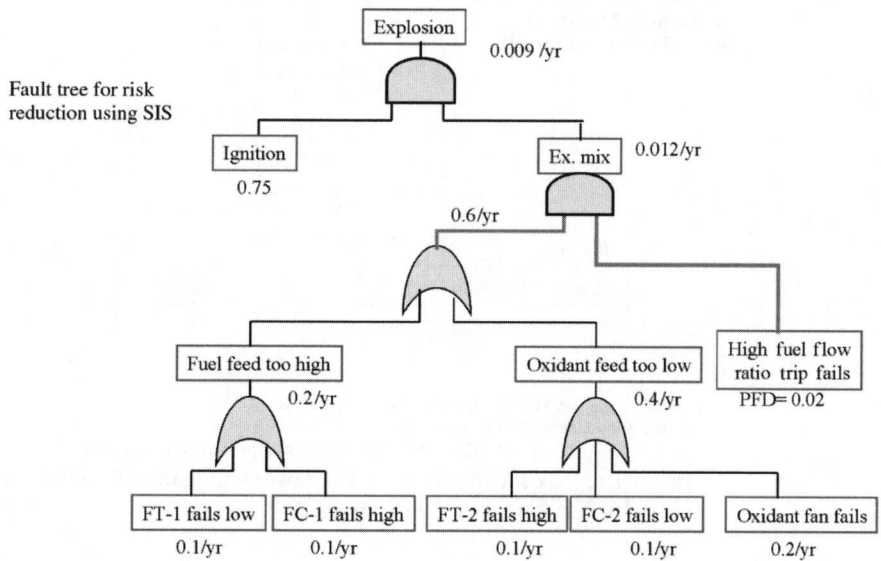

Figure An. 2
Fault tree with SIS added

Note the use of the AND gate to insert the SIS with its failure probability. This was described in Chapter 7.

Performance modeling for the trip system.

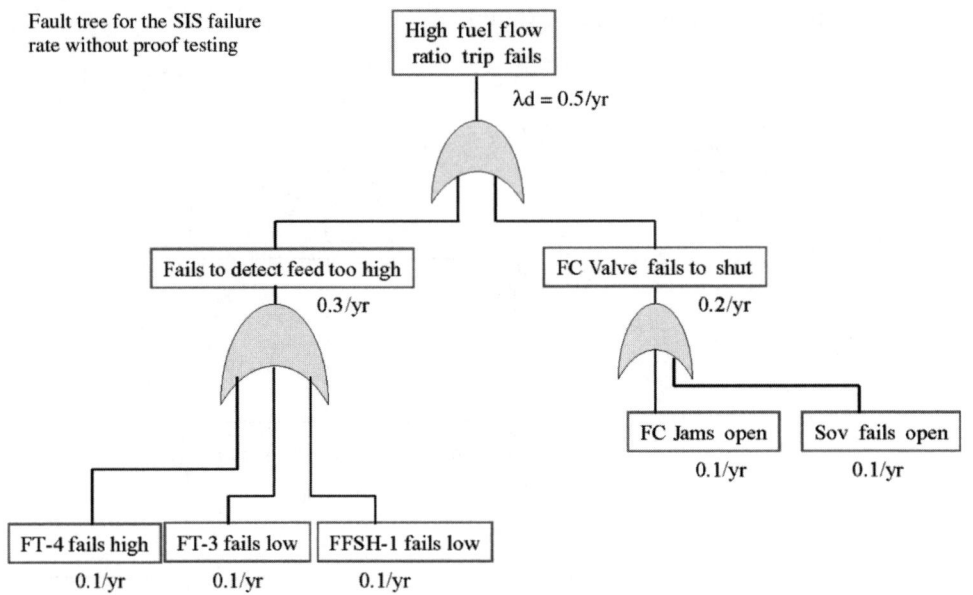

Figure An. 3
Probability of failure of trip system

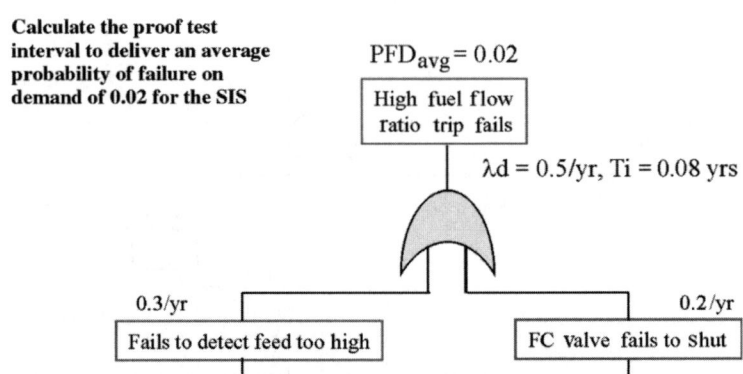

The PFD$_{avg}$ is given by the formula PFD$_{avg}$ = λd · Ti/2
Hence for a target of PFD$_{avg}$ = 0.02:
Ti = 0.02 x 2/5 = 0.08 years or approximately 4 weeks

(In practice look for components with lower failure rates than those assumed and try to avoid a Ti value lower than 12 weeks.)

Figure An. 4
Trip system PFD with the effect of proof testing

Index